PRIDE® METHODOLOGIES FOR IRM

(PRofitable Information by DEsign - through phased planning and control)

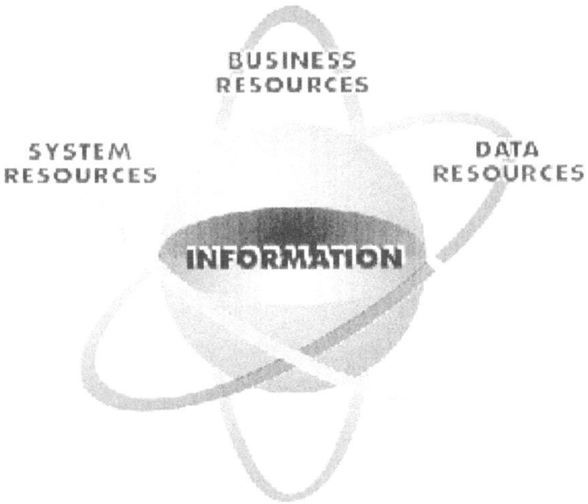

The science of Information Resource Management (IRM)

Tim Bryce

M&JB Investment Company
formerly M. Bryce & Associates (MBA)
P.O. Box 675
Palm Harbor, FL 34682-0675
United States
Tel: 727/786-4567
timb001@phmainstreet.com
http://www.phmainstreet.com/mba/

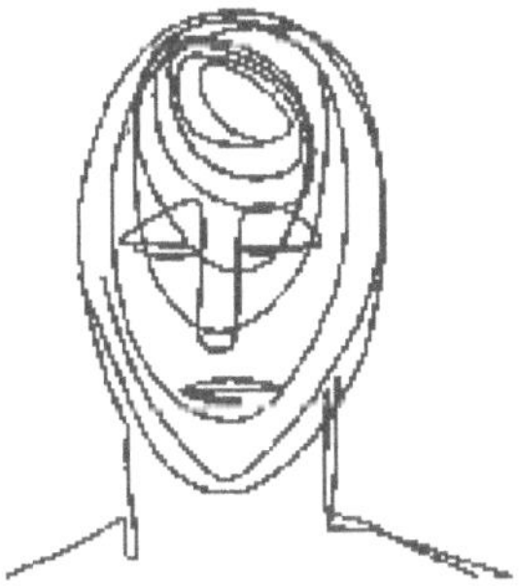

Since 1971:
Software for the finest computer - the Mind

MBA PRESS

Main entry under title:

PRIDE Methodologies for IRM

Includes index.

1. Information Resource Management 2. Management Science

I. Bryce, Tim

ISBN 0-9786182-2-X
ISBN-13 978-0-9786182-2-3

formerly M. Bryce & Associates (MBA),
P.O. Box 675
Palm Harbor, FL 34682-0675

Printed in the United States of America

DEDICATION

*To Milt Bryce, the Father of IRM Methodologies.
And to Jean, "The Boss," Milt's trusted
companion and confidant for 55 years.*

TABLE OF CONTENTS

ABOUT THE AUTHOR

TIM BRYCE

Tim is the Managing Director of M&JB Investment Company (M&JB), an international management consulting firm located in the Tampa Bay area of Florida.

Mr. Bryce graduated from Ohio University in 1976 with a Bachelor of Science degree in Communications (BSC) from OU's College of Communications, School of Communication Studies (formerly School of Interpersonal Communications). Upon graduation, he joined M&JB full time and served in a variety of capacities, including both sales and consulting. As Director, his responsibilities include product development, implementation, training and on-going support of all MBA customers on a worldwide basis. Because of this, he has traveled extensively providing training and consulting services at various levels of computer proficiency (novice to expert) on a variety of management and computer related subjects.

Tim is the principal author of M&JB's "PRIDE"-Enterprise Engineering Methodology (EEM) and the designer of the Computer Aided Planning (CAP) tool (a tool for calculating corporate priorities and performing an organization analysis) and Automated Systems Engineering (ASE), a tool used to generate system designs. He is also responsible for the development of M&JB's Automated Instructional Materials which includes instructions and documentation for all of the "PRIDE" methodologies for IRM,

A prolific writer, Tim has considerable experience in writing technical documentation (paper and on-line), help text, and web design. He has authored several papers on a variety of management and computer related subjects and was the co-author of the book, *The IRM Revolution: Blueprint for the 21st Century* (ISBN 0-9621189-0-7). He regularly maintains a blog which is read by thousands of I.T. professionals worldwide. His *"Management Visions"* commentaries are regularly broadcasted over the Internet.

Mr. Bryce has also made several presentations at computer trade related meetings and has given lectures at various universities. Tim also actively participates in various trade related associations and user groups, as well as community organizations. He is available to give presentations on subjects related to Information Resource Management (IRM).

Tim has also served on numerous Board of Directors of industry, non-profit, and fraternal organizations, as founder, President, and a variety of other capacities.

PREFACE

WHY THIS BOOK?

The intent of this book is to provide a uniform and consistent approach for the development of information resources. Over the last 50 years there has been a noticeable void in terms of standards for design and development. This has resulted in inconsistencies that have complicated the maintenance and upgrading of systems, software, and data bases. To overcome this problem, the "PRIDE" Methodologies for IRM are herein published as an attempt to bring standardization to the development process.

"PRIDE" represents a substantial body of work that has been field-tested by customers all over the world. The concepts and philosophies embedded in "PRIDE" have stood the test of time. We have also put considerable effort in articulating our concepts and terminology, thereby moving IRM from an art to a teachable science.

Over the last thirty years companies have taken a tool-oriented approach to solving their development problems. Not surprisingly, the tools and techniques have changed over the years, but the problems haven't: user information requirements are not satisfied, systems lack integration and adequate documentation, data redundancies plague companies, projects are delivered late and over budget, etc. Despite the fine tools and techniques available, the problems are essentially no different than they were 35 years ago. What is needed is not more tools, but rather how to orchestrate them and make our people more productive. In other words, a management-oriented approach; hence the publication of this book.

HOW TO USE THIS BOOK

The book is divided into six major parts:

An Introduction to IRM

Describing the concepts and philosophies of Information Resource Management. Included in a description of the concept of an "Information Factory." The section should be considered a prerequisite for the remainder of the book.

"PRIDE"-EEM

Describes the concepts and philosophies of the Enterprise Engineering Methodology, along with a description of the Phases and Activities.

"PRIDE"-ISEM

Describes the concepts and philosophies of the Information Systems Engineering Methodology, along with a description of the Phases and Activities.

"PRIDE"-DBEM

Describes the concepts and philosophies of the Data Base Engineering Methodology,
along with a description of the Phases and Activities.

"PRIDE"-PM

Describes the concepts and philosophies of Project Management, along with a description of PM activities, including: Planning, Estimating, Scheduling, Reporting, and Control.

Supplemental Narratives

Provides supportive tutorials on subjects such as Establishing a Technical Library, Quality Assurance, Establishing an IRM Repository, Forms, and a Glossary of Terms.

WHO SHOULD READ THIS BOOK?

This book will be of interest to a variety of people, including:

Chief Information Officers (CIO)
Development Management
Enterprise Engineers/Architects
Systems Analysts/Engineers
Software Engineers
Data Engineers
Data Base Administrators
End Users

It also complements a college curriculum for Information Resource Management, making it well suited for students.

EXAMPLES USED IN THIS BOOK

Pertinent examples are included throughout the book, primarily at the end of each section. Space prohibits the inclusion of full examples. As such, key deliverables are included herein. For more complete examples, see the Internet version as maintained at MBA's web site.

Throughout the methodologies, you will read about the preparation of estimates and schedules. Such samples can be found in the Project Management section and not the phase deliverables sections.

The examples contained herein include the names of fictional companies and people. Any relation to any organization or any person living or dead is strictly coincidental.

THE INTERNET VERSION

MBA maintains a separate Internet version of the methodologies. This includes additional materials not included herein, such as:

- Functional Matrices - quick summaries of the activities within each phase.
- Functional Descriptions - describing the duties and responsibilities of IRM specific functions; useful for developing job descriptions.
- Phase Review Checklists - defining the acceptance criteria when reviewing a phase deliverable.
- Supplemental Narratives - a variety of tutorials in support of the methodologies.
- More detail instructions of the various Activities, complete with advice on applying tools and techniques.
- More examples.

Adding all of the material as mentioned above would have easily doubled or tripled the size of this book. As such, this book complements the Internet version. Again, the Internet version as maintained at MBA's web site (see cover for address).

THE AUDIO BOOK VERSION

An abridged audio version of this book is also available from MBA and is suitable for the mobile manager of developer. It is also useful for the visually impaired.

CORPORATE PROFILE

M&JB Investment Company (fromerly M. Bryce & Associates) is the developer of the "PRIDE" Methodologies for IRM. M&JB is an international management consulting firm that was founded in 1971 in Cincinnati, Ohio. In 1985, the company relocated its offices to Palm Harbor, Florida (Tampa Bay area).

ABOUT OUR SLOGAN

"Software for the finest computer - the Mind."

Years ago when MBA was first starting to market "PRIDE," people would ask us what language the product was written in; was it COBOL? Fortran? It would surprise them when we told them it was written in "English." After all, "PRIDE" is not software for computers. It is a philosophy that transcends technical implementation. It is a way of thinking and managing, and affects our perspectives on development, which is why we coined the slogan, *"Software for the finest computer - the Mind."*

Since the inception of our company in 1971, the underlying theme in our methodologies and writings is the recognition of the vital role the human being plays in business. In the end, it is the human-being that matters most, not our technology.

- Tim Bryce
May 2006

Section 1

AN INTRODUCTION TO IRM

Productivity = Effectiveness X Efficiency
- Bryce's Law

Welcome to the "PRIDE" methodologies for Information Resource Management (IRM). This product includes three methodologies:

"PRIDE"-EEM (Enterprise Engineering Methodology) - used by business analysts to model the business, determine information requirements, and calculate an Enterprise Information Strategy.

"PRIDE"-ISEM (Information Systems Engineering Methodology) - used by Systems and Software Engineers to design enterprise-wide information systems.

"PRIDE"-DBEM (Data Base Engineering Methodology) - used by Data Base Engineers and DBA's to design an integrated corporate data base.

"PRIDE" also includes a Project Management (PM) system and an Information Resource Manager (IRM) to catalog and cross-reference information resources.

WHY WAS "PRIDE" CREATED?

"PRIDE" was created based on MBA's experiences in managing large IS/IT departments where it was necessary to bring uniformity and consistency to development efforts in order to improve productivity. The problems experienced by IT organizations today are essentially no different than they were in the early 1970's when "PRIDE" was first introduced:

• User information requirements are not satisfied. The development staff and end-users do not work together harmoniously; e.g., users lack confidence in the development staff, and the development staff is frustrated with the users.

• Little or no planning is performed; IT management can best be described as "reactive" as opposed to "active."

• Systems are poorly documented causing headaches in terms of maintenance and implementing changes.

• Systems lack integration thereby creating work redundancies.

• Data redundancy plaques corporate data bases. Consequently, inconsistent information is produced.

• Projects are rarely delivered on time and within budget.

• Development personnel are not working in a concerted manner (no standards). As a result, quality is lacking causing users to lose confidence in the IT staff and redundant development effort.

• Little time is spent in systems analysis with programming representing 85% of the development effort, thus causing programmers to re-write applications until they "get it right."

• Development personnel are constantly in a "firefighting mode" (maintenance) and not developing new systems.

• There is no effort to share and re-use information resources, causing developers to constantly re-write applications.

Understanding the enormity of the situation, MBA studied various disciplines to develop other types of products that are much more tangible than information resources. As such, MBA considered the **Five Basic Elements of Mass Production:**

1. **Division of Labor** - to break the production process into separate tasks performed by specialists or craftsmen.

2. **Assembly Line** - defining the progression and synchronization of work.

3. **Precision Tooling** - for mechanical leverage in the assembly line.

4. **Standardization of Parts** - for interchangeability and assembly by unskilled and semi-skilled workers.

5. **Mass Demand** - the impetus for mass production (as represented by a company's information requirements).

In studying production processes, MBA came to the conclusion that:

Productivity = Effectiveness X Efficiency

Most people fallaciously equate productivity with efficiency, which simply gauges how fast we can perform a given task. Effectiveness, on the other hand, validates the necessity of the task itself. There is nothing more unproductive than to do something efficiently that should not have been done at all. An industrial robot, for example, can efficiently perform tasks such as welding. However, if it welds the wrong thing or at the wrong time, then it is counterproductive. It therefore becomes important in the production of any product to define Who is to perform What work, When, Where, Why, and How (we refer to this as "5W+H").

From this, MBA devised a set of integrated engineering/manufacturing concepts for the development and control of information resources. These concepts became the foundation of the "PRIDE" methodologies. MBA then played the role of "Industrial Engineer" and detailed the production process (methodologies) in terms of the steps and sequencing of work, defined deliverables, and review points. By doing so, MBA defined the expert craftsman's rules for developing information resources. In other words, "PRIDE" does not do anything more than what a good IRM craftsman does already.

WHAT MAKES "PRIDE" UNIQUE?

The strength of "PRIDE" rests in its integration of concepts and philosophies, all of which are based on common-sense engineering and manufacturing concepts that have stood the test of time and have been applied to just about every field of endeavor imaginable around the world. This engineering/manufacturing approach is what distinguishes "PRIDE" from other CPA-based or programming related methodologies. The concepts embedded in "PRIDE" and its approach to development were also proven unique in a court of law (Trade Secret litigation: M. Bryce & Associates vs. Gladstone, et al). Because of its integration, a company can either use "PRIDE" in its entirety or in selective pieces and evolve into other parts of the product over time.

The emphasis in "PRIDE" is on design correctness and the production of a quality product. Each methodology consists of a work breakdown structure consisting of phases, activities, and operational steps (tasks), each specifying Who is to perform What work, When, Where, Why, and How (5W+H). The work breakdown structure is based on producing specific deliverables that can be reviewed and inspected to verify completeness, thereby assuring a quality product is produced. The deliverables can take many forms, be it a report, source code, data base structures, etc.

The structure and sequencing of the methodologies were not randomly devised such as the classic linear "water-fall" approach to programming. Instead, the "PRIDE" methodologies are based in the information resources affected; for example:

EEM - Phases map the logical and physical organizational structures.

ISEM - Phases map the system structure being developed.

DBEM - Phases map the logical/physical data base models.

Although "PRIDE" includes forms to document information resources, flowcharts, and to estimate time and costs, it is most definitely not a "forms driven" system. Instead, "PRIDE" is a way of thinking and looking at information resources. This is why we refer to it as "Software for the finest computer - the Mind." A "forms driven" methodology implies that if you fill out forms in a specific sequence, the desired work product, such as a system, will be automatically produced. We consider this a fallacious concept. "PRIDE" views documentation as a working tool and a by-product of the development process, just as blueprinting is to an architect. The processing of forms will not make a person a better Systems Engineer; the person must understand the process and the rationale by which it is based. Only then will they be able to channel their energies effectively and produce superior results.

"PRIDE" is intellectually honest. We have gone to great extremes to define our terms and concepts. We believe Information Resource Management (IRM) to be a science, one that can be taught. To do so, one must define their terminology and explain their concepts, which is why we have done so. One by-product from defining our terms is that companies can standardize and communicate with a common language, thereby avoiding the "Tower of Babel" effect that commonly plaques most application development organizations.

"PRIDE" recognizes development tools and techniques will change over time as technology changes. As such, it provides for a myriad of outside tools and techniques to be used throughout the methodologies, just as an assembly line allows such devices. Regardless of the marketing hyperbole of the computer industry, there is no single product on the marketplace that can do everything for everyone, in all situations. An overriding theme in "PRIDE" is that "if there is anything in life that it constant, it is change." Techniques and tools will come and go, but the assembly line process will remain constant.

"PRIDE" requires a change of focus and orientation. Instead of concentrating on programming, "PRIDE" is concerned with the overall architecture of a system. Due to its orientation towards up-front work, programmers become the beneficiaries of better specifications and, as such, improves their productivity. Due to the organization and discipline of "PRIDE", it can even make semi-skilled workers produce superior results. In the hands of a competent manager, we have never seen "PRIDE" fail.

CORPORATE PROFILE

M&JB Investment Company (M&JB) is an international management consulting firm specializing in Information Resource Management (IRM). The "PRIDE" methodologies have been used worldwide in every field of endeavor imaginable:

Countries Served:
Australia, Belgium, Brazil, Canada, Denmark, Japan, Korea, Mexico, New Zealand, Norway, Saudi Arabia, South Africa, Spain, United Kingdom, United States, Venezuela

Industries Served:
Aerospace, Automotive, Banking, Brewing, Chemical, Communications, Construction, Electronics, Energy, Engineering, Government (Federal/State/Local), Insurance, Manufacturing, Paper/Wood, Printing/Publishing, Public Utilities, Retail/Wholesale, Shipbuilding, Steel, Transportation

In particular, "PRIDE" products have been used extensively by firms throughout Japan, including several companies who have received the prestigious Deming Prize for quality.

"PRIDE" is a universally applicable approach that can be used in any company, on any information system, in any cultural environment, and is not restricted to any particular computer platform, data base, language, or programming tool/technique.

Since 1971, MBA has never failed to meet a customer implementation commitment, regardless of the customer's geographical location or training requirements.

Through the "PRIDE" product line, MBA has established several precedents and introduced many new concepts to the industry:

1. First to offer commercial methodologies for system design, data base design, and enterprise engineering.

2. First commercial repository for capturing, controlling, and re-using information resources.

3. First on-line methodology for accessing automated

instructional materials.

4. First methodology enacted into law (State of Minnesota, 1978).

5. Concepts and Techniques introduced: Information Driven Design, Structured Systems, Chronological Decomposition, Layered Documentation, Data Resource Management, 4 Data Base Models, Enterprise Decomposition, Priority Modeling, Bill of Material Processing for managing information resources, Mini-Project Manager, Estimate-to-do (versus percent complete), estimating by bill-of-materials.

While many of M&JB's competitors have come and gone, the "PRIDE" product line enters its fourth decade of use. This is a testament to the integrity and durability of the product and the company who produced it.

"With almost 1,500 installations worldwide and the fact MBA has existed in the methodology world for almost 20 years, it is fair to say that the company is well capable of managing its revenues and balancing them off with consistent research and development by investing about 80 percent of those revenues into R&D."

- DATAPRO

IRM: THE CONCEPT

"PRIDE" is an important part of an overall corporate program for Information Resource Management (IRM). The underlying philosophy behind IRM is to design, inventory and control all of the resources required to produce information. When standardized and controlled, these resources can be shared and re-used throughout the corporation, not just by a single user or application.

There are three classes of information resources:

- **BUSINESS RESOURCES** - Enterprises, Business Functions, Positions (Jobs), Human/Machine Resources, Skills, Business Objectives, Projects, and Information Requirements.

- **SYSTEM RESOURCES** - Systems, Sub-Systems (business processes), Administrative Procedures (manual procedures and office automation related), Computer Procedures, Programs, Operational Steps, Modules, and Subroutines.

- **DATA RESOURCES** - Data Elements, Storage Records, Files (computer and manual), Views, Objects, Inputs, Outputs, Panels, Maps, Call Parameters, and Data Bases.

INFORMATION RESOURCES

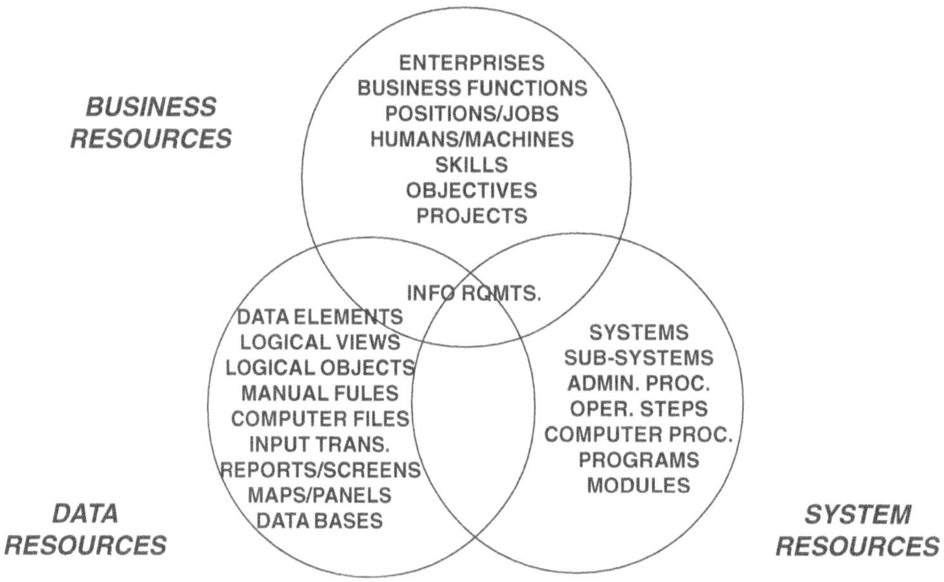

Integration is made possible through a Bill Of Materials Processor (BOMP)

These three classes of information resources provides the rationale as to why there are three complementary methodologies within "PRIDE".

• **ENTERPRISE ENGINEERING METHODOLOGY (EEM)** - for defining the mission and goals of the business and the development of an Enterprise Information Strategy synchronized with the business.

• **INFORMATION SYSTEMS ENGINEERING METHODOLOGY (ISEM)** - for designing and building enterprise-wide information systems (business processes crossing organizational boundaries). Software Engineering is considered a subset of ISEM.

• **DATA BASE ENGINEERING METHODOLOGY (DBEM)** - to design and develop the corporate data base, both logically and physically.

Each methodology consists of a series of defined phases, activities and operations. Laced throughout the methodologies are defined deliverables and review points to substantiate completeness and to provide an effective dialog between management and developers. The methodologies promote design correctness and the production of a quality product.

IRM/MRP ANALOGY

The concept of IRM is actually no different in intent than "Materials Resource Planning" (MRP) as used in manufacturing. Both are concerned with the efficient and cost effective use of resources. The classification and control of resources are the main objectives. Resources are classified to prove their uniqueness so that redundancy is not introduced and to promote sharing. Control is required to collect, inventory and retrieve resources as required by the business.

Whereas MRP is concerned with managing products and the parts required to produce them, IRM is concerned with managing information and the resources required to produce it.

One of the important by-products of cataloging and cross-referencing information resources is a model of the enterprise, including how it is organized and how it operates. Other benefits include:

• All information resources are controllable, permitting the ability to design integrated systems and perform an "impact analysis" of a proposed resource change.

3 METHODOLOGIES

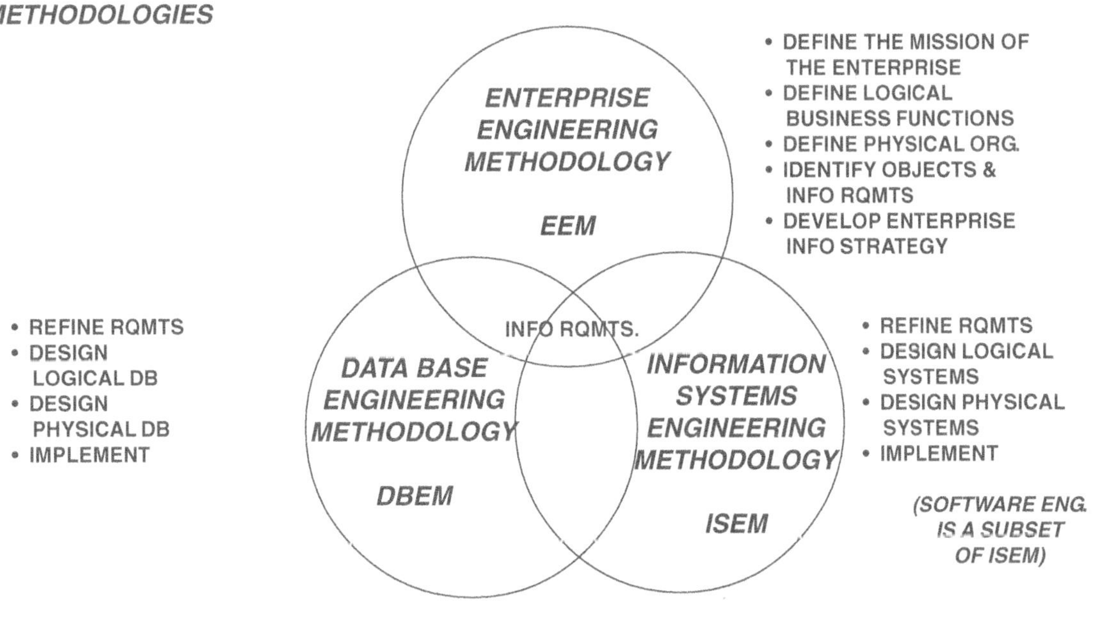

Each Methodology Consists of Phases, Activities, and Deliverables

IRM/MRP ANALOGY

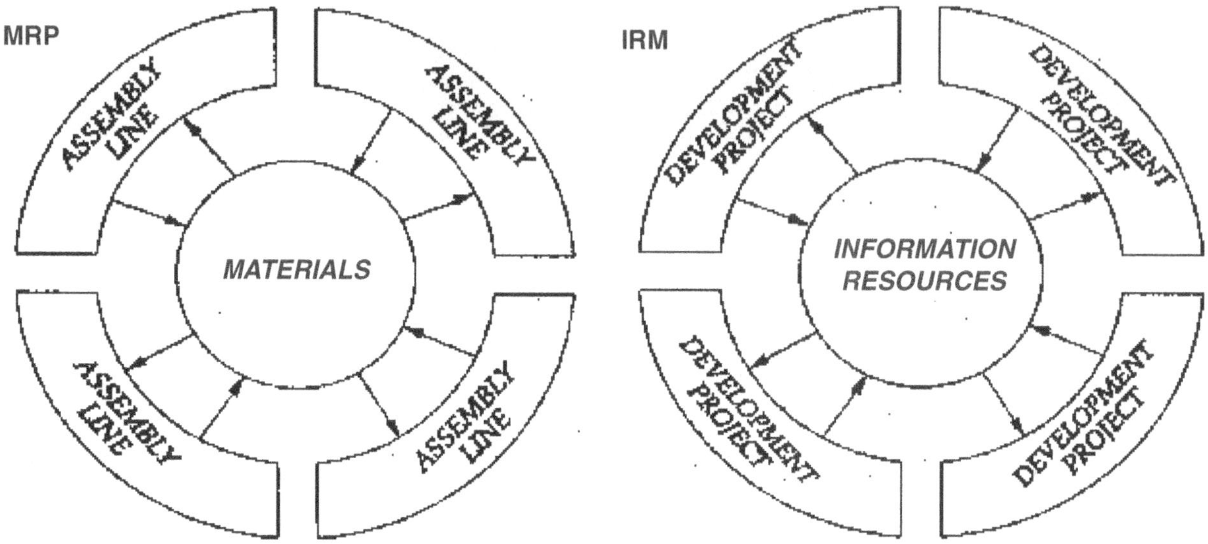

MISSION:
Standardize, Control, Share & Re-use

• Simplified search of information resources for reuse. Redundancy of resource definition is eliminated.

• Complete and current documentation of all information resources, in an organized and meaningful way.

• Communications within the organization is improved since developers and users would use standard and common definitions for information resources, all of which would be in standard business terminology.

INFORMATION FACTORY CONCEPT

There are several other engineering/manufacturing related practices that can be applied to the development of information resources. Just as it is possible to engineer a manufacturing environment to build products, it is entirely possible to engineer a manufacturing environment to build information resources.

There are three basic components within any manufacturing facility:

1. **ASSEMBLY LINES** - stages of work sequenced in such a way to develop products. Along the assembly line, a series of tools and techniques will be deployed, some implemented by the human being, others through automated assistance, such as robots.

2. **MATERIALS MANAGEMENT** - the business function concerned with standardizing parts so they may be re-used in various product assemblies. Further, it is concerned with collecting, storing and retrieving parts in the most efficient means possible (e.g., JIT - "Just In Time").

3. **PRODUCTION CONTROL** - oversees the assembly lines, looking for unanticipated delays or accelerations of production schedules. Consequently, corrective action can be taken as required to resolve problems.

These same disciplines can be utilized to develop and manage information resources:

1. **METHODOLOGIES** (Assembly Lines) - defines the work environment by establishing WHO is to perform WHAT, WHEN, WHERE, and WHY, thereby synchronizing the flow of work. Within the phases of the methodology, a variety of tools and techniques may be deployed which defines HOW the work is to be performed.

2. **RESOURCE MANAGEMENT** (Materials Management) - identifies and classifies information resources, thereby sharing and re-using resources. Also, ensuring they are collected, stored and retrieved in a timely manner.

3. **PROJECT MANAGEMENT** (Production Control) - is used to plan, estimate, schedule, report, and control project work.

FACTORY CONCEPT

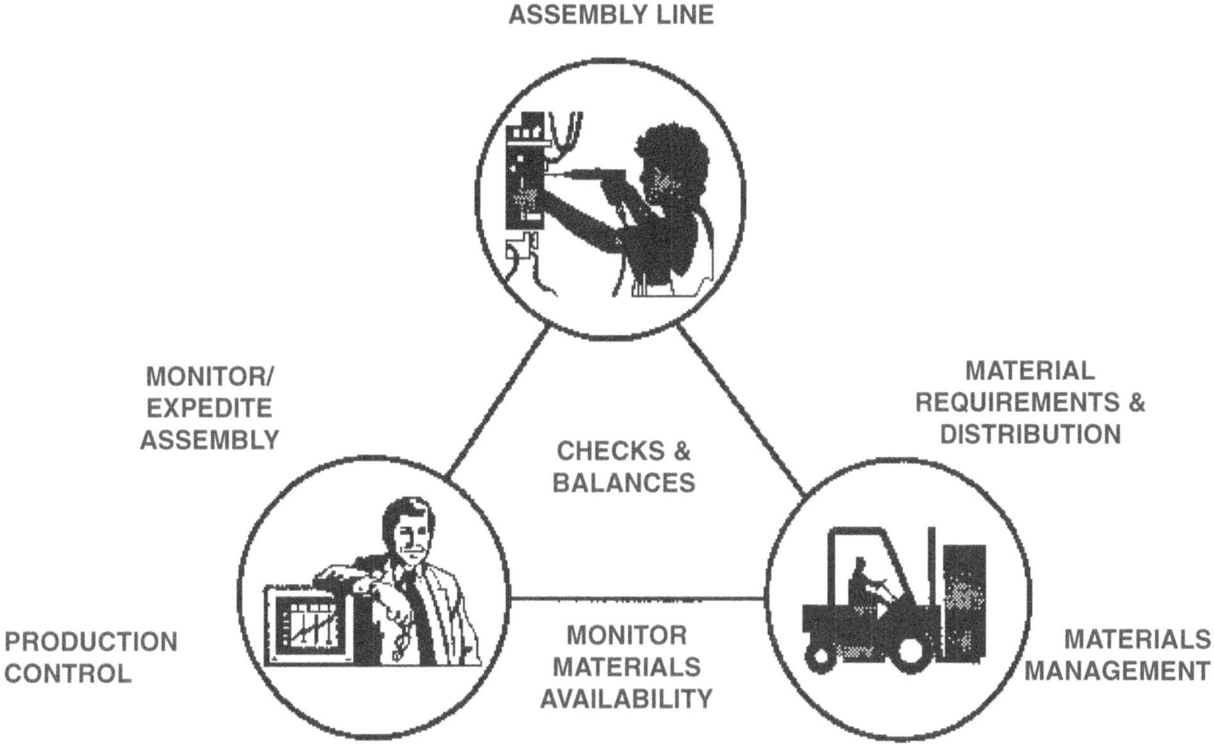

It is possible to apply the same discipline, organization and automation as used in an engineering/manufacturing facility.

Some people might argue that an "Information Factory" represents a much too rigorous environment that would stifle creativity. Far from it. An "Information Factory" is simply a recognition that the development of information resources can move from an art to a science. Ultimately, it represents organization, communications, and discipline. It also promotes teamwork and cooperation by sharing and re-using information resources.

With its emphasis on defined deliverables, validation/ acceptance criteria, and review points, "PRIDE" is compatible with a total quality assurance program, e.g., ISO-9000.

THE VOCABULARY OF "PRIDE"

The vendor has taken great care to keep the language of the methodologies clear, consistent and easy to understand. In fact, a comprehensive "Glossary of Terms" is included. Despite this, there are some common expressions used in "PRIDE" which the user will need to become familiar with. To illustrate:

RESOURCE refers to one of the many elements used to produce information. As developers proceed through the "PRIDE" methodologies, they will document (either manually or mechanically) each resource and link it to other resources, thereby creating a chain of relationships. Under "PRIDE," each resource is identified by number and name (consult the "Establishing a Repository" narrative under the "Supporting Narratives" section). Most resources, follow a "Part Number" numbering convention consisting of a two digit prefix denoting the type of resource, followed by an alphanumeric control number to identify the individual description. For example:

NUMBER	RESOURCE TYPE	USED TO DEFINE..
FE-XXXXX	Functional Entity	Enterprises, Business Functions
OE-XXXXX	Organizational Entity	Jobs/Positions
RE-XXXXX	Resource Entity	Employees, Consultants, Machines
SD-XXXXX	Skill Description	Knowledge, Experience, Talent, Feature (for equipment)
MI-XXXXX	Modification/Improve.	Business Objectives, Problems, Work Requests
PD-XXXXX	Project Description	Projects (any size or type)
IR-XXXXX	Information Rqmt.	Information related or non- info related (e.g., equipment specs)
MD-XXXXX	Module Description	Reusable program code (including subroutines)
DD-XXXXX	Data Description	Data Elements and Field Entries
FD-XXXXX	File Description	Physical Files (manual & computer), and Objects (Logical Files)
ID-XXXXX	Input Description	Screens, Documents, Verbal
OD-XXXXX	Output Description	Screens, Reports, Audio
RD-XXXXX	Record Description	Records, Views (Logical Records), Transactions, Maps, Panels, Messages, and Call Parameters.

These resource prefixes (DD, PD, RD, etc.) will become a natural part of the vernacular of the development organization.

Unlike the other resources, System resources follow a product structure numbering convention reflecting the four levels in a system/product hierarchy (a basic "PRIDE" concept):

XX	System number	LEVEL 1
XXXX	Sub-System number	LEVEL 2
XXXXXX	Procedure number (Administrative or Computer)	LEVEL 3
XXXXXXXX	Program or Operational Step number	LEVEL 4

NOTE: The XX-XX-XX-XX convention can be modified as required for installation standards; for example: XXX-X-XX-XX, XX-X-XXX-XX, etc.

The term **"IRM"** refers to the overall concept of Information Resource Management, but it is also used throughout "PRIDE" in reference to a **"repository"** where information resources are documented and inventoried (either manually or mechanically). Each resource is cataloged in the IRM by number and name in the manner mentioned above. For a more complete description of the organization of the IRM as a tool, consult the following section: Establishing a Repository.

METHODOLOGY:

Other commonly used expressions associated with the "PRIDE" methodologies include:

PHASE - consisting of a set of one or more activities of work. Phases represent significant milestones in the life of a project.

ACTIVITY - a unit of work consisting of one of more operations.

DELIVERABLE - a report, file, or some other object produced from a specific phase, activity or operation. Deliverables are sometimes referred to as "work products".

MOD/IMP - short for "Modification/Improvement"; refers to a business objective of any kind (small, medium, large; new development, maintenance activity, modification/improvements). The expression "Mod/Imp" is a recognition that most development activity is devoted to changing and enhancing existing information resources.

The layout and structure of the methodologies contained herein are uniform making it easy to learn and navigate. Once a person understands the structure of one "PRIDE" methodology, he/she will be able to easily navigate another.

END SECTION

Section 2

"PRIDE"-EEM ENTERPRISE ENGINEERING METHODOLOGY

Never embark on a journey without knowing your destination.
- Bryce's Law

BUSINESS PURPOSE

The "PRIDE"-Enterprise Engineering Methodology (EEM) is an important part of an overall philosophy of Information Resource Management (IRM) as defined by "PRIDE". This involves the development and control over all of the resources required to produce information. Planning is vital to the IRM philosophy. As such, EEM provides the organized framework to study a business and develop an Enterprise Information Strategy (EIS) that is synchronized with business objectives. Because of this, EEM precedes the other "PRIDE" methodologies which are oriented towards the development of systems and data resources:

The intent of any of the "PRIDE" methodologies, including EEM, is to define the business environment as to "Who" is to perform "What", "When", "Where" and "Why" (the "5W's"). As a result, it is used to convert a heterogeneous operating environment into a homogeneous environment. This improves communications and promotes cooperation and teamwork throughout an enterprise. Better organization and discipline also enhances the ability to build quality products and make effective use of resources. In addition, to the 5W's, the methodology provides "How" to perform the work by providing a variety of techniques and tools that are deployed throughout the methodology. A methodology, therefore, resembles an assembly line where work is performed in manageable stages.

EEM is a universally applicable approach that accommodates any type of company or institution, whether profit or non-profit, public or private. The methodology may be used to develop a total study of the enterprise or just a portion of it (e.g., Marketing, Administration, Manufacturing, etc.). The end result is an Enterprise Information Strategy (EIS) that specifies the priorities of the enterprise, along with a plan to implement them. As new objectives are introduced and others are implemented or deferred, the EIS is updated and maintained on a current basis.

CONCEPTS & PHILOSOPHIES

If you have observed your information systems over the years, you probably have noticed that they have been constantly changing, sometimes radically, sometimes subtly. For example, the payroll and billing systems you had thirty years ago do not resemble the systems you have today, nor will today's systems look the same thirty years from now. The day a system is installed is the day it begins to undergo change. Therefore, systems are built by evolution, not by revolution. No one has ever built the perfect system the first time, and no one ever will. This is because the company and its information requirements are constantly changing due to economics, competition, market demands, government regulation, politics, technology, etc.

If this is true, the more we know about our enterprise and the influences affecting it, the more effective we will be when developing information systems that will satisfy business needs. It is important, therefore, that before we build information systems, we must first understand the nature of the business they must serve. Developing systems that do not satisfy business needs is counterproductive.

Productivity = Effectiveness X Efficiency

One of the inherent problems associated with planning is that it is often regarded as unnecessary work. Many see it as an expensive luxury. Some people do not recognize that planning organizes and coordinates all work effort which, in turn, maximizes the productivity of development personnel and reduces costs. It assures that the relevant problems and opportunities of the business are addressed. One of the primary questions that Enterprise Engineering answers is, *"Are we doing the right things?"* (Effectiveness). This is an essential ingredient if we are to be productive. In contrast, Efficiency questions, *"Are we doing things right?"* (doing things faster and better). This may be an important issue but we must remember that there is nothing more unproductive than to build something efficiently that should not have been built at all. EEM, therefore, concentrates on the effectiveness issues of the business.

There have been numerous planning techniques developed over the years under a variety of names:

- "Corporate/Business Planning"
- "Strategic Systems Planning"
- "Business Systems Planning"
- "Information Engineering"
- "Strategic Information Planning"
- "Critical Success Factors"

Their objectives are all the same: to develop a plan for satisfying the information needs of an enterprise. Few, if any, have developed a common and cohesive approach addressing the IRM planning problem in its entirety. Most are simplistic techniques that address only a small portion of the overall picture. For example, some focus only on corporate objectives and priorities, others address only software, technology, and data base considerations. The reason for these varying viewpoints is the originators lacked an overall perspective of the IRM issue. The result:

- Corporate business plans and strategies are developed without consideration for information systems. As a result, systems are not synchronized with business plans.

- "Information Engineering" activities turn into esoteric data base projects.

- Systems are developed that do not support the information needs of users.

Perhaps the most serious failing is that planning is regarded as a one-time proposition; that it will not change over time. Developing a plan under the assumption that business conditions will not change is an exercise in futility and the plan will quickly become obsolete. Planning is an evolutionary process. As the needs of the enterprise change, so must the plans.

UNDERSTANDING THE ENTERPRISE

Enterprises take many forms, such as the conventional commercial venture, whether private or public, a government agency, a fraternal organization, even a simple household can be viewed as an enterprise. An enterprise is a defined business entity with a specific mission, whether profitable or non-profitable in intent. It exhibits the following characteristics:

A. An Enterprise has a mission or purpose -

For most companies, it is to make a profit through the sale of products and/or services. For non-profit organizations, such as government institutions, it is to regulate or administer laws and provide public services, e.g., police, transportation, fire fighting, etc.

B. An enterprise depends on other enterprises in order to survive -

Although an organization may be financially independent, it depends on the interaction with other enterprises, whether it be a parent company, customer, vendor, doctor, lawyer, competitor, employer, contractor, accountant, stockholders, government, etc. The identification of these external enterprises is an important consideration for Enterprise Engineering since these external entities represent significant forces influencing the enterprise. By addressing the information requirements of external enterprises, both strategic and tactical systems can be developed which provide a competitive advantage for a company.

C. An enterprise is a self-governing body -

Although it may be influenced by external enterprises, an enterprise usually operates autonomously, making its own decisions and taking its own actions. All enterprises have some form of internal government, regardless of whether it is a private or public institution. It may be governed by some form of president, managing director, or board of directors, whether elected or appointed. In addition, an enterprise is "owned" either privately or publicly, whether by a single owner or partnership, by shareholders, or by a membership. It is common for an enterprise to have a set of bylaws, articles of incorporation, or some other document outlining its legal "ownership" or government.

D. An enterprise is a self-contained unit -

It is able to produce or receive income, administer resources, and produce a product or service. All enterprises produce some form of product or service. Although a product is a tangible commodity that is easily understood, a service is highly intangible and may consist of such things as accounting, legal advice, banking, social work, government regulation, labor, etc. Like a product, a service is a billable or measurable commodity. The sale of products and services implies that an enterprise has clients it serves, whether they are a typical customer, the citizens within a government area, or the membership of a social organization. Some enterprises, such as insurance companies and banks, offer both services and products.

This leads to the idea that there is a simple and universal model for all enterprises:

UNIVERSAL ENTERPRISE MODEL

One area of the enterprise is concerned with producing income, which is typically the goal of a marketing function. Without financial income, all other work will quickly cease.

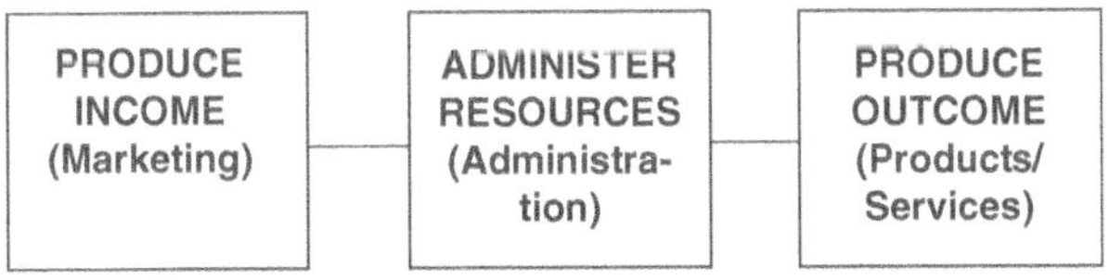

Another area of an enterprise administers resources, including human, financial, material, equipment and information resources. This includes the areas of accounting, finance, personnel, materials, and information resource management.

The last area of the company is concerned with producing a product and/or service. This represents the outcome of the enterprise and includes such things as manufacturing, consulting, training, or some other form of service, such as legal advice or accounting.

The boundaries of an enterprise can be typically defined by the set of "books" used for accounting purposes. These books, supported by a chart of accounts, describes how money is received and allocated through debits and credits. Ultimately, it defines the scope of the enterprise.

Perhaps the ultimate test for an enterprise is whether it can be spun off and made a separate operating unit. For example, a division within an automotive conglomerate could be considered an enterprise.

E. An enterprise has a structure, both logically and physically -

Each enterprise has a physical structure that reflects the organization in terms of human and machine resources. It also has a logical structure which reflects the organization of the underlying business functions. Functions are used to define the fundamental duties and responsibilities of the enterprise. Positions are used to define the physical positions or jobs implementing the functions. Resources, both human and machine, are used to implement positions. These resources have specific skills and proficiencies which qualify them for the work to be performed.

UNDERSTANDING FUNCTIONS

In order to fulfill the mission of the enterprise, certain functions must be performed. These functions define WHAT work must be performed and WHY. When organized into a hierarchy, functions represent the logical structure of the enterprise. A function, therefore, is defined as a logical grouping of one or more responsibilities for carrying out a specific portion of the mission of the enterprise.
Functions represent bodies of actions and decisions required to perform the duties and responsibilities of such things as marketing, accounting, manufactur-

ing, sales, customer service, shipping, receiving, inventory, finance, product development, etc. Since they are logical constructs, functions are relatively static. They will only change if the mission or nature of the enterprise changes, such as venturing into new business endeavors. Enterprises with common missions will have similar logical structures. For example, all life insurance companies are logically the same, all electric utilities are the same, automotive manufacturing, banking, etc. They will only differ if the products and/or services differ.

Functions exhibit the following characteristics:

A. Functions rely on other functions to form a whole-

An enterprise will consist of several functions which delineate specific areas of responsibility. Each function will rely on the other functions in order for the enterprise to operate effectively. For example, without an income producing function, such as sales, most companies will quickly go out of business regardless of how well the other functions are performing.

BASIC BUSINESS RESOURCES

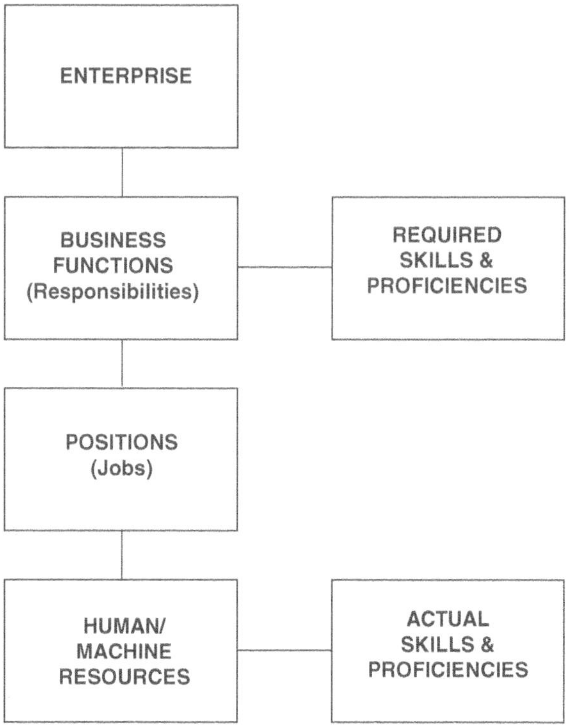

B. Functions denote the three basic levels of activity in an enterprise (policy, control and operations) -

Functions are organized into a hierarchy which denotes the types of actions and/or decisions involved. Policy functions refer to executive decisions where policy is made and objectives are formulated. Control functions relate to middle management actions and decisions to monitor day-to-day affairs and assure that executive decisions are met. Operational functions involve the routine activities or work of the

THE LOGICAL ENTERPRISE
Defines the business of the Enterprise

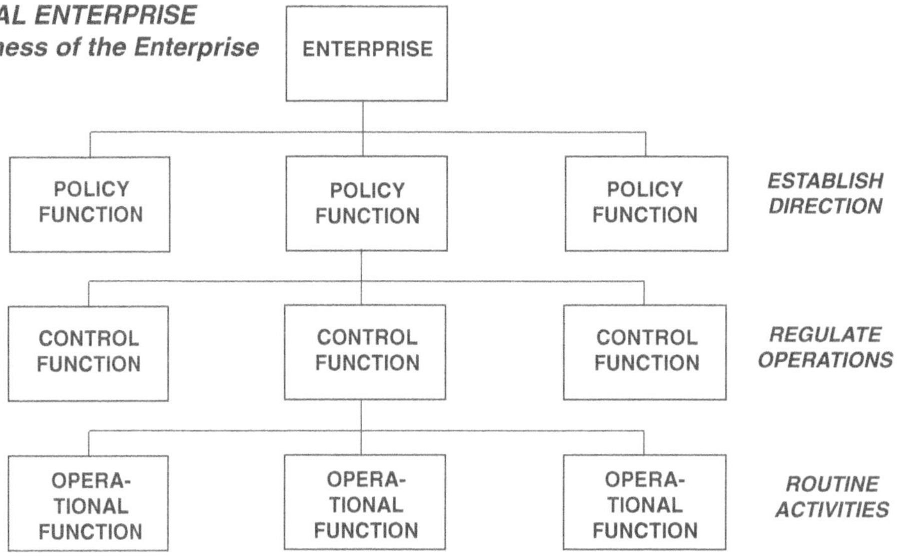

THE LOGICAL STRUCTURE REPRESENTS A STATIC VIEW OF THE ENTERPRISE; IT WILL ONLY CHANGE IF THE BUSINESS MISSION CHANGES

ENTERPRISES WITH COMMON BUSINESS MISSIONS WILL HAVE SIMILAR LOGICAL STRUCTURES

enterprise. These three levels can be charted as a hierarchy showing superior/subordinate/lateral functional relationships.

Information corresponds to the three levels. Policy information is used to establish the direction or policy for the enterprise. It includes such things as trend analysis, forecasts, profit and loss, etc. Control information is used to control operational activities and assure that policy decisions are implemented. Typical applications include production control, inventory control, accounts receivables, accounts payable, customer complaint analysis, error notification, progress/status reporting, etc. Operational information is used to perform the normal day-to-day operations of the enterprise, such as shipping, manufacturing, receiving, billing, payroll, processing customer orders/requests, etc.

The benefits from structuring the enterprise using logical functions are that it provides a guideline for establishing a chart of accounts for accounting purposes and can also provide the means to establish "profit centers" in the enterprise.

C. Functions deal with objects and require information -

Each function has at least one "object" it must deal with or manage. Objects represent facts and events required to operate and manage an enterprise. They may be as tangible as a product, employee or capital, or as intangible as a transaction, such as an order, debit or credit. In order to effectively manage these objects, functions require specific information about these objects in order to fulfill their mission. No single function has a monopoly on an object; they may be shared by many functions.

D. Functions communicate through information systems -

Information flows between functions and is the cement holding the enterprise together. Systems, with their inputs, outputs, files and processes represent the means by which functions interact with each other. They represent the vehicle by which duties and responsibilities are discharged.

SAMPLE ENTERPRISE MODEL

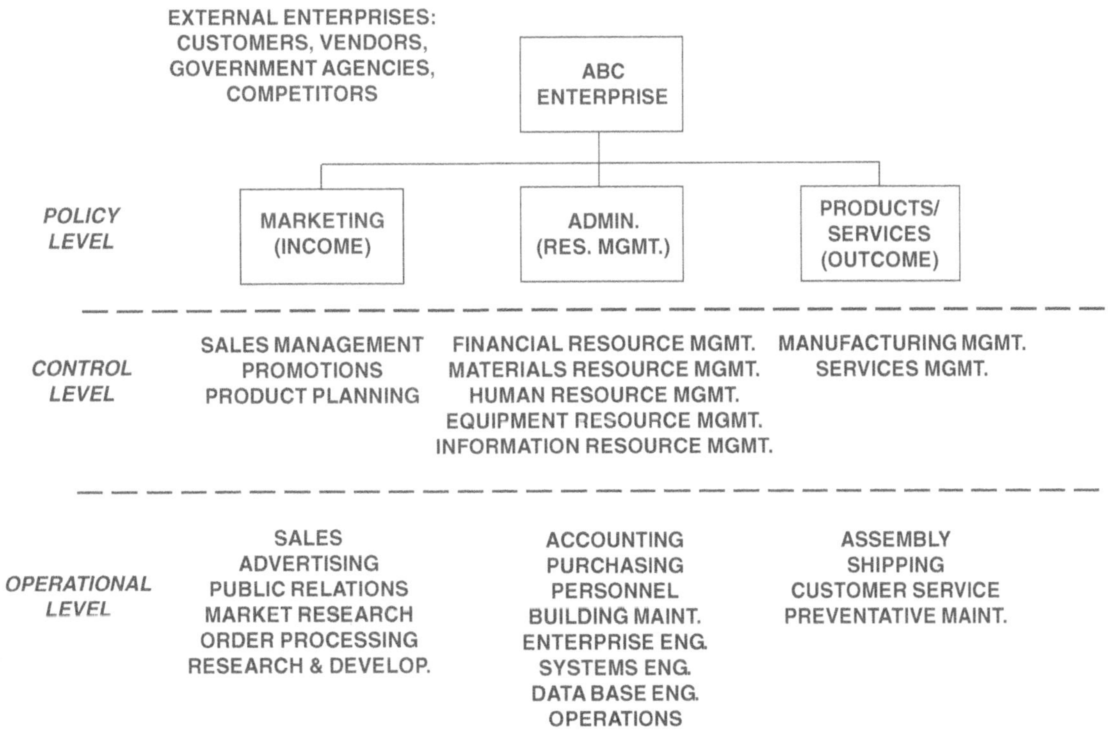

E. Functions require certain skills -

The talents required to perform one function may be totally different from those required to perform another. For example, the skills required to perform sales differs considerably with those required in manufacturing. This function/skills relationship becomes the basis for determining suitable positions to fulfill the functions and for recruiting resources with the appropriate skills. It also becomes the basis for determining if existing resources have the appropriate skills required to fulfill the functions.

F. Functions are implemented by positions -

Although enterprises with identical missions will have identical logical functions, they will differ considerably in how they are physically structured. Management will organize the enterprise physically in a manner they believe will be the most productive way to implement the logical functions. Whereas the logical functions represent a relatively static view of the enterprise, the physical positions are dynamic and will change more frequently.

Positions are organized into a hierarchy denoting administrative reporting relationships. Direct (internal positions) and dotted (external positions) are identified, as well as staff and line positions. This organization structure transcends management techniques for the use of resources, such as a matrix organization or project team.

THE PHYSICAL ENTERPRISE

Defines administrative relationships

There is not necessarily a one-to-one relationship between logical functions and physical positions; one function may be implemented by several positions, or several functions can be implemented by a single position. It is up to the discretion of management to determine the positions required to implement the logical functions. Each function must have at least one job to implement it, otherwise it will go unfulfilled.

Although the logical functions are organized into a simple three-level hierarchy, the physical positions can be organized into a complicated, multi-level structure.

One of the by-products of Enterprise Engineering is the identification of functions not being served by a position or functions with a position, but without resources. This means that a function is not being fulfilled and should be highlighted. It is also concerned with identifying excessive overlaps between positions and functions, both horizontally and vertically. This provides management with the ability to flatten the organization and develop an environment more aptly tuned to the needs of the enterprise.

Like functions, positions require information to discharge duties and responsibilities.

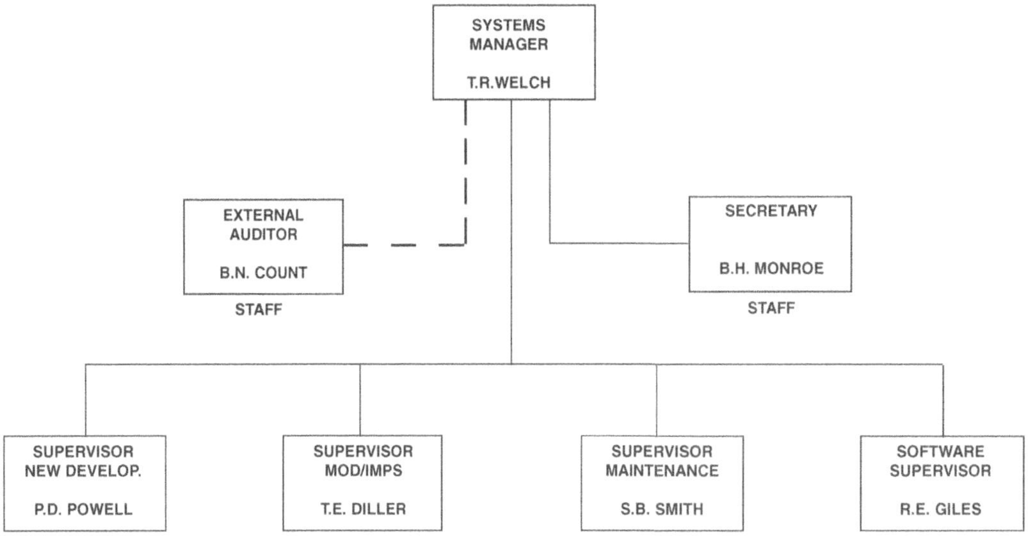

WHEREAS THE LOGICAL IS SOMEWHAT STATIC, THE PHYSICAL WILL CHANGE DYNAMICALLY

THE ORGANIZATION & OPERATION OF AN ENTERPRISE

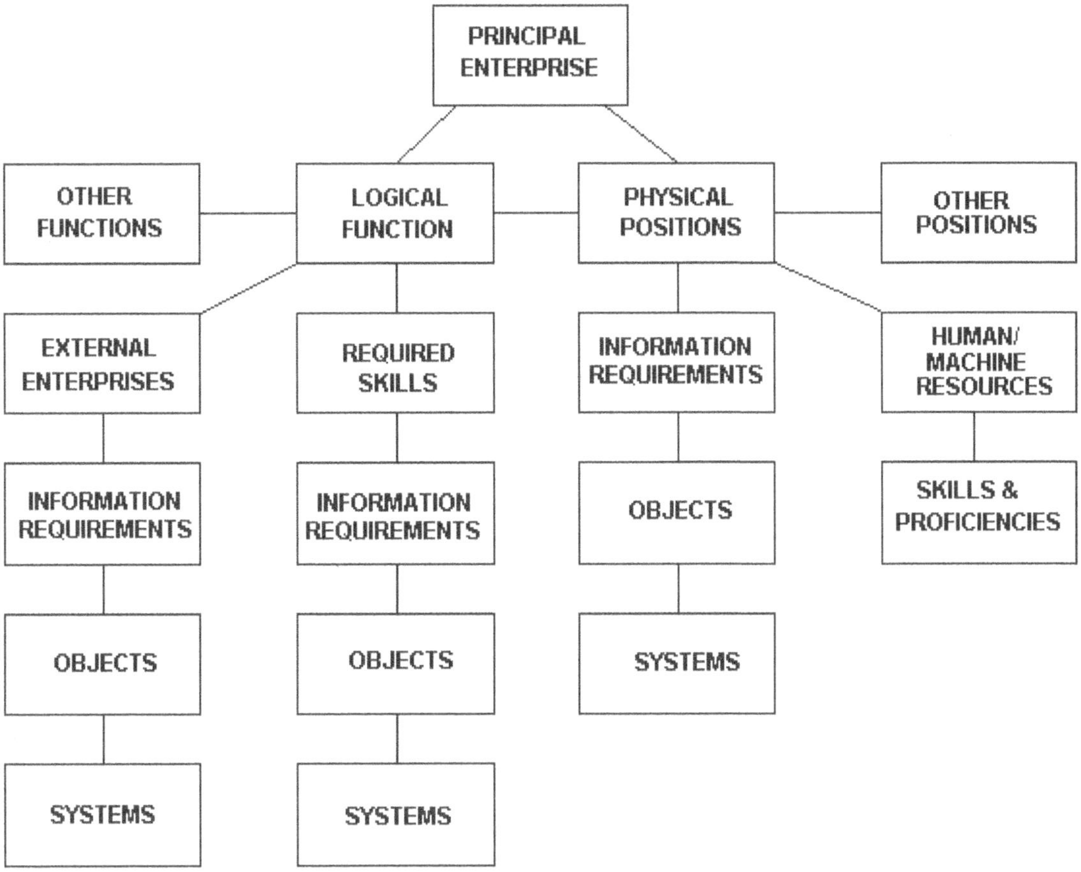

When we have defined and cross-referenced the resources, we now know how the enterprise operates, both logically and physically.

From this representation we can see the inner workings of the enterprise, including how it is structured, both logically and physically. We see the external enterprises influencing the enterprise, including their information requirements, objects, and systems required to support them. For the logical functions, we see the skills required to perform the function, its relationship to other functions, the information requirements and objects required to perform the function and the systems affecting the function. The physical positions show the use of human and machine resources with their pertinent skills and proficiencies, along with their relationships to other positions, and the information requirements, objects and systems associated with the position.

There are, of course, simpler ways of expressing these complicated relationships, such as through hierarchy charts and a series of matrices. In any event, by defining and cross-referencing these resources, we essentially know:

- The purpose and structure of the enterprise.
- The external enterprises influencing the enterprise.
- The objects required to manage and operate the enterprise.
- The information required to run the business.
- The systems used in running the business.

CALCULATING THE ENTERPRISE INFORMATION STRATEGY (EIS)

With this intelligence, we are now in a position to formulate an effective Enterprise Information Strategy (EIS), a plan to develop or modify information systems synchronized with business plans. There are basically eight steps required to develop the EIS:

A. Define the business plan, including both strategic and tactical objectives.

The strategies used to run an enterprise can be highly proprietary and sensitive, particularly in business. In some companies, business plans may be known only by a few key executives. A prospectus, precis or annual report typically will offer only a superficial description of corporate direction, usually for public dissemination. They offer more about what the company has done in the past, not necessarily what it plans to do in the future.

In order to formulate an effective information strategy, one that is synchronized with the direction of the enterprise, the business plan must be defined, even if it is to be maintained on a highly confidential basis. An information strategy that is not based on a business plan is unproductive. Development projects will tend to address the wrong problems/opportunities and will waste considerable resources, most notably time and money.

When defining the business plan, both tactical and strategic plans must be defined. Tactical plans are concerned with maintaining order within an enterprise, e.g., improve productivity, reduce costs, etc. Strategic plans are aimed at maintaining a market advantage (staying competitive). For example, entering a new market, introducing a new product or service, diversifications, mergers, acquisitions, etc.

B. Define the structure of the enterprise affected by the plan, both logically and physically.

As discussed, this includes a definition of the external enterprises influencing the enterprise, along with the internal "objects" used to manage and operate the business.

C. Define the information requirements required to run the enterprise.

D. Group information requirements based on commonality into objectives.

This grouping could be random, but this might result in objectives with an inordinate scope and complexity. If properly organized, the objectives will be more compact and easier to evaluate.

There is not necessarily a one-to-one relationship between information requirements and objectives. One objective may satisfy many information requirements, and one requirement may be satisfied by many objectives.

There are basically three considerations for grouping requirements:

1. By the functional area being served by the information (e.g., marketing, manufacturing, administration).

2. By information type (policy, control, and operational). Operational requirements are typically satisfied before control and policy requirements because operations represent the basic activities of the business and ultimately the core systems that collect data.

3. By requirements being implemented by existing systems, versus those that are not.

E. Perform a cost/benefit analysis on the information groupings. This will determine how the enterprise values the information.

F. List the objectives by priority ranking.

Based on the delivery date and cost/benefit analysis, each objective is assigned a priority "weight" value. The weight defines how much the enterprise values the objective and is based on a scale from 01 (high) to 99 (low). The arbitrary assignment of an objective "weight," without any specific rationale, will lead to inconsistent priority ratings. Standardization of priority weighting, based on precise and rational criteria will lead to more consistent results.

Based on their "weight" and due date, objectives are then ranked in priority sequence, using a scale from 0001 (high) to 9999 (low). This ranking becomes a part of the Enterprise Information Strategy.

G. Group objectives into projects based on commonality.

This grouping could also be random, but this may result in projects with an inordinate scope and complexity. If properly organized, the projects will be more compact and easier to evaluate and implement.

There is not necessarily a one-to-one relationship between objectives and projects. One project may

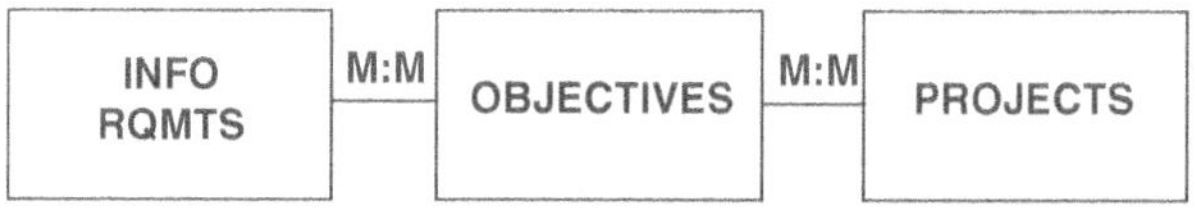

implement many objectives and one objective may be implemented by many projects. The resource relationships established so far include:

MANY-TO-MANY

There are fundamentally three considerations for grouping objectives:

1. Compatible work effort. For example, combine objectives dealing with modification/improvements to existing systems, maintenance (correction of defects), and new development. In all instances, the systems affected will be considered.

2. Compatible delivery due dates.

3. Common areas of the company affected by the objectives.

The underlying rationale behind the development of the Enterprise Information Strategy is how the enterprise values information. When the requirements were grouped into objectives, a cost/benefit analysis and delivery date was developed. This became the basis for the resulting projects and will ultimately become the criteria for ranking objectives and projects.

H. List the projects by priority ranking which is derived from the objective ranking.

The objective rankings are then used to determine the ranking of all projects. Here, the average ranking of all of the objectives that a project implements is used to establish the project ranking. For example:

In this small example, you see that the objective rankings have a direct effect on how the projects are ranked. What this means is that as information requirements and objectives change, the projects, in turn, will change. This concept promotes the fact that the Enterprise Information Strategy is a "living" document and is constantly undergoing change.

Both the proposed objective and project rankings are compared to existing priorities and the Business Plan. Adjustments to priority "weighting" and rankings are implemented accordingly.

A textual justification for the rankings is prepared for the EIS. The text explains the rationale for the objective and project rankings, what effect it will have on the existing Enterprise Information Strategy (changes, additions, deletions), and how it will accommodate the business plan.

OBJECTIVE NUMBER	RANKING	AVG OBJECTV. RANKING	PROJECT NUMBER	PROJECT RANK
00001	1	14	00320	3
00010	25			
00020	16			
00040	12	28	01216	4
00345	44			
00029	2	6	01011	1
00042	5			
00021	11			
00132	7	7	01211	2

"SAMPLE EIS OBJECTIVE RANKING"

RANK	OBJ. NUMB	NAME	WGHT	DATE REQSTED	DUE DATE	PROJ NUMB
0001	901	CUSTOMER ORDER SYSTEM	01	01/01/93	ASAP	PEND
0002	844	MATERIALS RESOURCE PLANNING	26	01/01/93	ASAP	132
0003	845	INVENTORY SYSTEM UPGRADE	25	01/01/93	04/01/88	132
0004	846	PARTS PICKING	25	01/01/93	04/01/88	132
0005	899	AUDIT ENGINEERING SYSTEM	55	12/15/93	02/10/88	133
0006	844	IMPLEMENT Q.A. PROGRAM	55	06/01/92	06/01/88	101
0007	007	PREP POLICIES & PROC MANUAL	75	01/01/93	ASAP	101
0008	014	SALES SYSTEM EVALUATION	75	01/01/93	ASAP	199
0009	222	CONSTRUCT WAREHOUSE	75	01/01/93	ASAP	557
0010	080	R.F.Q. ANALYSIS	75	01/01/93	03/01/88	162
0011	056	IMPLEMENT IRS CHANGES	75	01/01/93	04/15/88	200
0012	067	MODIFY SECURITY SYSTEM	75	12/15/92	OPEN	137

"SAMPLE EIS PROJECT RANKING"

RANK	PROJ NUMB	NAME	PROJECT STANDING	DATE REQUIRED	OBJ NUMB
0001	132	MATERIAL RESOURCE PLANNING	ACTIVE	ASAP	844 845 846
0002	133	ENGINEERING SYSTEM AUDIT	ACTIVE	02/10/93	899
0003	199	SALES SYSTEM	ACTIVE A	SAP	014
0004	101	REVISE QA/STANDARDS	ACTIVE	09/30/93	844 007
0005	557	WAREHOUSE PROJECT	ACTIVE	ASAP	222
0006	162	R.F.Q. ANALYSIS	ACTIVE	03/01/93	080
0007	200	IRS SYSTEM CHANGES	ACTIVE	04/15/93	056
0008	137	MODIFY SECURITY SYSTEM	ACTIVE	OPEN	067

The Enterprise Information Strategy represents a policy decision for management. It is ultimately charting corporate direction. The strategy specifies what objectives and projects are required to satisfy business plans.

If the EIS is acceptable as proposed, Information Resource Management is then instructed to implement the strategy accordingly. This includes initiation of Systems Engineering and Data Base Engineering related projects. It also includes the initiation of additional Enterprise Engineering related projects to maintain the evolution of the EIS.

GRAPHIC SUPPORT

There are fundamentally two types of graphics associated with Enterprise Engineering: hierarchy charts and matrices. The hierarchy charts are used to model the enterprise logically and physically. Matrices are used to describe influential factors affecting an enterprise, such as external enterprises and objects, and to develop the Enterprise Information Strategy.

HIERARCHY CHARTS

The purpose of any hierarchy chart is to show the superior, subordinate and lateral relationships between resources. In this way, it resembles a "tree" structure. With a hierarchy, a resource can have no more than one "parent" resource, but many "children." This is unlike a network diagram that allows multiple "parents."

SAMPLE HIERARCHY

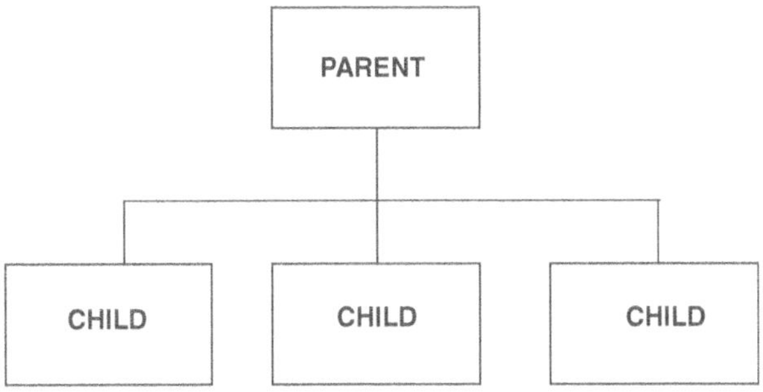

In Enterprise Engineering, two hierarchy charts are used: a function chart to model the logical enterprise, and; an organization chart to model the physical enterprise.

A function chart is a simple hierarchy using rectangular boxes to represent logical functions. The hierarchy shows the relationships between functions, for example:

SAMPLE FUNCTION CHART

The organization chart is similar in structure to the function chart. However, unlike the function chart, which shows some simple relationships, the organization chart may be more complicated due to direct/dotted and staff/line relationships. As previously shown, each rectangular box identifies a specific position along with the people who serve that position.

The physical structure of the enterprise is based on how management wishes to organize and operate the enterprise. It represents administrative reporting with levels of authority. Ultimately, it depicts the chain of command for an enterprise. Regardless of whether the enterprise operates with a "matrix" style of management or on a "project team" basis, there is always some form of hierarchical structure showing who works for whom. This is the intent of the organization chart.

MATRICES

The purpose of a matrix is to express elaborate resource relationships in a simple, easy-to-analyze format. It is drawn as a table with horizontal rows and vertical columns. For example:

COLUMNS

ROWS

	A	B	C	D	E	F	G	H	I	J	K	L	M	N	O	P	Q	R	S	T
1										X										
2		X																		
3		X																X		
4							X													
5													X							
6											X									
7															X					
8								X												
9				X																
10																	X			
11												X								

In the above example, relationships between columns and rows are expressed with an "X" in the coordinate cell.

METHODOLOGY CONSTRUCTION/NAVIGATION

The Enterprise Engineering Methodology (EEM) consists of an assembly of five phases detailing "Who" is to perform "What" work, "When", "Where" and "Why". Each phase consists of a defined set of activities (a total of 28); each activity consists of a series of operations or tasks to be performed. All phases, activities and tasks produce tangible deliverables that can be reviewed and checked for completeness. These deliverables substantiate adherence to the methodology and permits the measurement of progress. Both formal and informal review points are laced throughout the methodology which provides for an effective dialog between management and the enterprise engineers.

The first phase is essentially used to plan the EEM project; Phase 2 develops the logical model of the enterprise; Phase 3 develops the physical model; Phase 4 develops the Enterprise Information Strategy; and Phase 5 reviews the project.

PHASE 1 - "EEM Project Planning"

An EEM project may be initiated by the Information Resource Manager or as a result of another EEM project. During this phase the enterprise is defined in terms of its mission/charter, the external enterprises affecting it, and its business plan. The project may progress sequentially (Phases 2, 3, 4, 5) or may branch into parallel paths. It ultimately depends on the project scope and the level of management control required. Branching will typically occur in a major EEM project where the entire enterprise is being studied. In this event, branching is usually based on the three areas of the enterprise (marketing, administration, and products/services). Each area will branch into separate and parallel Phases 2 and 3.

The formal deliverable resulting from this phase is the "EEM Project Plan" which is reviewed with management for accuracy.

PHASE 2 - "Logical Enterprise Analysis"

Phase 2 is used to define the logical structure of the enterprise and how it operates. This includes a definition of the external enterprises influencing the functions, the "objects" used to manage and operate the enterprise and the information required to run the enterprise.

The formal deliverable resulting from this phase is the "Logical Enterprise Study" which is reviewed with management for accuracy.

PHASE 3 - "Physical Enterprise Analysis"

This phase is used to define the physical structure of the enterprise. It is similar to Phase 2 except it is used to also perform an organization analysis. During this activity, functions not being fulfilled are identified and overlaps in management are highlighted (both horizontally and vertically). This provides management with the ability to make more effective use of organization resources.

The formal deliverable resulting from this phase is the "Physical Enterprise Study" which is reviewed with management for accuracy.

PHASE 4 - "Develop Enterprise Information Strategy"

During this phase the Enterprise Information Strategy is developed and reviewed with management. The strategy specifies the objectives and projects the enterprise will follow.

The formal deliverable resulting from this phase is the "Enterprise Information Strategy" which is issued under executive management directive.

PHASE 5 - "EEM Evaluation"

This phase is used to initiate the Enterprise Information Strategy. Here, Information Systems Engineering and Data Base Engineering projects are initiated. Additional EEM projects are also initiated at this point. This means EEM is considered a "closed loop" methodology, and the Enterprise Information Strategy is constantly evolving.

Phase 5 is also used to audit the EEM project (proposed versus actual time and costs).

The formal deliverable resulting from this phase is the "EEM Project Evaluation Report."

WHO SHOULD PERFORM EEM?

As a high-level study of the business, Enterprise Engineering involves an analysis of organizational issues, policies and corporate directions. To effectively perform this type of work requires someone who is more in tune with the business as opposed to the technical detail of computers and software. Perhaps the best way to think of an "Enterprise Engineer" is more as a "business analyst" as opposed to a System/Software Analyst or someone from the Data Base organization. Their perspectives are totally different and are not typically adept for the type of work required to perform EEM. Ideally, a separate Enterprise Engineering group under a true IRM organization would be the most suitable arrangement. Unfortunately, many companies have yet to establish such an organization. Consequently, it is not uncommon to find managers from the various user departments (non-computer related) performing the function in conjunction with MIS management.

"Enterprise Engineers" require basic business skills. Fundamentally, they must be proficient in interpersonal relations/communications, possess good analytical/organizational skills, and knowledgeable about how the enterprise works.

OTHER FUNCTIONS PARTICIPATING IN EEM

Executives and end-users are interviewed and consulted during the course of an EEM project. As a result, their participation is critical. As in any other work effort, Project Managers are required to plan, estimate, schedule, and control an EEM project. Further, Quality Assurance personnel review and verify deliverables resulting from the various phases of the methodology. System Engineers and Data Engineers may also be consulted during "Current Systems Analysis" in Phase 1.

Where and when these other functions are involved during EEM is defined by the methodology.

BENEFIT$

The benefits of Enterprise Engineering are numerous:

- It assures that the company is "doing the right things," building information systems that support business objectives.
- The Enterprise Information Strategy is always synchronized with changing business conditions.
- Greater awareness and sensitivity to the external factors influencing the enterprise, along with the elements required to operate and manage it.
- Greater awareness of the mission and structure of the enterprise. This helps create a more productive and cooperative environment.
- Human resources have an improved understanding of the role they play in the enterprise.
- A more manageable organization structure.
- Improved communications and cooperation within the enterprise.
- Improved control over project planning and execution.
- An intelligent, efficient and confident approach to decision making; maximizing time and effort in planning.
- Greater awareness of the objectives and project priorities of the enterprise. This creates a spirit of teamwork.

END SECTION

Section 3

"PRIDE"-EEM PHASE 1 EEM PROJECT PLANNING

Organizations progress when the impact of good actions and decisions outweighs the impact of poor actions and decisions.
- Bryce's Law

The purpose of this phase is to initiate an EEM project. An Enterprise Engineering project may be used to study the entire enterprise or a portion of it (e.g., Manufacturing, Marketing, Administration); this is specified in the scope of the project. Several events occur during this phase:

- The scope of the EEM project is defined.
- A profile of the enterprise is defined.
- The current operating environment is defined.
- The enterprise's business plan is defined.
- A project plan is developed for the remainder of the project, including an Order-of-Magnitude Estimate and Schedule.
- A formal review with management is conducted to verify the phase findings.

Phase 1 requires considerable participation by Executive and User Management. Several interviews are required in order to properly define the mission of the enterprise, the current operating environment, and the business plan.

METHODOLOGY NAVIGATION

EEM projects are initiated based on a directive from Information Resource Management. At this time, Enterprise Resource Management assigns project personnel, including a Project Manager, and the project progresses according to the phases and activities of the methodology.

An EEM project may also begin as a result of a previous EEM project. Phase 5 of the methodology is used to initiate the Enterprise Information Strategy which includes a set of prioritized objectives and projects. At this time, Information Resource Management will initiate pertinent ISEM, DBEM and additional EEM projects. In this way, EEM is a "closed loop" methodology. It recognizes that the enterprise and its information requirements are constantly evolving. Change is constant.

Phase 1 will be used to evaluate the magnitude of the project, determine how much work is involved, and make decisions on how to proceed. Based on a formal review as conducted in Activity F, management may elect to:

1. Approve the project for proceeding to the next phase, "Logical Enterprise Analysis."

2. Request revisions to the findings produced in the Phase.

3. Discontinue the project.

If the intent of an EEM project is to study the entire enterprise, it is recommended that the Project be divided into multiple Phase 2's (one for each major area of the enterprise; (e.g., one Phase 2 to study Marketing, another to study Administration, and another to study Production). This will make the project more manageable. This is a decision which Project Management must make during Activity E.

GENERAL DISCUSSION

DEFINING THE ENTERPRISE

As a review of the key EEM concepts pertaining to this phase, the Enterprise Engineer must remember the fundamental attributes of an enterprise: An Enterprise...

- has a mission or purpose.
- is a self-contained unit.
- depends on other enterprises in order to survive.
- is a self-governing body.
- is a grouping of logical functions.

For a complete description of these key concepts, please refer to the narrative as provided in the methodology introduction section.

CURRENT SYSTEMS ANALYSIS

One of the principal deliverables from this phase is a definition of the enterprise. This includes a description of the products and/or services that the enterprise offers, the market it serves, how it is governed, external influences, etc. To effectively define this, an analysis must be performed of the current operating environment. This includes a high level review of the information systems of the enterprise. This review establishes the enterprise's "systems portfolio" thus providing executives with the means to grasp the magnitude of their systems.

Enterprise Engineering will also become familiar with the physical organization of the enterprise and how it operates. If desired, existing systems and data bases may be more fully defined in subsequent ISEM and DBEM projects.

BUSINESS PLANNING

The strategies used to run an enterprise can be highly proprietary and sensitive, particularly in business. In some companies, business plans may be known only by a few key executives. A prospectus, precis or annual report typically will offer only a superficial description of corporate direction, usually for public dissemination. They offer more about what the company has done in the past, not necessarily what it plans to do in the future.

In order to formulate an effective information strategy, one that is synchronized with the direction of the enterprise, the business plan must be defined, even if it is to be maintained on a highly confidential basis. An information strategy that is not based on a business plan is an exercise in futility. Development projects will tend to address the wrong problems/opportunities and will waste considerable resources, most notably time and money.

In defining the business plan, both tactical and strategic plans must be defined. Tactical plans are concerned with maintaining order within an enterprise, e.g., improve productivity, reduce costs, etc. Strategical plans are aimed at maintaining a market advantage (staying competitive). For example, entering a new market, introducing a new product or service, diversifications, mergers, acquisitions, etc.

ORDER-OF-MAGNITUDE

Following the early analytical activities, a project plan is developed which defines the remaining phases of the project, along with assigned personnel. From this, an Order-of-Magnitude (OOM) estimate and schedule can be prepared which defines the time and costs required to complete the project. Management will consider the OOM estimate and schedule when determining whether the project should be continued, revised or discontinued.

DESCRIPTION OF PHASE ACTIVITIES

Activity A - Develop Project Scope

Information Resource Management submits an Objective to Enterprise Resource Management who initiates an EEM project. This includes the assignment of appropriate Project Management and Enterprise Engineering personnel. Project Management prepares a Project Scope which defines the project's objectives and the area(s) of the enterprise to be studied. Based on the scope, Project Management and Enterprise Engineering prepare a Detail estimate and schedule for the phase. This is reviewed with Enterprise Resource Management for approval.

Activity B - Enterprise Definition

Enterprise Engineering analyzes and/or revises the definition of the enterprise. This is based, in part, on interviews with key Executive Management personnel. The enterprise is defined in terms of its mission (purpose), government, products and/or services offered, and the external enterprises that influence it.

Activity C - Current Systems Analysis

A system study is performed to evaluate the information systems that currently support the enterprise. The intent is to inventory the existing systems and how they are used in the enterprise. The level of detail required is at the system and sub-system levels only; existing procedures and programs may be defined in a subsequent ISEM project. This type of analysis is used to establish a "System Portfolio" which lists the systems and sub-systems used by the enterprise.

Existing logical "objects" (files) are defined and related to the systems. The result, a series of matrices which shows how existing systems and objects are used in the enterprise. From this, Enterprise Engineering becomes intimate with the strengths and weaknesses of existing systems.

Although this activity is primarily performed by the Enterprise Engineer, Systems Engineers and Data Engineers may be consulted to provide assistance in locating/defining systems and "objects."

Activity D - Business Planning

Enterprise Engineering interviews Executive Management to define and analyze business plans. Both tactical and strategic plans are evaluated. Tactical plans are principally concerned with improving the organization and control of enterprise operations. Strategic plans are concerned with maintaining a market advantage (competitive edge). A formal statement of the current business plan is prepared as a result of this activity. The plan includes a description of the information systems, both current and proposed, that are intended to satisfy the plan.

Activity E - Project Planning

Based on the analysis performed so far, Project Management prepares a project plan that specifies the phases required to complete the project. From Phase 1, the project may branch into parallel Phase 2's, or may simply progress to a single Phase 2. Based on the plan, an Order-of-Magnitude Estimate and Schedule is prepared for the remainder of the project.

Project Management and Enterprise Engineering then assemble the deliverables produced in the phase and packages them into a single phase review document which is evaluated by Quality Assurance prior to conducting the formal phase review with Executive Management.

Activity F - Phase 1 Review

Project Management conducts a formal review of the Phase 1 deliverables with Executive Management, Information Resource Management, Enterprise Resource Management, and possibly User Management. The review is used to evaluate the work performed thus far and revise as required. At this time, management will review the formal Phase 1 "EEM Project Plan" consisting of:

- Phase Cover Page
- Project Scope
- Enterprise Profile Report
- Business Plan
- Current System Analysis
- System Portfolio
- Position/System Matrix
- Organization Chart
- Project Plan
- Project Estimate/Schedule Recap
- Phase Review Checklist

Based on this report and subsequent review meeting, management may elect to revise parts of the report, discontinue the project, or approve it for continuation to Phase 2 of EEM.

SAMPLE PHASE 1 DELIVERABLES

OD-90233	ENTERPRISE PROFILE	"PRIDE"-EEM
AUG 4, 2004	RELEASE 1.2.0	PAGE 1

****
* FE-00100 TRUNDLE MANUFACTURING CORP. *
****

BUSINESS PURPOSE/CHARTER

The Trundle Manufacturing Corporation (TMC) was founded in 1971 as a manufacturer and retailer of trundle related products. A trundle is a castor, wheel or roller used to support and move furniture, trucks, and portable machines.

PRODUCTS AND/OR SERVICES

Currently, TMC markets products on a domestic basis only. Stock items include two product lines consisting of various models of trundles and trundle brackets. When requested by a client, TMC will engineer and manufacture trundles specifically tailored to the customers requirements. TMC warrants all products to be free of defects for ten years from the date of purchase. To this end, TMC provides a maintenance service for customers. The company is a relative newcomer to the fiercely competitive trundle market, having been founded by the C.E.O. and Owner, Rupert J. Jones, in 1971. They have maintained their niche by offering quality products and service at a competitive price.

FORM OF MANAGEMENT/GOVERNMENT

TMC is a privately held company. Its management structure centers around the Chief Executive Officer (CEO), Chief Operating Officer (COO), Chief Financial Officer (CFO), and Chief Information Officer (CIO), all of whom are responsible for determining corporate policies and strategies. The COO is supported by the Vice Presidents of Marketing, Manufacturing, and Administration, all of which participate in the development of policy decisions affecting the operations of the company. The CFO is supported by the Treasurer and Controller. The executives are supported by subordinate managers and supervisors who are responsible for implementing management decisions. Consult the TMC Function and Organization Charts for additional detail. Currently, there are over 200 employees in the company.

Executive and middle management consists of exempt employees who are paid monthly salaries. Workers are non-exempt employees that are paid an hourly wage. Sales reps are paid on a commission basis. There are no union contracts at TMC. All employees are eligible to participate in the company's pension and profit sharing plans. Also, the company provides compensation in the form of dental and health insurance. Consult the TMC Personnel Manual for additional details.

With the exception of some regional sales offices, all corporate operations, including manufacturing, are performed centrally at the company's headquarters.

OTHER CORPORATE CONSIDERATIONS

* The work week is Monday-Friday (Sat-Sun off).
* There are eight (8) hours in the business day.
* Standard holidays are observed by the company.

For additional details concerning corporate policy, consult the TMC Policy Manual.

* * * END OF OUTPUT * * *

SAMPLE PHASE 1 DELIVERABLES

OD-90251	BUSINESS PLAN	"PRIDE"-EEM
AUG 11, 2004	RELEASE 1.2.0	PAGE 1

AS DEFINED ON PD-00447 CORPORATE DEVELOPMENT PROJECT

* * WARNING: CONFIDENTIAL! * *

This document contains the business objectives of TMC. It is not to be divulged to parties outside of the corporation.

OVERVIEW

As the trundle market expands, TMC continues to grow at a rate of 20% per year. This growth has provided TMC with a surplus of cash which management wishes to invest.

STRATEGIC PLANS

I. To promote TMC's domination of the trundle market, management has defined three strategic objectives:

A. To acquire the ABC Company, the number 3 trundle company. This would clearly show TMC's dominance of the market. It is anticipated that this would be a "friendly" takeover, but would involve administrative adjustments to the TMC corporate structure.

B. To enter foreign markets, particularly Asia, where there is an absense of strong competition.

C. To move into the office furniture business, specifically tailored for executives. This would lead into additional peripheral office equipment.

TACTICAL PLANS

I. Product Specific Considerations - To have TMC products clearly recognized as the world's best trundle products, TMC must ...

A. Reduce production costs in order to price products competitively.

B. Improve speed of delivery of products to customers, particularly custom made products.

C. Improve quality of products. This will lead to greater customer satisfaction and a larger share of the trundle market. To do this, management wishes to invest in manufacturing automation, "just in time" (JIT) material management techniques, and new approaches to shipping.

AS OF: August 11, 2004

* * * END OF OUTPUT * * *

SAMPLE PHASE 1 DELIVERABLES

OD-90343	SYSTEM PORTFOLIO	"PRIDE"-EEM
AUG 11, 2004	RELEASE 1.2.0	PAGE 1

SYSTEMS & PROCESSES SUBORDINATE TO FE-00100 TRUNDLE MANUFACTURING CORP.

SYS/S-S	NAME	FREQ	OFFS	RESP	EFFECTIVE
AP	ACCOUNTS PAYABLE SYSTEM				MAY 01, 1988
AR	ACCOUNTS RECEIVABLE SYSTEM				MAY 01, 1988
BL	BILLING SYSTEM				JUN 01, 1992
CS	CUSTOMER SERVICE SYSTEM				JUN 01, 1976
GL	GENERAL LEDGER SYSTEM				MAY 30, 1977
HR	HUMAN RESOURCE SYSTEM				MAY 03, 1982
IC	INVENTORY CONTROL SYSTEM				DEC 31, 1974
MN	MANUFACTURING SYSTEM				MAY 15, 1972
OP	TRUNDLE ORDER PROC. SYS (TOPS)				NOV 20, 1988
OP-00	CUSTOMER ORDER INQUIRY	R		15S	NOV 30, 1988
OP-04	WEEKLY ORDER STATUS	1W	1D	1H	NOV 30, 1988
OP-05	DAILY ORDER PRODUCTION	1D	7H	1H	NOV 30, 1988
OP-06	CREDIT CHANGE REQUEST	R		15S	NOV 30, 1988
OP-07	CUSTOMER RECORD MANAGEMENT	R		1H	NOV 30, 1988
OP-08	ORDER-PROC FILE MAINTENANCE	R		15S	NOV 30, 1988
PA	PAYROLL SYSTEM				JAN 12, 1975
SH	SHIPPING SYSTEM				DEC 31, 1973
TS	TRUNDLE SALES SYSTEM				JUN 01, 1992
TS-00	REQUEST SALES INFO	R		10S	AUG 06, 1992
TS-02	WEEKLY SALES REPORTS	1W	5D	2H	AUG 06, 1992
TS-03	MONTHLY SALES REPORTS	1M	30D	4H	AUG 06, 1992
TS-05	MAINTAIN SALES FILES	R		10S	AUG 06, 1992

* * * END OF OUTPUT * * *

END SECTION

Section 4

"PRIDE"-EEM
PHASE 2
LOGICAL ENTERPRISE ANALYSIS

Enterprises with identical missions will also be identical in terms of their logical structure.
- Bryce's Law

The purpose of this phase is to develop the logical model of the enterprise, which is represented using functions. Several events occur during this phase:

- Business functions are defined and organized into a hierarchy.
- Information requirements and "objects" needed to perform the business functions are defined and re lated to the logical model.
- The set of skills required to perform the business functions are defined.
- The project's Order-of-Magnitude estimate and schedule is updated.
- A formal review with management is conducted to verify the phase findings.

Phase 2 requires considerable participation by User Management. Interviews are required to properly define business functions and to specify "objects" and information requirements.

METHODOLOGY NAVIGATION

Depending on the magnitude of the enterprise, this phase can either be executed for the entire enterprise or, as is more common, for a specific area (e.g., Marketing, Administration, Manufacturing, etc.). This breaks the project work into smaller, more manageable pieces. In other words, it is not unusual for an EEM project to branch from Phase 1 to parallel Phase 2's. This decision is left to the discretion of the Project Manager who must judge the project's complexity.

Phase 2 will be used to define the enterprise logically and to specify the "objects" and information requirements needed to perform the business. Based on a formal review as conducted in Activity H, management may elect to:

1. Approve the project for proceeding to the next phase, "Physical Enterprise Analysis."

2. Request revisions to the findings produced in the Phase.

3. Discontinue the project.

Following Phase 2, the EEM project will proceed to Phase 3, "Physical Enterprise Analysis." Again, depending on the scope of the project and enterprise, there may be either one Phase 3 or parallel Phase 3's which will correspond with the various Phase 2's.

GENERAL DISCUSSION

Understanding logical business functions can be a very abstract concept for some people to comprehend. People tend to assimilate physical structures, such as jobs/positions, more readily than logical structures. It is the Enterprise Engineer's responsibility to clarify this when interviewing users and executives. In fact, it is probably easier for the user to explain the physical organization structure first and then let the Enterprise Engineer interpret the business functions.

As a review of the key concepts pertaining to this phase, the Enterprise Engineer must remember the fundamental attributes of a business function: Business Functions...

• are a scope of responsibilities that define "what" is to be accomplished and "why"; not "who" and "how."

• rely on other functions to form a whole enterprise.

• can be separated into three high-level basic areas in any enterprise:

1. produce/generate income
2. administer resources
3. produce product/service

• denote the three basic levels of activity in an enterprise (Policy, Control and Operations). The three levels can be charted as a hierarchy that shows superior/subordinate/lateral functional relationships.

• deal with objects and require information. Each function has at least one "object" that it must deal with or manage.

• communicate through information systems.

• require certain skills.

• are implemented by positions.

For a complete description of these key concepts, please refer to the narrative as provided in the methodology introduction section

DESCRIPTION OF PHASE ACTIVITIES

Activity A - Detail Estimate & Schedule

Enterprise Engineering prepares a Detail estimate and schedule for the activities of the phase, which is reviewed with Project Management for approval.

Activity B - Interview Management

Enterprise Engineering conducts interviews with Executive Management and User Management to determine areas of responsibilities and actions/decisions that are made within that area. Also, discussed are the "objects" and information requirements that are used to operate and manage the enterprise, along with a description of the external enterprises that influence actions/decisions.

Activity C - Prepare Functional Model

Enterprise Engineering prepares a logical model of the enterprise using functions. Each function defines a scope of activity with specific responsibilities to be performed. In essence, the function defines "What" must occur and "Why"; however, it does not specify "How" (which is what systems are used for). The functions are organized into a hierarchical model denoting the areas of responsibility, the three basic levels of activity in the enterprise (Policy, Control and Operational), and dependencies between functions. The functional model (Function Charts and Profiles) is reviewed with User Management to verify the functions and with Quality Assurance to substantiate compliance with definition standards.

Activity D - Prepare Objects & Requirements

Enterprise Engineering defines the "objects" (which are represented using logical files) and Information Requirements for each function. Data Engineering provides assistance in locating existing objects and establishing new ones. At this stage, a basic definition of the object is all that is required. Additional descriptions about the object, using a comprehensive data model, will be defined in succeeding DBEM projects. Each requirement is analyzed to determine what existing Information Systems, if any, are used to implement the requirement. Both the objects and the information requirements are attached to pertinent functions.

Activity E - Prepare Functional Matrices

Enterprise Engineering prepares a series of matrices which show the relationships between functions, systems, objects and information requirements. These matrices will be reviewed by management for accuracy during the phase review.

Activity F - Prepare Functional Skills

Enterprise Engineering defines the skills required to perform each function. These skills will be used for performing an organization analysis in the succeeding phase.

Activity G - Update Project Plan

Project Management reviews and revises as required the Project Plan along with the Order-of-Magnitude estimate and schedule for the remainder of the project.

Project Management and Enterprise Engineering then assemble the deliverables resulting from the phase and packages them into a single phase review document which is evaluated by Quality Assurance prior to conducting the formal phase review with management.

Activity H - Phase 2 Review

Project Management conducts a formal review of the Phase 2 deliverables with Information Resource Management, User Management, Executive Management, and Enterprise Resource Management. At this time, management will review the formal Phase 2 "Logical Enterprise Study" consisting of:

- Phase Cover Page
- Function Chart
- Function Profile Report
- Enterprise/Function Matrix
- Function/Object Matrix
- Enterprise-Function/Object Matrix
- Enterprise-Function/Information Matrix
- Information Requirements
- Function/Skill Matrix
- Project Plan
- Project Estimate/Schedule Recap
- Phase Review Checklist

Based on this report and subsequent review meeting, management may elect to revise parts of the report, discontinue the project, or approve it for continuation to Phase 3 of EEM.

The next few pages include sample deliverables resulting from this phase.

SAMPLE PHASE 2 DELIVERABLES

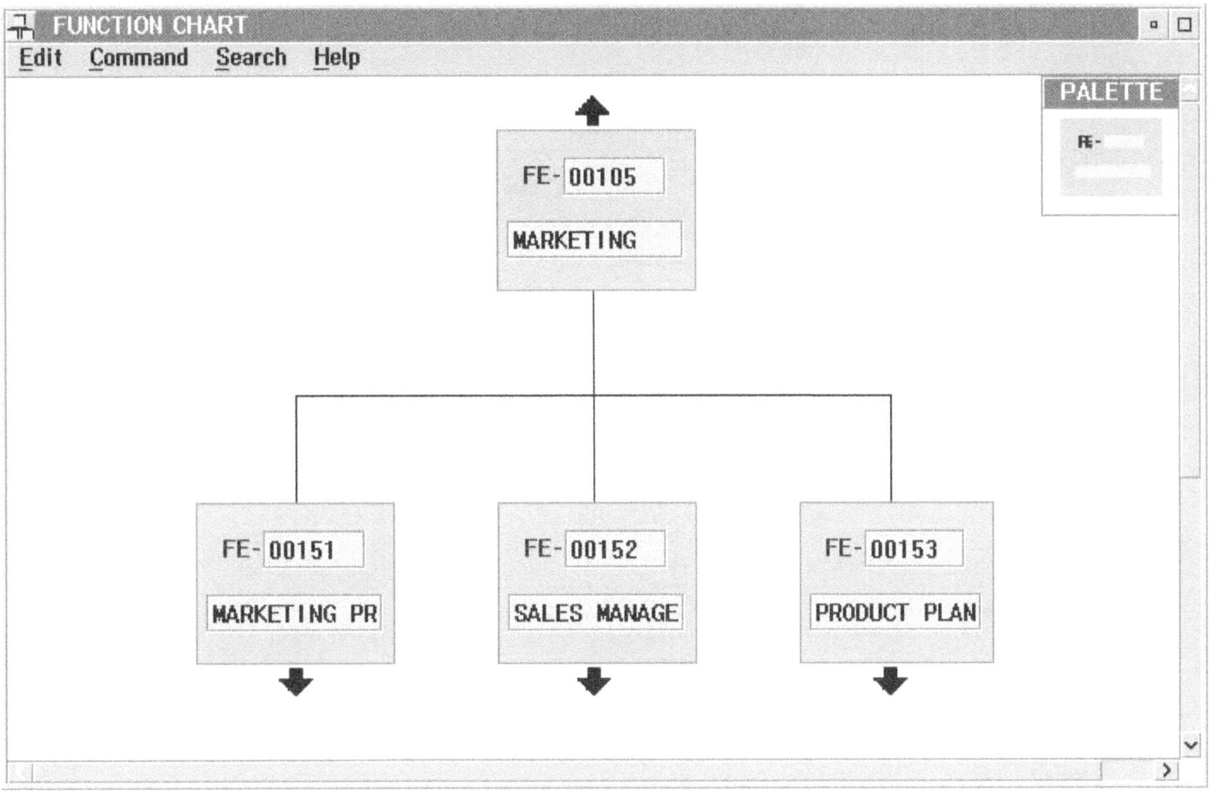

OD-90234	FUNCTION PROFILE	"PRIDE"-EEM
AUG 30, 2004	RELEASE 1.2.0	PAGE 1

```
*******************************************
*        FE-00105 MARKETING               *
*******************************************
```

INTERNAL/EXTERNAL- INTERNAL FUNCTION CODE- MK

SCOPE OF FUNCTION

This is a Policy level function. It has primary MANAGEMENT responsibility for producing income for the corporation.

RESPONSIBILITIES

1. Directs and oversees all Marketing activities for TMC.
2. Reports on all Marketing activities to the corporation.
3. Establishes a Marketing budget for the necessary financial expenditures for each fiscal year.
4. Controls all Marketing Service related financial expenditures.
5. Establishes and provides for adequate Marketing facilities and equipment.
6. Puts into effect all Marketing related contract commitments.
7. Establishes and maintains all Marketing related policies.
8. Establishes and maintains all Marketing related functional and organizational descriptions.
9. Establishes Marketing objectives and strategies to satisfy corporate needs.
10. Recruits and manages personnel for managerial positions in Marketing.

SAMPLE PHASE 2 DELIVERABLES

SUBORDINATE FUNCTIONS:

FE-00151	MARKETING PROMOTION
FE-00152	SALES MANAGEMENT
FE-00153	PRODUCT PLANNING

ORGANIZATIONAL ENTITIES:

OE-00401	CHIEF OPERATING OFFICER
OE-00405	VICE PRESIDENT, MARKETING

SUPERIOR FE COMPONENTS:		TYPE OF FE
FE-00100	TRUNDLE MANUFACTURING CORP.	INTERNAL ENTERPRISE
FE-00301	FEDERAL GOV'T - POST OFFICE	EXTERNAL ENTERPRISE
FE-00307	STATE GOV'T - CONSUMER AFFAIRS	EXTERNAL ENTERPRISE
FE-00320	TMC CORPORATE LAWYERS	EXTERNAL ENTERPRISE
FE-00324	TMC ADVERTISING MEDIA	EXTERNAL ENTERPRISE
FE-00325	TMC INDUSTRY CONTACTS	EXTERNAL ENTERPRISE
FE-00326	TMC STOCK CUSTOMER	EXTERNAL ENTERPRISE
FE-00327	TMC SPECIAL PRODUCT CUSTOMERS	EXTERNAL ENTERPRISE
FE-00328	TMC PRODUCT COMPETITION	EXTERNAL ENTERPRISE
FE-00332	CITY CHAMBER OF COMMERCE	EXTERNAL ENTERPRISE

INPUTS/OUTPUTS USED:

ID-00500	MARKETING SYSTEM WINDOW
OD-00456	MARKETING STATUS SUMMARY
OD-00890	SALES TRENDS/FORECASTS

FILES USED:

FD-01001	SALES REGION OBJECT
FD-01004	CUSTOMER OBJECT
FD-01006	PRODUCT OBJECT
FD-01007	PART OBJECT
FD-01009	VENDOR OBJECT
FD-01010	GOVERNMENT AGENCY OBJECT
FD-01011	COMPETITOR OBJECT
FD-01016	ORDER OBJECT
FD-01017	PURCHASE ORDER OBJECT

SYSTEMS USED:		RESPONSIBILITY
MK	MARKETING SYSTEM	APPROVAL
MK04	MARKETING STATUS QUERY	EXECUTION

INFORMATION REQUIREMENTS:		INFO FLOW
IR-00150	ANNUAL CORPORATE REPORT	ORIGINATOR
IR-00164	CORPORATE OBJECTIVE DESCRIPTN.	RECEIVER
IR-00165	WEEKLY OBJECTIVE ANALYSIS	RECEIVER
IR-00229	POLICY DESCRIPTION	RECEIVER
IR-00151	MONTHLY MARKETING STATUS	RECEIVER
IR-00191	CORPORATE POLICY REVISIONS	ORIGINATOR

SKILLS USED:		REQ'D PROFICIENCY
SD-00082	TARGET MARKET ANALYSIS	6
SD-00084	DEFINING PRODUCT MARKETS	6
SD-00085	PRODUCT PRICING ANALYSIS	6
SD-00086	STRATEGIC MARKET PLANNING	6
SD-00087	SALES ANALYSIS	6
SD-00088	MARKETING FORECASTING	6

*** END OF OUTPUT ***

OD-90189	INFORMATION REQUIREMENTS	"PRIDE"-ISEM
SEP 01, 2004	RELEASE 1.2.0	PAGE 1

****
* IR-00101 CUSTOMER ORDER *
****

ENTITY(S) REQUIRING INFORMATION:	NUMBER	INFO. FLOW
1. CUSTOMER SERVICE	FE-00118	ORIGINATOR
2. ORDER PROCESSING	FE-00127	RECEIVER
3. TMC STOCK CUSTOMER	FE-00326	ORIGINATOR
4. CUSTOMER SERVICE REP	OE-00453	RECEIVER
5. CUSTOMER SERVICE REP	OE-00453	ORIGINATOR

TYPE REQUIREMENT: INFORMATION INFORMATION TYPE: OPERATIONAL

FREQUENCY:	OFFSET:	RESPONSE TIME:
UPON REQUEST	N/A	15 SECOND(S)

BUSINESS PURPOSE

The purpose of this requirement is to interact with a customer to place or revise an order for corporate products.

ACTIONS AND/OR BUSINESS DECISIONS

Order Processing will...

1. Check product availability and advise the customer of anticipated delivery dates.

2. Create, revise or delete an order.

   NOTE: Deletion is only permitted if the order has not yet been processed by Customer Service.

3. Notify Customer Service to take necessary actions to fill the order.

4. Request changes, as required, to the customer's credit rating (i.e., increase, decrease) to Customer Credit.

BENEFITS

1. Improved communications/relations with the customer, thereby enhancing customer satisfaction.

2. Improved communications internally, particularly with Customer Service and Customer Credit. Such communications improves control over customer credit, thereby preventing unauthorized shipments.

OBJECTS/FILES AFFECTED:

* PRODUCT OBJECT (FD-00100)
* CUSTOMER OBJECT (FD-00101)
* ORDER OBJECT (FD-00102)

REQUIRED DATA ELEMENTS:

* PRODUCT NUMBER (DD-00003)
* NAME (DD-00018)

(continued on next page)

SAMPLE PHASE 1 DELIVERABLES

(continued from previous page)

* UNIT PRICE (DD-00111)
* ORDER NUMBER (DD-00012)
* DATE (DD-00016)
* QUANTITY ORDERED (DD-00036)
* TERMS (DD-00044)
* STATUS (DD-00041)
* QUANTITY ON HAND (DD-00508)
* CONTACT NUMBER (DD-00042)
* DELIVERY PRIORITY (DD-00050)
* CUSTOMER NUMBER (DD-00001)

IMPLEMENTED BY INPUTS/OUTPUTS:

* PRODUCT AVAILABILITY (OD-00010)
* CUSTOMER ORDER (OD-00011)
* CUSTOMER PROFILE (OD-00012)

SUPPORTED BY SYSTEM RESOURCES:

* TRUNDLE ORDER PROC. SYS (TOPS) (OP)
* CUSTOMER ORDER INQUIRY (OP-00)
* ORDER-PROC FILE MAINTENANCE (OP-08)

* * * END OF OUTPUT * * *

SAMPLE PHASE 2 DELIVERABLES

OD-90203 E N T E R P R I S E / F U N C T I O N M A T R I X "PRIDE"-EEM

SEP 12, 2004 RELEASE 1.2.0 PAGE 2.1.1

ENTERPRISES > / FUNCTIONS V	FE-00300 STATE GOV'T - I.R.S.	FE-00301 STATE GOV'T - POST OFFICE	FE-00302 STATE GOV'T - SOCIAL SECURTY	FE-00303 COUNTY GOVT - LABOR DEPT.	FE-00304 COUNTY GOVT - OSHA
FE-00101 ADMIN.	X	X	X	X	X
FE-00102 MANUFACTURING		X			
FE-00103 FINANCIAL RESOURCE MGMT	X	X	X	X	X
FE-00104 MATERIAL RESOURCE MGMT		X			
FE-00105 MARKETING		X			
FE-00106 HUMAN RESOURCE MANAGEMENT		X	X	X	X
FE-00107 INFORMATION RESOURCE MGMT		X			
FE-00108 OFFICE RESOURCE MGMT		X			
FE-00109 ACCCCOUNTS/ RECEIVABLES		X			
FE-00110 PARTS INVENTORY					

SAMPLE PHASE 2 DELIVERABLES

OD-90205 F U N C T I O N / O B J E C T M A T R I X "PRIDE"-EEM

SEP 12, 2004 RELEASE 1.2.0 PAGE 2.1.1

FUNCTIONS > / OBJECTS V	FE-00101 ADMINISTRATION	FE-00102 MANUFACTURING	FE-00103 FINANCIAL RESOURCE MANAGEMENT	FE-00104 MATERIAL RESOURCE MANAGEMENT	FE-00105 MARKETING
FD-01001 SALES REGION OBJECT EL					X
FD-01002 SALES TERRITORY OBJECT EL					X
FD-01003 EMPLOYEE OBJECT EL	X		X		
FD-01004 CUSTOMER OBJECT EL	X	X	X	X	X
FD-01006 PRODUCT OBJECT EL	X	X	X	X	X
FD-01007 PART OBJECT EL	X	X	X	X	X
FD-01008 EQUIPMENT/MACH. OBJECT EL	X				
FD-01009 VENDOR OBJECT EL	X		X	X	X
FD-01010 GOV'T AGENCY OBJECT EL	X		X		X
FD-01011 COMPETITOR OBJECT EL					X

EL - ENTERPRISE LOGICAL

SAMPLE PHASE 2 DELIVERABLES

OD-90206 E N T E R P R I S E - F U N C T I O N / S Y S T E M M A T R I X "PRIDE"-EEM

SEP 12, 2004 RELEASE 1.2.0 PAGE 2.1.1

FUNCTIONS > SYSTEMS V	FE-00118 CUSTOMER S ERVICE	FE-00119 PRODUCTION CONTROL	FE-00120 PRODUCT MA INTENANCE	FE-00121 PRODUCT AS SEMBLY	FE-00122 PRODUCT SH IPPING ING
AP ACCOUNTS PAYABL E SYSTEM					
AR ACCOUNTS RECEIV ABLE SYSTEM	INFO				
BL BILLING SYSTEM					
CS CUSTOMER SVC SYSTEM	APPROVAL		APPROVAL		APPROVAL
GL GENERAL LEDGER SYSTEM					
HR HUMAN RESOURCE SYSTEM					
HR INVENTORY CONT ROL SYSTEM					
MN MANUFACTURING SYSTEM		APPROVAL		APPROVAL	
PA PAYROLL SYSTEM					
SH SHIPPING SYSTEM		APPROVAL			X

NOTE: THE CELL ENTRY REPRESENTS THE 'RESPONSIBILITY' FIELD ENTRY. IF A NULL ENTRY, 'X' IS ENTERED.

SAMPLE PHASE 2 DELIVERABLES

OD-90207 E N T E R P R I S E - F U N C T I O N / I N F O M A T R I X "PRIDE"-EEM

SEP 12, 2004 RELEASE 1.2.0 PAGE 2.1.1

FUNCTIONS > INFO RQMTS V	FE-00330 TMC MORTGAGE COMPANY	FE-00331 COUNTY GOV'T - EMERGENCY SVCS.	FE-00332 CITY CHAMBER OF COMMERCE ANAGEMENT	FE-00101 ADMINISTRATION	FE-00102 MANUFACTURING
IR-00107 REVIEW CREDIT VIOLATION					
IR-00110 SALES PERFORMANCE					
IR-00107 SALES PERFORMANCE					
IR-00111 SALES REP ASSIGNMENT					
IR-00112 MONTHLY SALES ANALYSIS					
IR-00113 WEEKLY SALES COMMISSIONS					
IR-00150 ANNUAL CORPORATE REPORT				O	O
IR-00151 MONTHLY MARKETING STATUS					
IR-00152 MONTHLY ADMINISTRATION STATUS				R	
IR-00153 MONTHLY MANUFACTURING STATUS					R
IR-00154 WEEKLY MARKETING STATUS					R

INFORMATION FLOW KEY: O = ORIGINATOR, P = PARTICIPATOR, R=RECEIVER IF A NULL ENTRY, 'X' IS ENTERED.

SAMPLE PHASE 2 DELIVERABLES

OD-90388 F U N C T I O N / S K I L L M A T R I X "PRIDE"-EEM

SEP 12, 2004 RELEASE 1.2.0 PAGE 2.1.1

FUNCTIONS > SKILLS V	FE-00101 ADMINISTRA TION	FE-00102 MANUFACTUR ING	FE-00103 FINANCIAL RESOURCE M ANAGEMENT	FE-00104 MATERIAL R ESOURCE MA NAGEMENT	FE-00105 MARKETING
SD-01001 PRESENTATION SKILL	5				5
SD-01002 CONTRACT WRITING	7		7		7
SD-01003 EMPLOYEE RELATIONS	X		X		
SD-01004 CUSTOMER RELATIONS	X	X	X		X
SD-01006 PRODUCT PRICING	X	X	X		X
SD-01007 MRP CONCEPT	X	X	X	X	X
SD-01008 EQUIPMENT/ MACHINE SKILL	X				
SD-01009 VENDOR RELATIONS	X		X	5	X
SD-01010 GOVERNMENT ANALYSIS	X		X		3
SD-01011 COMPETITION EVALUATION					6

REQUIRED PROFICIENCY IS EXPRESSED IN THE CELL WHEN AVAILABLE

END SECTION

Section 5

"PRIDE"-EEM PHASE 3 PHYSICAL ENTERPRISE ANALYSIS

All companies have a culture. In order for employees to function and succeed, it is essential they understand and believe in the culture.
- Bryce's Law

The purpose of this phase is to develop the physical model of the enterprise, which is primarily represented using positions (jobs). Several events occur during this phase:

- The administrative reporting structure of the enterprise is organized into a hierarchy.
- Information requirements and "objects" needed to perform the business functions are defined and related to the physical model.
- An organization analysis is performed which compares the logical business functions against the physical organization structure. Also, the corporate culture is analyzed.
- The project's Order-of-Magnitude estimate and schedule is updated.
- A formal review with management is conducted to verify the phase findings.

Phase 3 requires considerable participation by User Management. Interviews are required to properly define positions and relationships, and to specify the use of information resources.

METHODOLOGY NAVIGATION

Depending on the magnitude of the enterprise or project, this phase can either be executed for the entire enterprise or, as is more common, for a specific area (e.g., Marketing, Administration, Manufacturing, etc.). This breaks the project work into smaller, more manageable pieces. In other words, it is not unusual for an EEM project to have multiple Phase 3's, particularly if the project had already branched into multiple Phase 2's earlier. In this situation, a Phase 3 follows each Phase 2. This decision is left to the discretion of the Project Manager who must judge the project's complexity.

Phase 3 will be used to define the enterprise physically and to specify the "objects" and information requirements needed to perform the business. Based on a formal review as conducted in Activity G, management may elect to:

1. Approve the project for proceeding to the next phase, "Enterprise Information Strategy."

2. Request revisions to the findings produced in the Phase.

3. Discontinue the project.

GENERAL DISCUSSION

Whereas functions are used to describe the enterprise in logical terms (What and Why), positions are used to describe the physical enterprise (Who). Positions, therefore, are used to physically implement the logical functions of the enterprise. Although enterprises with common missions will have identical functional definitions, it is unlikely two enterprises will be physically organized the same way. This is because management greatly influences how the enterprise is organized.

As a review of the key concepts pertaining to this phase, the Enterprise Engineer must remember the fundamental attributes of an organizational entity: A Position...

- is a prescribed set of duties and responsibilities; another name is "job." A position is usually based on a position title, such as President, Vice President, Director, Manager, Supervisor, Clerk, Secretary, Laborer, etc. The relationships between positions represent administrative reporting.

- describes the work to be performed and the functions to be fulfilled.

- is used to process data and/or use information.

For a complete description of these key concepts, please refer to the narrative as provided in the methodology introduction section.

DESCRIPTION OF PHASE ACTIVITIES

Activity A - Detail Estimate & Schedule

Enterprise Engineering prepares a Detail estimate and schedule for the activities of the phase, which is reviewed with Project Management for approval.

Activity B - Interview Management

Enterprise Engineering conducts interviews with Executive Management and pertinent User Management to determine the positions that are used to fulfill functions. Superior/Subordinate/Lateral relationships between positions are noted, along with job titles, if any. Both human and machine resources (such as computers) are also identified and related to the positions they serve. The fundamental skills of the resources are also reviewed.

Activity C - Prepare Position Model

Enterprise Engineering prepares a physical model of the enterprise using OE, RE, SD descriptions in the IRM. Documentation produced includes Human/Machine Resource Profiles, Resource/Skill Matrices, Position Descriptions, and Organization Charts. The position model is reviewed with User Management to verify relationships and with Quality Assurance to substantiate compliance with definition standards.

Activity D - Prepare Position Matrices

Enterprise Engineering prepares a series of matrices which show the relationships between functions, positions, systems, objects and information requirements. These matrices will be reviewed by management for accuracy during the phase review.

Activity E - Prepare Organization Analysis

Enterprise Engineering prepares an analysis of the current organization. The relationship between functions, positions, resources and skills are analyzed to compare the physical organization to the logical. Several organizational considerations may be identified, such as: Functions that are not being fulfilled by a Position; an inordinate amount of Positions serving a Function (excessive overlaps); Functions being served with inappropriately skilled resources; etc. In addition, Enterprise Engineering defines the Corporate Culture as it exists today. All of this is packaged into an "Organization Analysis" report for management to study during the phase review. From these observations, management may elect to make some organizational changes.

Activity F - Update Project Plan

Project Management reviews and revises as required the Project Plan along with the Order-of-Magnitude estimate and schedule for the remainder of the project.

Project Management and Enterprise Engineering then assembles the deliverables resulting from the phase and packages them into a single phase review document which is evaluated by Quality Assurance prior to conducting the formal phase review with management.

Activity G - Phase 3 Review

Project Management conducts a formal review of the Phase 3 deliverables with Information Resource Management, User Management, Executive Management, and Enterprise Resource Management. At this time, management will review the formal Phase 3 "Physical Enterprise Study" consisting of:

- Phase Cover Page
- Organization Chart
- Position Description Report
- Function/Position Matrix
- Position/Information Matrix
- Position/Object Matrix
- Position/System Matrix
- Organization Analysis
- Human/Machine Resource Profiles
- Resource/Skill Matrix
- Project Plan
- Project Estimate/Schedule Recap
- Phase Review Checklist

Based on this report and subsequent review meeting, management may elect to revise parts of the report, discontinue the project, or approve it for continuation to Phase 4 of EEM.

The next few pages include sample deliverables resulting from this phase.

SAMPLE PHASE 3 DELIVERABLES

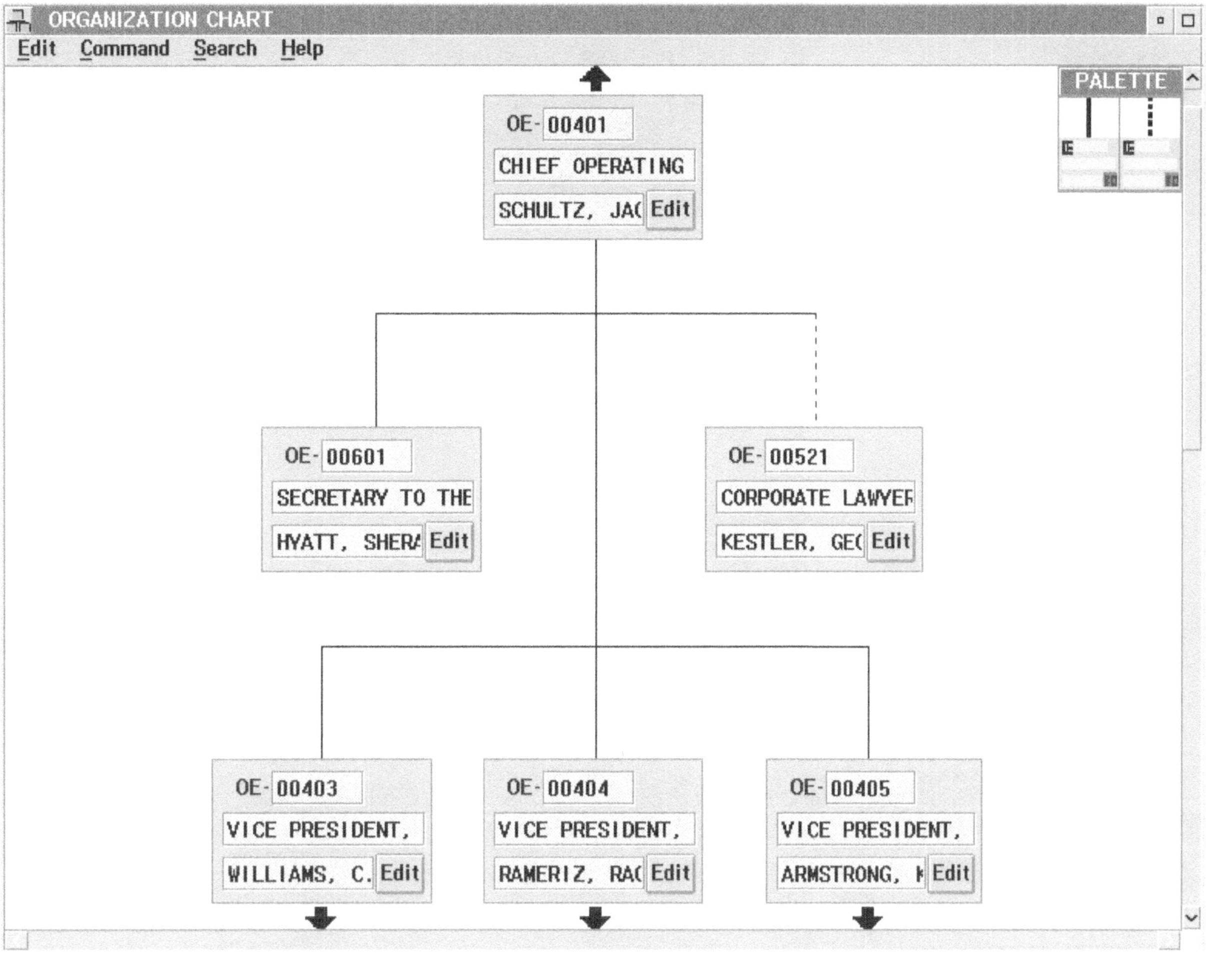

SAMPLE PHASE 3 DELIVERABLES

OD-90235 POSITION DESCRIPTION "PRIDE"-EEM

SEP 15, 2004 RELEASE 1.2.0 PAGE 1

```
***************************************************************
* OE-00452          NAT'L SALES MANAGER                       *
***************************************************************
```

INT/EXT- INTERNAL FUNCTION CODE- MKTG ORG CODE-
TELEPHONE NO- 513/862-1212 LOCATION- MKTG DEPT.

SCOPE OF ORGANIZATIONAL ENTITY

Has primary MANAGEMENT responsibility for implementing all Sales related functions. This specifically includes Management and Control of all corporate sales activities at a National level.

SPECIFIC DUTIES AND RESPONSIBILITIES

1. Manages all Sales related activities.

2. Reports to the Marketing Director on all Sales activities.

3. Establishes a departmental operating budget for the necessary expenditures for each fiscal year. The budget is to be approved by the Marketing Director.

4. Controls all departmental related expenditures.

5. Manages all Sales related facilities and equipment.

6. Advises and recommends to the Marketing Director any policies as they pertain to Sales activities. Only the Marketing Director can set Policy. Entity is responsible for the implementation of all corporate and departmental related policies.

7. Creates, updates and deletes all Sales related functional, organizational, resource and skill descriptions.

8. Recruits and manages all personnel within the Sales Department. This includes handling personnel problems.

9. Works with the Marketing Director and the other corporate Managers to determine Sales objectives and strategies. Entity is then responsible for meeting these goals within time and cost constraints.

10. Assures that all sales related records are accurately and efficiently processed and maintained. 11. Assures that an effective line of communication is developed and maintained between the corporation and its market.

12. Advises other parts of the company, such as Manufacturing, on sales related trends.

13. Assures that all sales programs are effectively developed, maintained and implemented

REQUIRED KNOWLEDGE/SKILLS/EXPERIENCE

1. In-depth knowledge about the workings of the corporation, including its policies and procedures.

2. In-depth knowledge of all corporate legal obligations (i.e., contracts, invoicing, shipment, production, etc).

3. In-depth knowledge of all corporate products and the market.

(continued on next page)

SAMPLE PHASE 3 DELIVERABLES

(continued from previous page)

4. Familiarity with the capabilities of all corporate equipment and facilities, particularly those that relate to the Sales Department.

5. The experience and ability to make critical decisions regarding all Sales related policies and procedures.

6. The experience and ability to prepare a budget and to control costs.

7. The experience and ability to critically evaluate the performance of Sales functions, the organization, and related resources.

8. The experience and ability to manage the activities of others.

9. The experience and ability to recruit and release personnel for Sales Department positions.

10. The experience and ability to manage department facilities and equipment.

11. The experience and ability to execute corporate legal obligations as they pertain to Sales activities.

12. The experience and ability to assume responsibility for performing assigned tasks and meeting objectives within time and cost constraints.

13. Possess good communication skills.

14. Possess good analytical and problem solving skills.

15. Must be results oriented.

EVALUATION OF PERFORMANCE

The following list is intended as a guide to evaluate the performance of the resource(s) related to this organizational entity:

1. Departmental expenditures have not exceeded budget.

2. Resource(s) enforces all corporate and departmental related policies and procedures as established by the Marketing Director.

3. Resource(s) has a thorough understanding and knowledge of all Sales functions and responsibilities.

4. Resource(s) has a thorough understanding and knowledge of all corporate policies, procedures, functions and products.

5. Departmental personnel conduct themselves in a professional manner.

6. The Marketing Director is aware of all prominent departmental activities.

7. Resource(s) assumes responsibility for performing assigned tasks and achieves them within time and cost constraints.

8. All departmental related facilities and equipment are well maintained and in good functioning order.

9. All functional, organizational, resource and skill descriptions are current and accurate.

10. There is adequate staff personnel required to perform all departmental functions and are all well trained.

(continued on next page)

SAMPLE PHASE 3 DELIVERABLES

(continued from previous page)

11. Resource(s) do not commit a substantial number of errors or mistakes.

12. Resource(s) works well with co-workers.

13. Records and files are accurately and efficiently maintained.

14. An effective line of communications exists between the company and its customers and with the general market.

15. Adequate technical assistance is provided to corporate Product Development Departments.

ORGANIZATIONAL ENTITY RELATIONSHIPS

Subordinate to the Marketing Director. Maintains a lateral working relationship with other Marketing Departments including Promotion and Customer Services.

SUBORDINATE POSITIONS:		RELATIONSHIP
OE-00456	EASTERN SALES MANAGER	DIRECT/LINE
OE-00458	WESTERN SALES MANAGER	DIRECT/LINE
OE-00615	SECRETARY TO NAT'L SALES MGR	DIRECT/STAF

RESOURCES IMPLEMENTING POSITION:

RE-00121	HOLBROOK, A.F.

AFFECTED FE COMPONENTS:		TYPE OF FE
FE-00152	SALES MANAGEMENT	INTERNAL FUNCTION
FE-00156	SALES REPRESENTATION	INTERNAL FUNCTION
FE-00127	ORDER PROCESSING	INTERNAL FUNCTION

SUPERIOR POSITIONS:		RELATIONSHIP
OE-00405	VICE PRESIDENT, MARKETING	DIRECT/LINE

INPUTS/OUTPUTS USED:

OD-00700	SALESMAN APPRAISAL REPORT
OD-00701	SALES REP ASSIGNMENT
OD-00702	WEEKLY SALES COMMISSIONS
OD-00703	MONTHLY SALES FORCE ANALYSIS
OD-00704	MONTHLY PRODUCT SALES ANALYSIS
OD-00014	WEEKLY ORDER STATUS ANALYSIS
OD-00017	CUSTOMER DETAIL SHEET
ID-00125	WEEKLY ORDER STATUS REQUEST
ID-00128	CUSTOMER DETAIL REQUEST
ID-00100	ORDER PROCESSING WINDOW

FILES USED:

FD-00201	MASTER SALES FILE
FD-00220	SALES PROSPECT MANUAL FILE

SYSTEMS USED:		RESPONSIBILITY
TS	TRUNDLE SALES SYSTEM	APPROVAL AUT
OP	TRUNDLE ORDER PROC. SYS (TOPS)	APPROVAL AUT
OP04	WEEKLY ORDER STATUS	APPROVAL AUT
OP07	CUSTOMER RECORD MANAGEMENT	APPROVAL AUT
OP08	ORDER-PROC FILE MAINTENANCE	APPROVAL AUT

(continued on next page)

SAMPLE PHASE 3 DELIVERABLES

(continued from previous page)

OP0403	PREPARE ORDER STATUS REQUEST	EXECUTION
OP0405	CONVERT ORDER STATUS REQUEST	EXECUTION
OP0409	BURST/DISTRIBUTE ORDER STATUS	EXECUTION
OP0411	ANALYZE WEEKLY ORDER STATUS	EXECUTION
OP0703	PREPARE CUSTOMER DETAIL REQST.	EXECUTION
OP0707	REVIEW/REVISE CUSTOMER DETAILS	EXECUTION
OP0803	ORDER-PROC FILE UPDATE SESSION	EXECUTION

INFORMATION REQUIREMENTS:		INFO FLOW
IR-00110	SALES PERFORMANCE APPRAISAL	RECEIVER
IR-00111	SALES REP ASSIGNMENT INFO.	RECEIVER
IR-00110	SALES PERFORMANCE APPRAISAL	ORIGINATOR
IR-00111	SALES REP ASSIGNMENT INFO.	ORIGINATOR
IR-00112	MONTHLY SALES ANALYSIS	RECEIVER
IR-00112	MONTHLY SALES ANALYSIS	ORIGINATOR
IR-00113	WEEKLY SALES COMMISSION INFO	RECEIVER
IR-00113	WEEKLY SALES COMMISSION INFO	ORIGINATOR
IR-00102	CUSTOMER PROFILE	ORIGINATOR
IR-00104	ORDER STATUS ANALYSIS	RECEIVER
IR-00107	CUSTOMER SETUP/MODIFICATIONS	RECEIVER

*** END OF OUTPUT ***

OD-90204 F U N C T I O N / P O S I T I O N M A T R I X

SEP 19, 2004 RELEASE 1.2.0 PA

FUNCTIONS > POSITIONS V	FE-00101 ADMINISTRA TION	FE-00102 MANUFACTUR ING	FE-00103 FINANCIAL RESOURCE M ANAGEMENT	FE-00104 MATERIAL R ESOURCE MA NAGEMENT	FE-00105 MARKETING	FE-00106 HUMAN RESO URCE MANAG EMENT	FE-00107 INFORMATIO N RESOURCE MANAGEMNT	FE-00108 OFFICE RES OURCE MANA GEMENT	FE-00109 ACCOUNTS/R ECEIVABLES	FE PA NT
OE-00224 SHIFT SUPERVISO R										
OE-00400 CHIEF EXECUTIVE OFFICER	X	X			X					
OE-00401 CHIEF OPERATING OFFICER	X	X			X					
OE-00402 SAFETY INSPECTO R										
OE-00403 VICE PRESIDENT, ADMINISTRATION				X		X		X		
OE-00404 VICE PRESIDENT, MANUFACTURING		X								
OE-00405 VICE PRESIDENT, MARKETING					X					
OE-00406										

SAMPLE PHASE 3 DELIVERABLES

OD-90242 P O S I T I O N / I N F O M A T R I X

OCT 10, 2004 RELEASE 1.2.0 P

POSITIONS > / INFO RQMTS V	OE-00001 PRESIDENT	OE-00002 SOFTWARE ENGINEERING MANAGER	OE-00003 DIRECTOR, MKTG/CUST-SVCS	OE-00004 SOFTWARE ENGINEER	OE-00005 DIRECTOR, PRODUCTION	OE-00006 SECRETARY/ TREASURER	OE-00008 DIRECTOR, PRODUCT DEVELOPMENT	OE-00010 VICE PRESIDENT	OE-00012 SOFTWARE E SUPERVISOR	OE- CUS
IR-30004 IMPLEMENTATION MATERIALS			O R							
IR-30007 CUSTOMER REFERENCE INFO			O R							
IR-30008 CUSTOMER RELEASE INFORMATION			P		R		O			
IR-30010 SHIPPING INFORMATION			O		R					
IR-30011 DIRECT MAIL INSTRUCTIONS			O R							
IR-30013 RECEIPT INFORMATION						O R				
IR-30015 BILLING INFORMATION						O R				
IR-30016						O				

OD-90208 P O S I T I O N / O B J E C T M A T R I X

SEP 19, 2004 RELEASE 1.2.0 PA

POSITIONS > / OBJECTS V	OE-00224 SHIFT SUPERVISOR	OE-00401 CHIEF OPERATING OFFICER	OE-00402 SAFETY INSPECTOR	OE-00403 VICE PRESIDENT, ADMINISTRATION	OE-00404 VICE PRESIDENT, MANUFACTURING	OE-00405 VICE PRESIDENT, MARKETING	OE-00411 OFFICE SUPPLIES CLERK	OE-00421 MANAGER, MATERIALS	OE-00431 PURCHASING SUPERVISOR	OE- DIR PRO
FD-30001 SALES REGION OBJECT AL						X				
FD-30002 SALES TERRITORY OBJECT AL						X				
FD-30004 CUSTOMER OBJECT AL		X		X	X	X				
FD-30006 PRODUCT OBJECT AL	X	X		X	X	X			X	
FD-30007 PART OBJECT AL	X				X			X	X	
FD-30009 VENDOR OBJECT AL				X			X			
FD-30010 GOVERNMENT AGENCY OBJECT AL		X	X	X	X					
FD-30011										

SAMPLE PHASE 3 DELIVERABLES

OD-90209 P O S I T I O N / S Y S T E M   M A T R I X

AUG 16, 2004 RELEASE 1.2.0

POSITIONS > / SYSTEMS V	OE-00224 SHIFT SUPERVISOR	OE-00401 CHIEF OPERATING OFFICER	OE-00402 SAFETY INSPECTOR	OE-00403 VICE PRESIDENT, ADMINISTRATION	OE-00404 VICE PRESIDENT, MANUFACTURING	OE-00405 VICE PRESIDENT, MARKETING	OE-00411 OFFICE SUPPLIES CLERK	OE-00421 MANAGER, MATERIALS	OE-00431 PURCHASING SUPERVISOR	(I F
CS CUSTOMER SERVICE SYSTEM										
HR HUMAN RESOURCE SYSTEM		APPROVAL	X	APPROVAL	X					
IC INVENTORY CONTROL SYSTEM		APPROVAL		APPROVAL	APPROVAL			APPROVAL	APPROVAL	
MN MANUFACTURING SYSTEM	INFO	APPROVAL			APPROVAL					
PA PAYROLL SYSTEM										
SH SHIPPING SYSTEM		APPROVAL			APPROVAL					
TR CURRENT TRUNDLE SALES SYSTEM						APPROVAL				
TS										

OD-90390 R E S O U R C E / S K I L L   M A T R I X "

SEP 19, 2004 RELEASE 1.2.0 PA

RESOURCES > / SKILLS V	RE-00002 JONES, ROBERT	RE-00004 SMITH, GEORGE	RE-00018 DEMARCO, JOHN	RE-00021 CONDU, CONNIE	RE-00025 FRANCIS, BETTY	RE-00036 THORNTON, FRED	RE-00043 SIMPSON, R.J.	RE-00049 WELCH, LUCILLE	RE-00063 CONSTANTINE, K.K.	RE-C KREE JOSE
SD-00001 ASSEMBLER PROGRAMMING SKILL					X		5	5	5	
SD-00002 BASIC PROGRAMMING SKILL					X		5	5		
SD-00003 COBOL PROGRAMMING SKILL			X		5		5	5	5	
SD-00004 MICROSOFT PROGRAMMRS WORKBENCH			3		5		5	5	5	
SD-00006 DATA ENTRY SKILL						5				
SD-00009 FILING SKILL						5				
SD-00011 FORTRAN PROGRAMMING SKILL					1		5	5		
SD-00014 RPG PROGRAMMING					9		5	5	5	

SAMPLE PHASE 3 DELIVERABLES

OD-90389	ORGANIZATION ANALYSIS	"PRIDE"-EEM
SEP 19, 2004	RELEASE 1.2.0	PAGE 1

AS DEFINED ON PD-00002 TRUNDLE EEM PROJECT

An analysis of the TMC logical/physical organization was recently performed. The study revealed several interesting findings&colon.

1. There appears to be excessive levels of management required to perform the Production Control function and Sales Management function. For both functions, three levels of management are used to control a modest size staff.

2. The human resources implementing the Sales Representation function indicates a low proficiency rating and may reflect the need to improve the skills of the sales force.

For further information, consult the attached Organization Analysis.

CORPORATE CULTURE

A statistical survey was recently conducted within the manufacturing area; this included Production, Materials, and Distribution. Although no real unethical problems surfaced, the survey does indicate that there are strong departmental loyalties and competitiveness between the departments. This has created a political situation where the departments compete for recognition as opposed to a spirit of cooperation in this vital part of the business.

A statistical summary is attached to this report which reflects these findings.

*** END OF OUTPUT ***

END SECTION

Section 6

"PRIDE"-EEM PHASE 4 ENTERPRISE INFORMATION STRATEGY (EIS)

There is nothing more unproductive than to build something efficiently that should not have been built at all.
- Bryce's Law

The purpose of this phase is to develop an information strategy that is synchronized with the mission and goals of the enterprise. Several events occur during this phase:

- The current Enterprise Information Strategy (EIS) is evaluated for accuracy and currency.
- Information Requirements (IR) are grouped into objectives (MI) based on commonality.
- Objectives (MI) are grouped into projects (PD) based on commonality.
- A proposed new Enterprise Information Strategy is calculated. This includes a priority ranking of all objectives (MI) and projects (PD).
- A formal review with executive management is conducted to confirm/revise the EIS.

METHODOLOGY NAVIGATION

This phase cannot be started until the last Phase 3 of the project has been completed. Normally, there is only one Phase 4 per EEM project.

Phase 4 will be used to define the Enterprise Information Strategy. This will be a high-level policy decision for the business. Based on a formal review as conducted in Activity F, management may elect to:

1. Approve the project for proceeding to the next phase, "EEM Evaluation."

2. Request revisions to the findings produced in the Phase.

3. Reject the proposed EIS (existing priorities remain in effect).

Following Phase 4, the EEM project will proceed to Phase 5.

GENERAL DISCUSSION

All enterprises have objectives that they aspire to reach in order to survive or succeed in their mission. These objectives are implemented through projects which must satisfy the objectives within time and cost constraints.

Objectives are supported by information which is used to make decisions. Therefore, information requirements must be specified when defining objectives. The criteria for grouping requirements into objectives may be varied:

1. Grouping requirements within a particular functional area (e.g., Marketing, Administration, Manufacturing).

2. Grouping requirements by information type (Policy, Control and Operational).

3. Grouping requirements by whether they are currently implemented by an existing information system or not implemented by any system.

4. Or combinations of the above.

If anything is constant, it is change. Enterprise objectives constantly change primarily due to external factors, such as market conditions, economics, competition, government regulation, etc. Because of this, information requirements, jobs and even the logical functions may change. Objectives, priorities and projects must be reviewed routinely to assure that the enterprise is moving in the right direction.

Objectives have the following characteristics:

1. **OBJECTIVES ARE TIME DEPENDENT.** One of the attributes that makes objectives volatile, is that they all have maximum delivery dates, whether it be expressed in days, weeks, months, or years. Even the expression "As Soon As Possible" (ASAP) refers to certain mandatory time constraints. The required time frame could be as precise as April 15th (the date when U.S. income tax returns are due) or as imprecise as the end of the decade.

2. **OBJECTIVES PERTAIN TO CERTAIN TYPES OF WORK EFFORT.** There are three basic types of work effort: Maintenance, which is the correction of problems or errors; New Development, which represents new areas of interest to the enterprise; and Modification/Improvements which represents changes and enhancements to existing things.

3. **THE COMPLETE DESCRIPTION FOR AN OBJECTIVE INCLUDES THE FOLLOWING ELEMENTS:**

   1. Statement of PROBLEM AND/OR OPPORTUNITY to be addressed; including a description of the "objects" involved.

   2. AREAS AFFECTED - both internal functions and external enterprises, directly and indirectly.

   3. MEASUREMENT CRITERIA which substantiates when the objective has been met. This determination must be quantifiable. There is little point in establishing objectives if you cannot verify their completion.

4. **OBJECTIVES SHOULD NOT BE DEPENDENT ON OTHER OBJECTIVES.** They should be defined as independent entities for simplicity. Objectives can be grouped into projects as required. Dependencies can then be expressed through projects.

5. **OBJECTIVES HAVE A PRIORITY.** Priorities are used to identify the urgency or importance of the objective to the enterprise, and dictate what objectives should be turned into projects.

6. **OBJECTIVES MAY BE GROUPED WITH OTHER OBJECTIVES INTO PROJECTS.** An objective is implemented through one or more projects. This implies that there is not a one-to-one relationship between objectives and projects. One objective may be implemented through several projects. Conversely, one project may help implement several objectives. Although the criteria for grouping objectives can be arbitrary, it should be based on the common characteristics of objectives, including:

   1. Compatible work effort: Modification/Improvements, New Development, and Maintenance.

      a. For Modification/Improvements and Maintenance, note the common system resources affected by the objective.

      b. Compatible due dates.

c. Common areas of the company affected by the objectives.

Great care should be taken when combining an objective with other objectives that are currently being implemented by project work (especially those projects nearing completion).

PROJECTS & METHODOLOGIES

A project is a scope of work consisting of one or more phases of work. It is an application of resources to a specific objective through the execution of a prescribed sequence of events. This sequence is governed by a methodology.

Since projects implement objectives, and objective priorities are constantly changing, project priorities are constantly changing. This means that project priorities can be derived from the priorities of the objectives that the project implements. As in the case of ranking objectives, projects can also be ranked from 1 to X. This is calculated simply by taking the CURRENT ranking of all objectives implemented by a project and calculating the average. The average sets the ranking precedence. Objective due dates and dates submitted are used to break ties in priorities.

Like the objective ranking, the project ranking is a "living" document that is constantly evolving. It should be monitored and analyzed on a routine basis.

Projects employ methodologies in order to provide organization and to denote precedent relationships, both internal to the project and external to other projects.

Methodologies consist of one or more phases of work. Each phase represents a key event in a project and consists of one or more activities. Each activity represents sub-events that are used to accomplish the overall key event. Each activity consists of a series of operational steps that represent specific tasks within each phase.

Sample methodologies include EEM, ISEM, and DBEM.

ENTERPRISE INFORMATION STRATEGY (EIS)

An EIS is a plan to satisfy the information needs of an enterprise. It is not something that is permanently set; rather, it is a "living" document that is constantly evolving and staying in tune with the enterprise. It must accommodate ever changing objectives and projects (new, defer, complete, cancel). The EIS details the following items:

1. Enterprise objectives, in priority ranking - this includes the type of work effort involved and the parts of the enterprise affected.

2. Enterprise projects, in priority ranking - this includes projects for developing information systems, as well as non-systems related projects. It also includes a description of the parts of the enterprise affected.

3. A justification for the ranking.

Although the Enterprise Engineering function assists in the development of the EIS, it is Executive Management and Information Resource Management that determines the final plan.

For a complete description of these key concepts, please refer to the narrative as provided in the methodology introduction section.

DESCRIPTION OF PHASE ACTIVITIES

Activity A - Detail Estimate & Schedule

Enterprise Engineering prepares a Detail estimate and schedule for the activities of the phase, which is reviewed with Project Management for approval.

Activity B - Group Requirements into Objectives

Enterprise Engineering evaluates the Information Requirements as they pertain to this EEM project. Requirements are analyzed in terms of the types of information involved (Policy, Control and Operational), whether they are being implemented by an existing system, the functions and positions they serve, etc.

Enterprise Engineering then groups Information Requirements into Objectives (MI) based on commonality. A cost/benefit analysis of the information requirements is also performed when grouping requirements. This grouping will be proposed to management for decision during the phase review.

Activity C - Group Objectives into Projects

Enterprise Engineering groups Objectives (MI) into

proposed Projects (PD). This grouping will also be reviewed with management as part of the Enterprise Information Strategy.

Activity D - Prepare Enterprise Information Strategy

Enterprise Engineering prepares a preliminary Enterprise Information Strategy (EIS) for review in the next activity. The EIS lists objectives and projects in priority ranking, along with a narrative justifying the rankings. This is packaged as a formal report which is checked by Project Management and Enterprise Resource Management as a preliminary review, and Quality Assurance to verify that it conforms with standards.

Activity E - Phase 4 Review

Enterprise Resource Management conducts a formal review of the Phase 4 deliverables with Information Resource Management and Executive Management. Enterprise Engineering and Project Management also participate in the review to answer questions. At this time, management reviews the proposed "Enterprise Information Strategy" as prepared in the previous activity, consisting of:

- Phase Cover Page
- Current Objective Ranking Report
- Current Project Ranking Report
- Business Plan
- Various matrices to show relationships between IRM resources
- Proposed Objective Ranking Report
- Proposed Project Ranking Report
- Enterprise Information Strategy (textual justification)
- Phase Review Checklist

This review results in a final determination of the Enterprise Information Strategy so that Information Resource Management can implement accordingly.

The next few pages include sample deliverables resulting from this phase.

SAMPLE PHASE 4 DELIVERABLES

OD-90236 CURRENT OBJECTIVE RANKING REPORT "PRIDE"-EEM

OCT 12, 2004 RELEASE 1.2.0 PAGE 1.1

THE FOLLOWING OBJECTIVES PERTAIN TO: FE-00100 TRUNDLE MANUFACTURING CORP.

RANK	MI NO.	OBJECTIVE NAME	WGHT	DATE REQUIRED	REQUEST STANDING	PROJECT NUMBER
0001	MI-00112	AUTOMATE PRDCT ASSEMBLY SCHDL	09	ASAP	APPROVED	PD00012
0002	MI-00103	IMPRV CSTMR SRVC RESP. TIME	21	ASAP	REVIEW	
0003	MI-00116	TMC ENTERPRISE ANALYSIS	21	2004/11/30	ACTIVE	PD00002
0004	MI-00111	CENTRALIZ HUMAN RSRC MGMT SYS	21	2005/11/30	ACTIVE	PD00015
0005	MI-00108	AUTOMATE PARTS REORDERING	29	ASAP	REVIEW	PD00012
0006	MI-00104	REDUCE PRODUCT SHIPPING COSTS	29	ASAP	APPROVED	PD00014
0007	MI-00109	MODIFY PARTS INVENTORY SYSTEM	29	2005/01/01	REVIEW	PD00012
0008	MI-00113	MODIFY ACCTS RECV. SYSTEM	29	OPEN	REVIEW	PD00014
0009	MI-00101	DEFINE NEW MARKETS/PRODUCTS	55	2005/05/01	APPROVED	PD00013
0010	MI-00115	MODIFY PAYROLL SYSTEM	59	ASAP	APPROVED	PD00015
0011	MI-00102	TRACK ADVRT TO SALE GENERATED	59	OPEN	REVIEW	PD00010
0012	MI-00114	INTGRT ACCTS PAY W/VENDOR SYS	79	OPEN	REVIEW	PD00010

* * * END OF OUTPUT * * *

OD-90237 CURRENT PROJECT RANKING REPORT "PRIDE"-EEM

OCT 12, 2004 RELEASE 1.2.0 PAGE 1.1

THE FOLLOWING OBJECTIVES PERTAIN TO: FE-00100 TRUNDLE MANUFACTURING CORP.

RANK	PD NO.	PROJECT NAME	PROJECT STANDING	DATE REQUIRED	MI NUMBER
0001	PD-00002	TRUNDLE EEM PROJECT	APPROVED	2004/11/30	MI-00116
0002	PD-00012	INTEGRATED MRP SYSTEM	ACTIVE	ASAP	MI-00109 MI-00108 MI-00112
0003	PD-00015	CNTRL HUMAN RSRC SYS (CHRiS)	APPROVED	2005/04/30	MI-00111 MI-00115
0004	PD-00014	TRUNDLE ORDER SHIPPING SYSTEM	ACTIVE	OPEN	MI-00113 MI-00104
0005	PD-00013	NEW PRODUCT DEVELOPMENT SYS	APPROVED	OPEN	MI-00101
0006	PD-00010	VENDOR INFO/PAYMT RPRTG-VIPR	ACTIVE	2005/12/31	MI-00102 MI-00114

* * * END OF OUTPUT * * *

SAMPLE PHASE 4 DELIVERABLES

OD-90391	ENTERPRISE INFORMATION STRATEGY	"PRIDE"-EEM
OCT 11, 2004	RELEASE 1.2.0	PAGE 1

AS DEFINED ON PD-00002 TRUNDLE EEM PROJECT

* * WARNING: CONFIDENTIAL! * *

The Enterprise Information Strategy consists of a priority ranking of objectives and projects, both of which have been attached to this report.

The E.I.S. reflects the direction of the corporation as stated in the TMC Business Plan. Because of management's desire for tighter control over cash flow and to improve quality and delivery, the corporation will concentrate on Sales and Manufacturing related priorities. This specifically includes creating a new Sales System and enhancing the MRP system.

- B. Karpoff
Information Resource Manager
October 11, 2004

* * * END OF OUTPUT * * *

OD-90262	E.I.S. PROPOSED MI RANKINGS		"PRIDE"-EEM
OCT 11, 2004	RELEASE 1.2.0	DATE OF LAST RANK - 2002/05/29	PAGE 1

THIS RANK	LAST RANK	OBJECTIVE NUMBER & NAME	STANDING	PROJECTS
0001	NEW	MI-00901 SALES SYSTEM EVALUATION	APPROVED	PD-00901
		WEIGHT=05 DATE REQUIRED=2002/09/30 DATE REQUESTED=2002/05/15		
0002	0001	MI-00112 AUTOMATE PRDCT ASSEMBLY SCHDL	APPROVED	PD-00012
		WEIGHT=09 DATE REQUIRED=ASAP DATE REQUESTED=2001/12/14		
0003	0002	MI-00103 IMPRV CSTMR SRVC RESPONSE TIME	REVIEW	
		WEIGHT=21 DATE REQUIRED=ASAP DATE REQUESTED=2005/05/18		
0004	0003	MI-00116 TMC ENTERPRISE ANALYSIS	ACTIVE	PD-00002
		WEIGHT=21 DATE REQUIRED=2004/06/01 DATE REQUESTED=2003/12/07		
0005	0004	MI-00111 CENTRALIZE HUMAN RSRC MGMT SYS	ACTIVE	PD-00015
		WEIGHT=21 DATE REQUIRED=2005/11/30 DATE REQUESTED=2004/06/19		

(continued on next page)

SAMPLE PHASE 4 DELIVERABLES

(continued from previous page)

0006	0005	MI-00108 AUTOMATE PARTS REORDERING	REVIEW	PD-00012
		WEIGHT=29 DATE REQUIRED=ASAP DATE REQUESTED=2004/01/03		
0007	0006	MI-00104 REDUCE PRODUCT SHIPPING COSTS	APPROVED	PD-00014
		WEIGHT=29 DATE REQUIRED=ASAP DATE REQUESTED=2004/05/18		
0008	0007	MI-00109 MODIFY PARTS INVENTORY SYSTEM	REVIEW	PD-00012
		WEIGHT=29 DATE REQUIRED=2005/01/01 DATE REQUESTED=2002/04/15		
0009	0008	MI-00113 MODIFY ACCTS RECV. SYSTEM	REVIEW	PD-00014
		WEIGHT=29 DATE REQUIRED=OPEN DATE REQUESTED=2002/05/03		
0010	0009	MI-00101 DEFINE NEW MARKETS/PRODUCTS	APPROVED	PD-00013
		WEIGHT=55 DATE REQUIRED=2005/05/01 DATE REQUESTED=2002/05/18		
0011	0010	MI-00115 MODIFY PAYROLL SYSTEM	APPROVED	PD-00015
		WEIGHT=59 DATE REQUIRED=ASAP DATE REQUESTED=2003/01/07		
0012	0011	MI-00102 TRACK ADVRT TO SALE GENERATED	REVIEW	PD-00010
		WEIGHT=59 DATE REQUIRED=OPEN DATE REQUESTED=2004/05/18		
0013	0012	MI-00114 INTGRT ACCTS PAY W/VENDOR SYS	REVIEW	PD-00010
		WEIGHT=79 DATE REQUIRED=OPEN DATE REQUESTED=2002/05/03		

* * * END OBJECTIVE RANKING * * *

SAMPLE PHASE 4 DELIVERABLES

```
--------------------------------------------------------------------------------------------------
OD-90263                      E.I.S. PROPOSED PD RANKINGS                          "PRIDE"-EEM
--------------------------------------------------------------------------------------------------
OCT 11, 2004             RELEASE 1.2.0      DATE OF LAST RANK - 2002/05/29          PAGE 1

THIS    LAST
RANK    RANK    PROJECT NUMBER & NAME                       STANDING        DATE REQ'D

0001    NEW     PD-00901 TRUNDLE SALES SYSTEM PROJECT       ACTIVE          ASAP

                OBJECTIVES      RANKING
                MI-00901 0001          OBJECTIVE AVERAGE RANK=0001

0002    0001    PD-00002 TRUNDLE EEM PROJECT                APPROVED        2004/06/30

                OBJECTIVES      RANKING
                MI-00116 0004          OBJECTIVE AVERAGE RANK=0004

0003    0002    PD-00012 INTEGRATED MRP SYSTEM              ACTIVE          ASAP

                OBJECTIVES      RANKING
                MI-00108 0006
                MI-00109 0008
                MI-00112 0002          OBJECTIVE AVERAGE RANK=0005

0004    0003    PD-00015 CNTRL HUMAN RSRC SYS (CHRiS)       APPROVED        2005/04/30

                OBJECTIVES      RANKING
                MI-00111 0005
                MI-00115 0011          OBJECTIVE AVERAGE RANK=0008

0005    0004    PD-00014 TRUNDLE ORDER SHIPPING SYSTEM      ACTIVE          OPEN

                OBJECTIVES      RANKING
                MI-00104 0007
                MI-00113 0009          OBJECTIVE AVERAGE RANK=0008

0006    0005    PD-00013 NEW PRODUCT DEVELOPMENT SYS        APPROVED        OPEN

                OBJECTIVES      RANKING
                MI-00101 0010          OBJECTIVE AVERAGE RANK=0010

0007    0006    PD-00010 VENDOR INFO/PAYMT RPRTG (VIPR)     ACTIVE          2004/12/31

                OBJECTIVES      RANKING
                MI-00114 0013
                MI-00102 0012          OBJECTIVE AVERAGE RANK=0012

                        * * * END PROJECT RANKING * * *
```

SAMPLE PHASE 4 DELIVERABLES

OD-90228	WORK REQUEST/OBJECTIVE	"PRIDE"-EEM
OCT 10, 2004	RELEASE 1.2.0	PAGE 1.1.1

```
*********************************************************************
* MI-00901           SALES SYSTEM EVALUATION       *
*********************************************************************
```

REQUIRED DATE- ASAP   USER PRIORITY WEIGHT- 15   TYPE- MODIFICATION
DATE RECEIVED- 2004/10/18   REQUEST STANDING - ACTIVE

DESCRIPTION-

The purpose of this request is to investigate problems with the current sales system and make recommendations for improvement. The current system appears to be very labor intensive, requiring manual extraction and manipulation of data which could be prone to error, causing false information.

The Sales System directly affects the Sales Department and indirectly affects the following departments: Customer Services, Production, Shipping, and Materials.

SEVERITY/PRIORITY

The Sales System is key to the company in that it is the primary means to control sales activities, evaluate income which influences cash flow, and make recommendations to Production for scheduling.

PROJECT(S) IMPLEMENTING OBJECTIVE:

PD-00452   TRUNDLE SALES SYSTEM PROJECT

REQUIREMENTS AFFECTED-

IR-00110   SALES PERFORMACE ANALYSIS
IR-00111   SALES REP INFO
IR-00112   MONTHLY SALES ANALYSIS
IR-00113   WEEKLY SALES COMMISSIONS

INPUTS/OUTPUTS AFFECTED-

(NONE SPECIFIED)

FILES AFFECTED-

(NONE SPECIFIED)

SYSTEM RESOURCES AFFECTED-

TR CURRENT TRUNDLE SALES SYSTEM

* * * END OF OUTPUT * * *

SAMPLE PHASE 4 DELIVERABLES

OD-90187 PROJECT SCOPE "PRIDE"

OCT 10, 2004 RELEASE 1.2.0 PAGE 2.1.1

```
**********************************************************
*              PD-00437 FINANCE PROJECT                  *
**********************************************************

PROJECT STANDING- ACTIVE              TYPE- DIRECT
DATE REQUIRED- ASAP                   PRIORITY- 0003
PLANNED DATE- 2005/06/01   A          S OF- 2004/05/01
```

DESCRIPTION-

The purpose of this project is to implement several changes to the Finance System, resulting from new government regulations. The IRM Department will execute the project. Finance and Accounting will act as the user and also perform the Project Management function.

THIS PROJECT IMPLEMENTS THE FOLLOWING OBJECTIVE(S):

MI-00019 FEDERAL PROCUREMENT CHANGES
MI-00052 STATE TAXATION

REQUIREMENTS AFFECTED-

IR-00100 FEDERAL REPORTING
IR-00201 MONTHLY STATE TAXATION

SYSTEM RESOURCES AFFECTED-

FS FINANCE SYSTEM

*** END OF OUTPUT ***

END SECTION

Section 7

"PRIDE"-EEM PHASE 5 EEM EVALUATION

A project will only be accomplished if the individuals performing the work want to do it.
- Bryce's Law

The purpose of this phase is to evaluate project results and initiate pertinent ISEM/DBEM/EEM projects. Two events occur during this phase:

- ISEM, DBEM and EEM projects are started in accordance with the Enterprise Information Strategy.

- An evaluation of the EEM project is performed included an analysis of proposed versus actual estimated time, costs and schedules.

METHODOLOGY NAVIGATION

This represents the concluding phase of the EEM project. It is performed either immediately after Phase 4, or if a project has been discontinued earlier on (such as after Phase 1, 2 or 3).

GENERAL DISCUSSION

With the Enterprise Information Strategy in place, Information Resource Management begins to implement project plans. Systems Resource Management, Data Resource Management, Enterprise Resource Management and Data Processing Operations are briefed on the strategy and their role in its implementation. Objectives and Project Scopes are reviewed and tactics discussed. Following this meeting, Information Systems Engineering projects, Data Base Engineering projects and additional Enterprise Engineering projects are initiated.

Project Management performs an evaluation of the EEM project. Order-of-Magnitude estimates and schedules as prepared in Phases 1, 2 and 3 are compared to actual figures. An OOM Estimate/Actual report is used to show actual time, costs and schedules. Actual figures are compared and analyzed to the original estimates. This proposed versus actual comparison is then reviewed with project personnel, Enterprise Resource Management and Information Resource Management. Based on the findings of this analysis, it may be desirable to update estimating guidelines. Project Management then prepares concluding comments about the project which is included in the formal phase deliverable.

DESCRIPTION OF PHASE ACTIVITIES

Activity A - Initiate Strategy

Project Management prepares a Detail estimate and schedule for the activities of the phase.

Information Resource Management initiates the Enterprise Information Strategy by establishing pertinent ISEM, DBEM and EEM projects. Systems Resource Management, Data Resource Management, Enterprise Resource Management, and Data Processing Operations are briefed on the EIS and required project work.

Activity B - Prepare EEM Evaluation

Project Management prepares an "EEM Project Evaluation Report" which provides an analysis of proposed versus actual project time, estimates and schedules. It is based on the project plans, estimates and schedules as developed in Phases 1, 2 and 3. Concluding comments about the project are prepared by Project Management and packaged with the OOM Estimate/Actual analysis as part of the final phase report. This report is then reviewed by the members of the project, Enterprise Resource Management and Information Resource Management.

OD-90392	PROJECT EVALUATION	"PRIDE"-EEM
OCT 16, 2004	RELEASE 1.2.0	PAGE 1.1

AS DEFINED ON PD-00002 TRUNDLE EEM PROJECT

The corporate planning project was recently and successfully concluded. During this time, TMC's logical and physical sides of the business were documented and analyzed. From this analysis an Enterprise Information Strategy (EIS) was calculated and set into motion by Executive Management. This has triggered several other projects for systems and data base design.

Although the project was successfully completed, the learning curve used to implement the methodology resulted in a 9% overrun on the Phase 1 Order-of-Magnitude Estimate, for a total project cost of $55,750. However, the software tools accompanying the methodology allowed us to complete the project approximately six weeks ahead of schedule. The entire project took approximately two and a half months to perform. Consult the attached OOM Estimate/Actual report for details on estimated versus actual figures.

LESSONS LEARNED

1. The "priority modeling" tool provided by "PRIDE" is capable of automatically recalculating priorities on request. It is suggested that the Enterprise Resource Manager monitor priorities on a weekly basis. A formal review of all priorities should be conducted on a quarterly basis.

2. Due to time constraints of the project, detailed descriptions of jobs within the Financial Resource Management area were not completed as thoroughly as desired. It is recommended that the Personnel Department complete this work.

3. User cooperation significantly expedited project development for which we give our thanks. 4. All user departments can now query their objective and project priorities through a query facility.

- Elaine Nardo
Project Manager
October 13, 2004 *

* * END OF OUTPUT * * *

END SECTION

Section 8

"PRIDE"-ISEM INFORMATION SYSTEMS ENGINEERING METHODOLOGY

An information system is a product that can be engineered and manufactured like any other product.
- Bryce's Law

The "PRIDE"-Information Systems Engineering Methodology (ISEM) is an important part of an overall philosophy of Information Resource Management (IRM) as defined by "PRIDE". This involves the development and control over all of the resources required to produce information. Whereas the "PRIDE"-Enterprise Engineering Methodology (EEM) is principally concerned with developing an Enterprise Information Strategy, methodologies such as ISEM and the "PRIDE"-Data Base Engineering Methodology (DBEM) are concerned with actually building the system and data resources needed to produce information.

The intent of any of the "PRIDE" methodologies, including ISEM, is to define the business environment as to "Who" is to perform "What," "When," "Where" and "Why" (the "5W's"). As a result, it is used to convert a heterogeneous operating environment into a homogeneous environment. This improves communications and promotes cooperation and teamwork throughout an enterprise. Better organization and discipline also enhances the ability to build quality products and make effective use of resources. In addition, to the 5W's, the methodology provides "How" to perform the work by providing a variety of techniques and tools that are deployed throughout the methodology. A methodology, therefore, resembles an assembly line where work is performed in manageable stages.

ISEM is a generic and universally applicable approach for building any type of information system, regardless of industry, type of application, or software language/technique. It is based on tried and proven approaches that are so fundamental to sound system design that tailoring to individual development requirements is not only unnecessary but highly undesirable. It has been used around the world to develop applications in every field of endeavor imaginable, including banking, communications, education, government, insurance, manufacturing, retail, transportation, utilities, etc.

ISEM is concerned with building major systems in their entirety; this includes integrated business processes, complete with administrative and computer processing. As such, ISEM provides the high-level systems architecture required by software engineering to develop computer programs. Because of this, software engineering is considered a subset of ISEM. Although ISEM provides the ability to design any type of system, it does not mandate the use of any particular programming language or design technique. Consequently, the programming phases of ISEM (4-II, 5 and 6) allow for the use of any pertinent software engineering tool or technique.

CONCEPTS & PHILOSOPHIES

"It must be remembered that there is nothing more difficult to plan, more doubtful of success, nor more dangerous to manage than the creation of a new system. For the initiator has the enmity of all who would profit by the preservation of the old institution and merely lukewarm defenders in those who would gain by the new ones."

- Machiavelli, "The Prince" (1513)

INFORMATION = DATA + PROCESSING

The processing of data is surely as important as the data to be processed. Whereas the data represents "what" is to be processed, processing (or systems) represents "how" it is to be processed, using formulas, algorithms or calculations. An invalid calculation is just as misleading as invalid data; both will produce erroneous information. From this perspective, both data and processing must be carefully designed and controlled as resources for producing information. Whereas Data Base Engineering is concerned with the integrity of the data, Information Systems Engineering is concerned with the integrity of systems.

An information system is not an amorphous mass. It has specific characteristics that can be precisely defined. For example:

- It has a purpose; to produce information to serve business needs.

- It operates routinely in a predictable manner. In fact, it runs in time frames required to make specific decisions and take actions.

- It is a grouping of one or more elements which form a whole; this defines the scope of the system.

These three characteristics represent the basic elements of any system, information or otherwise.

With the advent of the computer many in the field have come to regard an information system as nothing more than a collection of programs or "jobs." From this perspective, there is little consideration for processing through non-computer means. As a result, non-computer processing and data are treated as irrelevant, and users are left with the awkward task of synchronizing administrative processing with computer processing.

Those biased with programming suffer from a narrow perspective of the system. Anything outside of computer software and hardware is considered meaningless. This is one of the primary reasons why information systems lack integration; some developers tend to ignore the overall picture and concentrate only on the computer. Many developers are more concerned with programming than they are with engineering the overall system. The analogy here is that there are many carpenters on the job, but very few architects. This is why systems development is usually regarded as an art rather than a science.

A SYSTEM IS A PRODUCT

The fact of the matter is, an information system is a product that can be engineered and manufactured like any other product. With a product orientation, a system consists of several levels of abstraction, from general to specific, that define the processing.

Information System - Represents the highest level of the product and the scope of the system. An information system is arbitrarily defined by an organization; it is based on the total number of sub-systems that an organization desires to satisfy business needs. It is possible for a company to operate with a single system consisting of hundreds of sub-systems. It is also conceivable to have several systems. It ultimately depends on how the organization perceives the system. In any case, an information system is defined as one or more sub-systems that satisfy information requirements in particular time frames.

Sub-Systems - Represents a business process that exists within a unique time frame. It is a logical process that dictates "what" data is to be processed and "when." Each sub-system consists of one or more administrative procedures and one or no computer procedures. As an example, a payroll application may have the following sub-systems:

- Daily time posting by non-exempt employees.
- Weekly time posting by exempt employees.
- Payroll/Employee adjustments.
- Query employee time history.

- Weekly paychecks for non-exempt employees.
- Monthly paychecks for exempt employees.
- Monthly government reporting.
- Quarterly government reporting.
- Year end government reporting.

Administrative Procedures - Physically defines "who" is to perform "what" operations from an administrative perspective, e.g., manual processing, office automation processing, data entry, etc. The administrative procedure consists of one or more operational steps.

Operational Steps - Represent specific actions to be performed or decisions made.

Computer Procedure - Like the administrative procedure, this procedure defines "what" operations the computer will perform. The computer procedure consists of one or more programs.

Programs - These are comparable to the operational steps of the Administrative Procedure. Programs represent executable tasks to be performed by the computer.

There is little difference between an administrative procedure and a computer procedure; the only difference is the "actor" who has to perform the work. Whereas administrative procedures will deal with manual and non-computer processing, the computer procedure deals with computer processing. Of the two types of procedures, administrative procedures are the most commonly overlooked and omitted. Systems fail more for the lack of administrative procedures than they will for well written computer procedures. If people do not know how to use the systems, the systems will not be used.

This product structure provides the architectural framework required to engineer an information system. All information systems, that function, have this structure. In manufacturing terms, it can be described as a "four-level bill of materials" which shows the product from the general to the specific. Any information system can be modeled using this standard system structure. Omission of any of these resources will highlight problem areas in a system.

STANDARD SYSTEM STRUCTURE
A System is a Product

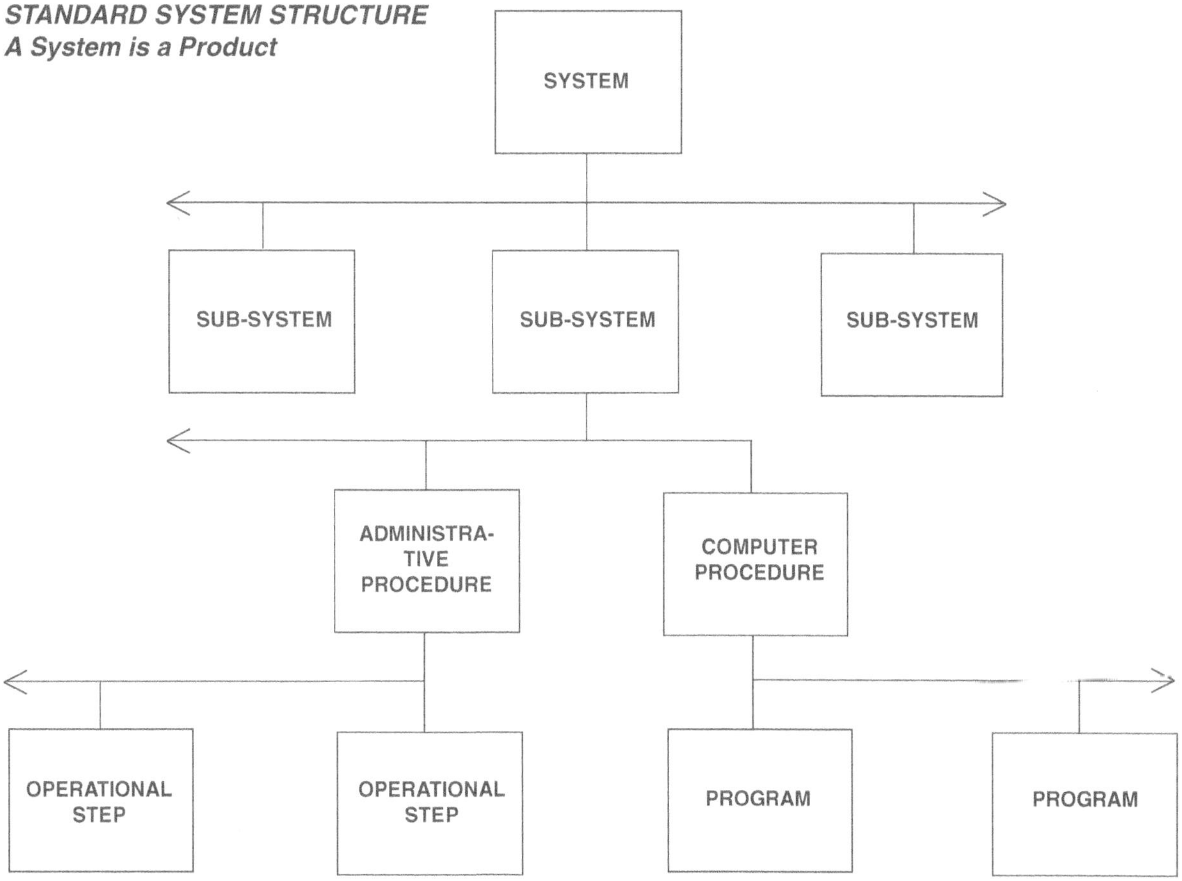

During system design, the product structure is decomposed or "exploded" downward until the smallest elements are defined. The product structure is then "imploded" upward during testing and implementation. This "explosion/implosion" technique is common to engineering/manufacturing.

SYSTEMS INTEGRATION

The only way systems communicate, internally or externally to other systems, is through data. It is the only way that a system interfaces with other systems, a sub-system to other sub-systems, an administrative procedure to a computer procedure, program to program; through data. Data represents the adhesive element that holds systems together. It should be defined and standardized so it can be re-used in as many applications that require it. However, because many people in the field view systems as nothing more than isolated programs, they feel it is more convenient to define data on an application-by-application basis. This is the primary reason for data redundancy.

Under the engineering/manufacturing philosophy, data is treated as a "materials management" issue where it is standardized and controlled for maximum efficiency and cost effectiveness. This scenario calls for data to be defined in the same manner as a part to a product: it is classified, inventoried and re-used. Data base, therefore, is treated as a management issue concerned with controlling all of the data required to produce information, regardless of where used or how stored. It is not a technical problem that is concerned only with data physically stored on the computer and managed by a DBMS.

INFORMATION DRIVEN DESIGN

Since the intent of an information system is to produce information, the more we understand about information, the better we can accommodate its implementation. This is why we refer to this concept as "Information Driven Design," a system design based on the characteristics of information.

INFORMATION DRIVEN DESIGN CONCEPT

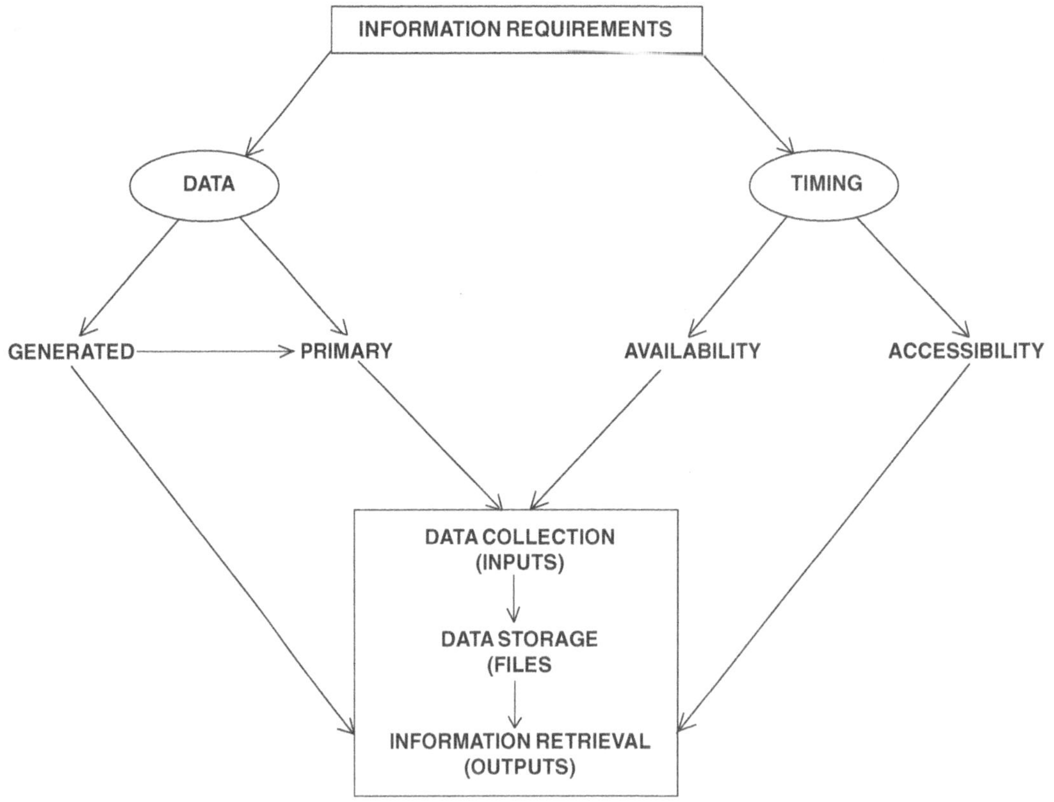

INFORMATION SYSTEMS

Information is used to support specific actions and/or decisions of the business. There are actually three types of information:

• **Policy Information** - To implement executive decisions.

• **Control Information** - To monitor policy and manage operations.

• **Operational Information** - To implement the routine day-to-day activities of the business.

Information is a dynamic and perishable commodity. It only has value at the time it is required. Whereas the definition of data is constant, information requirements can change for a variety of reasons, such as politics, government, competition, economics, people, etc. Ultimately, corporate survival depends on providing the users with accurate and timely information.

TIMING

In order to properly specify information requirements, it is not sufficient to merely determine what data is necessary; it is also necessary to define the timing of the information. This timing will ultimately dictate how data will be collected, stored, and retrieved to produce information.

Failure to recognize timing as an important element of design will result in the data base being out of synchronization with the application. For example, consider a situation where data is collected on a routine weekly basis. Daily analysis of the data will be inappropriate since the data will remain constant until the next weekly update. To resolve the conflict, data collection should be changed to daily basis.

All information systems operate in time frames, such as instantaneous, daily, weekly, monthly, quarterly, etc. If this is true, why not make use of this timing consideration during system design as opposed to discovering it afterwards while trying to correct the data base design?

There are three aspects to timing: frequency, offset and response time.

• **FREQUENCY** defines "how often" the information is required, e.g., upon request, hourly, twice daily, once a week, monthly, quarterly, semi-annually, etc.

• **OFFSET** defines when processing should begin, such as the beginning of the week, end of the month, etc. However, if the frequency is 'upon request,' then there is no scheduled offset; this is because the information can be requested at any point in time.

• **RESPONSE TIME** defines how fast the information must be delivered to the user. For example, five seconds, two hours, one day, etc. This should not be confused with computer 'response time' or 'throughput' which is concerned with machine speed. Rather, response time is concerned with the maximum amount of time that will transpire between the request and delivery of the information, so the user can make the necessary decisions and/or take action. This implies that if the response time is exceeded, it is no longer information, only historical data.

The data required to produce information must be defined in terms of what primary data must be supplied externally to the system by a user, and what generated data must be calculated internal to the system. Data relationships can be extensive. For example, the data element "Net Pay" may be based on a complicated computation: NET PAY = GROSS PAY - FICA - CITY TAX - UNION DUES - (ETC.)

Other data elements used in the formula may also be generated, such as: GROSS PAY = HOURS WORKED X PAY RATE

What this means is that in order to arrive at the correct value for "Net Pay," we must be able to reach all of the primary values, such as "Hours Worked" and "Pay Rate," in a timely manner. If we cannot do this, "Net Pay" will be incorrect.

Defining these data dependencies has typically defaulted to the programmer who redefines the relationships with each application and buries it in the program source code, making maintenance and change considerably difficult. (It is not unusual to find "Net Pay" defined differently in multiple applications throughout a company.)

The basic operations that can be performed on data include "create," "update" and "reference" ("delete" is the opposite of "create"). In programming terminology, a "create" represents a "write," an "update" represents a "read/write," and a "reference" represents a "read" only.

The timing and data specifications resulting from the

information requirements will ultimately dictate the type of system to be created. For example, if information is needed upon request and within a matter of seconds, this will probably result in an "interactive" type of process. However, if the information is required upon request but within a few hours, this will probably result in "batch" type processing (it may even be processable manually). These specifications are the basic building blocks for all systems and software design.

Producing an information system design that correctly satisfies requirements is a vital part of information driven design. The resulting system will only be as good as the specifications from which it is based. With this approach, the emphasis is on business analysis as opposed to technical detail. No amount of elegant programming or technology will solve a problem if it is not defined or understood correctly.

CHRONOLOGICAL DECOMPOSITION

This is the technique used to implement the concept of "Information Driven Design." As the name indicates, timing is an essential part of chronological decomposition. Based on information requirements, chronological decomposition breaks a system into its logical business processes (sub-systems) based on timing with their specific inputs, outputs and files. It synchronizes all of the data required to support the system into an application logical data base.

Essentially, the technique works backwards, from the receiver of the information, to the source of the primary data; from outputs to inputs. In this way, it forces the systems engineer to consider all of the data required for the application. This eliminates the problem of forgetting required data. It also emphasizes the need for sharing data. As a design develops, consideration is given to using data from other applications. After all, why create new files and processes if they already exist in the data base?

Perhaps the most significant benefits derived from chronological decomposition, as reported by users, is that it organizes the data and simplifies processing.

With the logical design defined, consideration is then given to the most appropriate way to physically process the data, either manually or computer assisted. This is where the "work flow" is designed, from start to end; from input to output. Here is where functional decomposition and data driven design excels. For software engineering, the characteristics of the data, its structures and what functions the computer must perform (e.g., create, update and reference) dictates the required programs. These specifications are the result of chronological decomposition. The physical characteristics of the data defines its validity. The data structures denote input, file and output relationships. The functional requirements determine how the data will be read and written in a program, whether sequentially, iteratively or selectively.

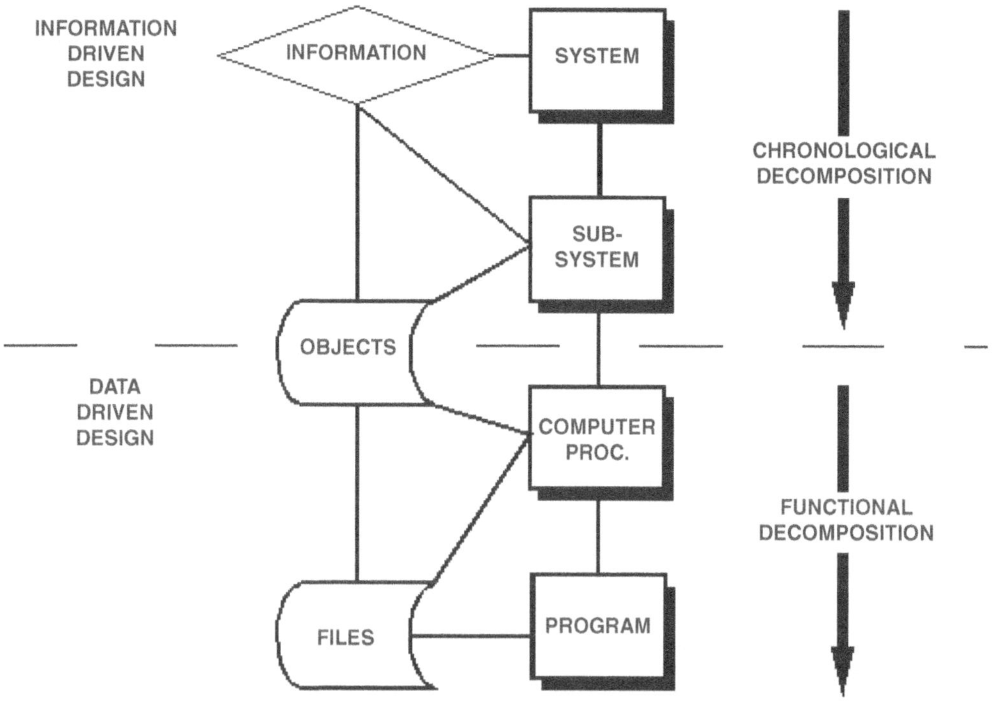

In no way should chronological decomposition be confused with functional decomposition. Both have distinctly different purposes and objectives. Whereas the former is concerned with logical system design based on the characteristics of information, the latter is concerned with how data is to be processed physically. They are not conflicting techniques. In fact, they are complementary as long as they are placed in the proper perspective. To try to use one in place of the other is unproductive. They simply have entirely different roles.

METHODS OF PROCESSING

Regardless of the jargon of the day, all processing is based on "transactions," whether it is to request an output or to collect data. Buzzwords like "on-line," "real-time," "interactive," and "batch" only cloud the issue.

A transaction represents some form of operation. For output reporting, "request," "display," "print," "extract," "search," etc. are common transactions. For input, "new," "add," "change," "delete," "update," "charge," "credit," "debit," "deposit," etc. are typical transactions.

Transactions can be processed either one at a time (as in "interactive") or in groups ("batch"). What is then important is the volume of transactions and the speed they must be processed (as specified by timing). This will determine the physical constraints of the equipment to be used. "Batch" processing has the advantage of processing high volumes of transactions within a relatively short period of time per transaction. "Interactive" processing has the advantage of processing individual transactions quickly.

The tendency to be more concerned with the techniques of processing than with the problems to be solved can cause systems to be physically implemented in ways resulting in high operating costs. The governing issues should be, *"What is the business problem to be solved?"* and, *"How effective is the processing method concerning the value and timing of the information to be produced?"* Programming techniques have little value in these areas. Consider, for example, that batch processing will always be a viable solution for information systems design regardless of the direction technology moves. This method of processing is analogous to many manufacturing operations where products are more economically manufactured in groups or batches - as opposed to one at a time. The point is, this situation will always be true regardless of how close the costs become between the two methods. For example, can you imagine preparing the corporate payroll interactively (one check at a time)? Technology should implement the logical business solutions to the problem and not be a platform to display technical elegance.

FORMS OF PROCESSING

In any processing, computer or otherwise, there are three basic forms of processing that can be used: sequence, iteration, and choice.

SEQUENCE *- This type of processing represents a continuous series of steps, for example:*

ITERATION *- Represents repetition until a certain condition is met, for example:*

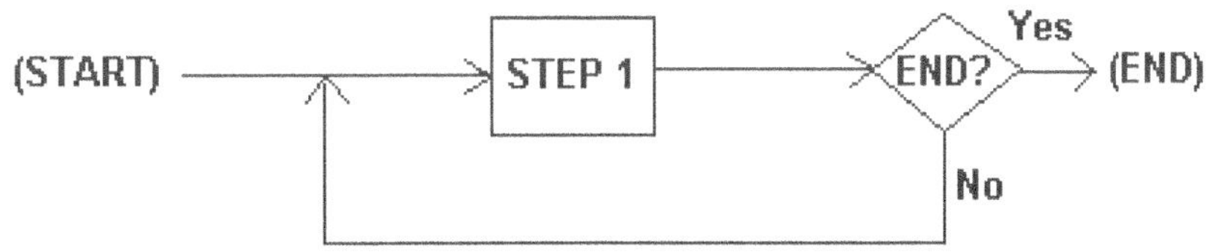

CHOICE *- Permits the selection of processing paths based on prescribed criteria. It also allows for parallelism; for example:*

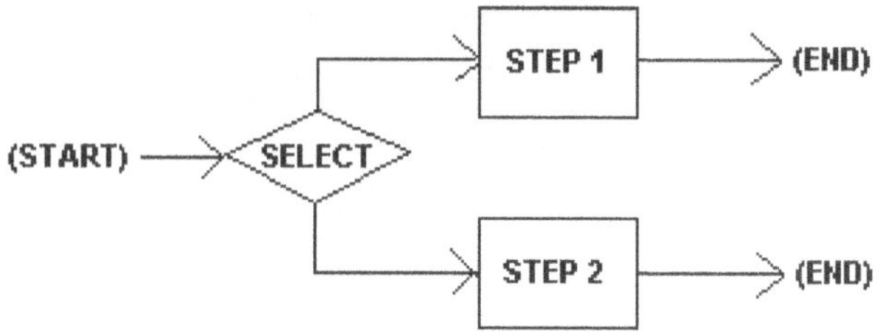

Obviously, these processing constructs can be combined in many different ways to process data. All procedures, programs, and steps consist of combinations of these constructs.

Again, timing has a role to play here. Frequency, offset, and response time are used to synchronize processing. It dictates how often processing must occur, when each process is to start, and how fast it must process data. Without timing, processing will be awkward and cumbersome to perform.

LAYERED DOCUMENTATION

Whereas system design moves a system from logical specifications to ultimate physical implementation, documentation, which is a reflection of the design, should do likewise. Documentation is a working tool and a by-product of design. This is comparable to blueprinting, where a product is designed to various levels of detail, from the general to the specific. In addition to being used as a tool to refine designs, this "layered" approach to documentation is used to communicate the various levels of abstraction to users and designers.

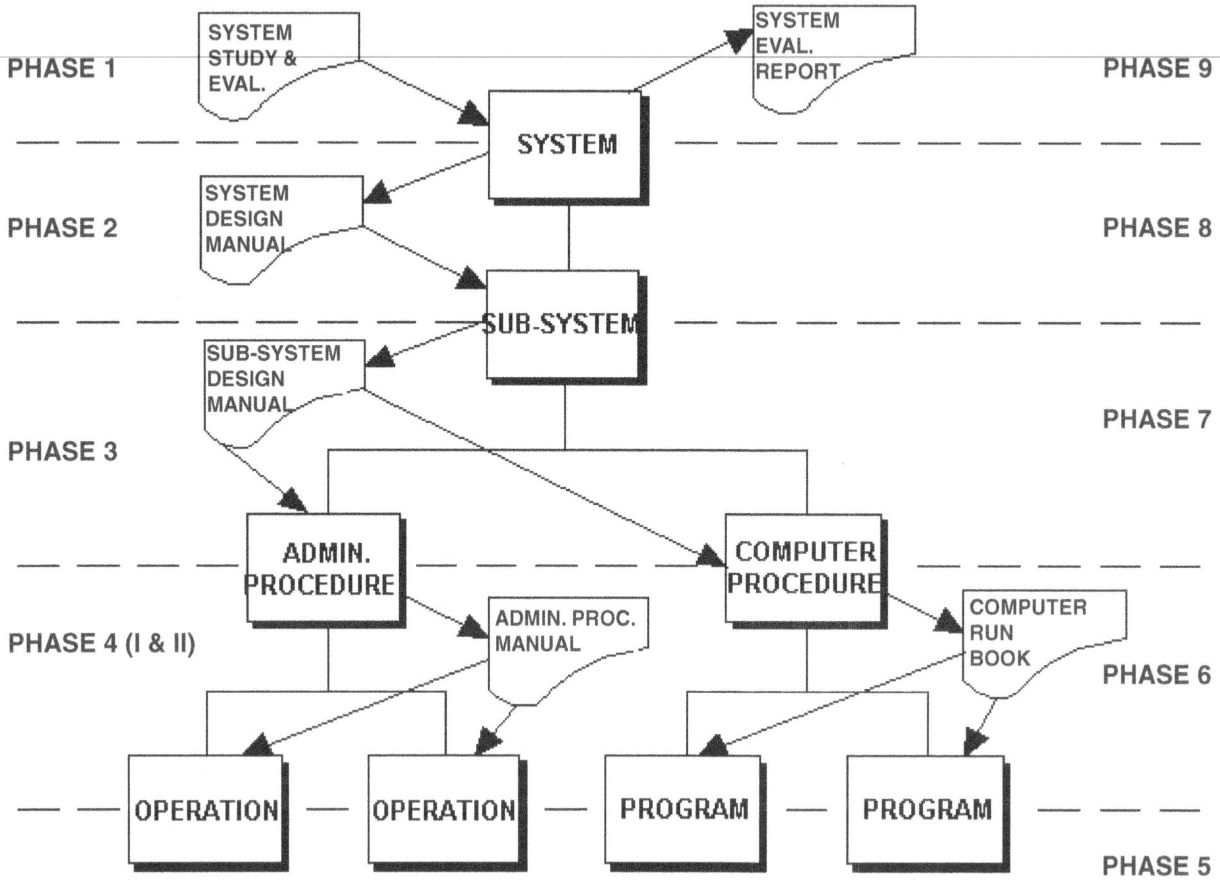

Each Level Provides Specifications for the Next

The major deliverables resulting from the design phases of ISEM include:

1. **System Study & Evaluation Report (Phase 1)** - This manual represents the "feasibility study" where information requirements are defined and a proposed system solution developed. An information flow diagram and system concept diagram are included in this manual.

2. **System Design Manual (Phase 2)** - This manual defines the specifications for all of the sub-systems in the application. A system flowchart is maintained in this manual which depicts the sub-systems within the system.

3. **Sub-System Design Manual (Phase 3)** - This manual defines the specifications for the procedures in a single sub-system. This includes a sub-system flowchart showing the procedures within the sub-system.

4. **Administrative Procedure Manual (Phase 4-I)** - This manual defines the specifications of the operational steps in the procedure. This represents the manual used by end-users when operating the system. Unlike the computer procedure flowchart that shows programs, the administrative procedure is written using "playscript," a manual procedure language developed by Les Matthies. "Playscript" is based on a script to a play, with "actors" and "actions."

5. **Computer Run Book (Phase 4-II)** - This manual defines the specifications for the programs in a computer procedure. This is where a computer procedure flowchart is maintained showing the programs within the procedure. In addition to being used for program design, this manual will become the documentation used by computer operations personnel. For each program, a Software Structure Diagram will be prepared expressing the internal processing logic.

This "blueprinting" approach to documentation shows how it is used as a working tool during design. As modifications are made to the system, the appropriate level in the documentation incorporates the changes. This simplifies maintaining documentation and keeps it current with the system. In addition to flowcharts and supporting narratives, test requirements are incorporated into the documentation for use during the testing and implementation phases of ISEM.

GRAPHICS

The basic types of graphics associated with ISEM include:

Information Flow Diagram (IFD) - This diagram illustrates how information is used by an enterprise for a single system. It defines the boundaries of the system as to who originates, receives and participates in the flow of information in an application. NOTE: The IFD should not be confused with "Data Flow Diagrams" as used in programming.

INFORMATION FLOW DIAGRAM

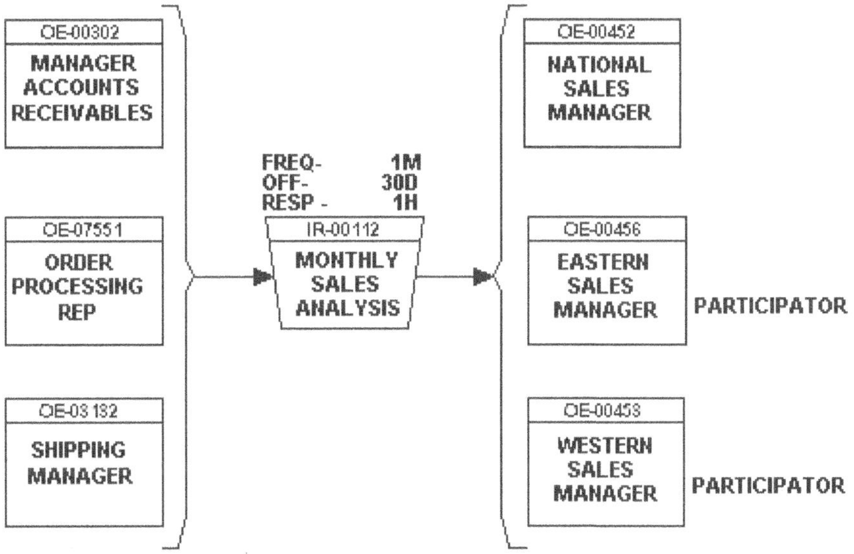

The IFD can also be drawn with Functional Entities (FE)

The **System Concept Diagram** provides a free-form schematic of the overall system. Like an architect's "rendering," this diagram is used to paint a picture for the user to easily visualize the end product. Like the Information Flow Diagram, it denotes the boundaries of the application.

SYSTEM CONCEPT DIAGRAM
Free-Form Schematic

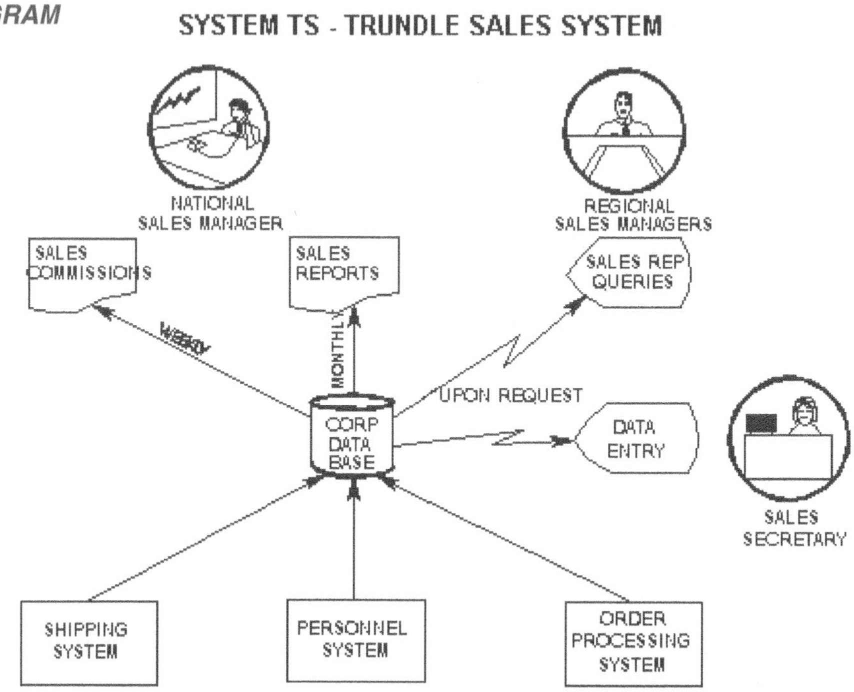

The **System Flowchart** shows the second level of detail in the system hierarchy: sub-systems. Each sub-system is drawn with its inputs, outputs, files and timing. A Table of Files section at the end of the flowchart shows how all of the files in the application are used by the various sub-systems (create/update/reference).

SYSTEM FLOWCHART
Uses ANSII Standard Symbols

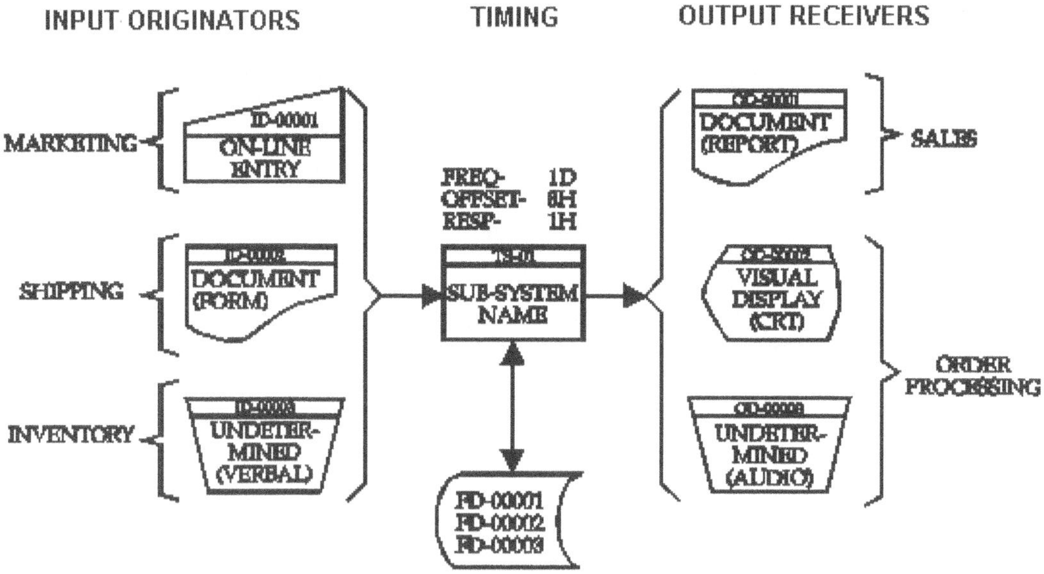

ARROWS INDICATE DIRECTIONAL PROCESSING FLOW (LEFT-TO-RIGHT)
GRAPHIC RELATES TO SUB-SYSTEM NARRATIVE

The **Sub-System Concept Diagram** is a free-form schematic illustrating the general processing flow in a sub-system. It is drawn in a similar manner as the System Concept Diagram.

SUB-SYSTEM CONCEPT DIAGRAM
Free-Form Schematic

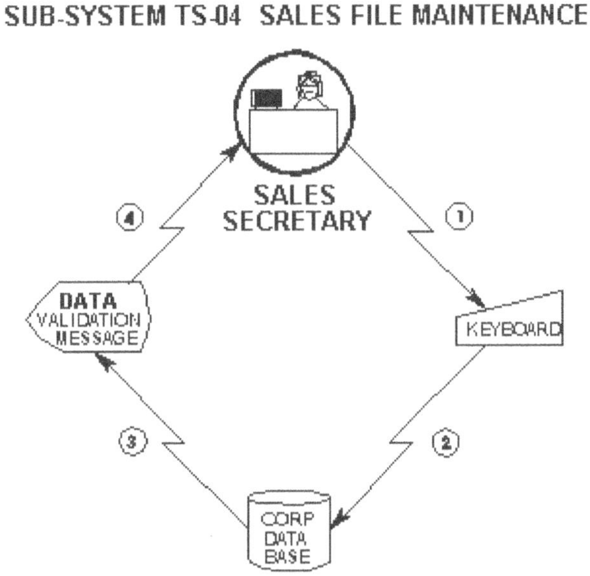

Used to describe the physical processing of the sub-system

The **Sub-System Flowchart** shows the third level of detail in the system hierarchy: procedures - both administrative and computer. This type of flowchart represents a "process diagram" showing the flow of work in the sub-system. The precedent relationships between procedures are drawn which specify sequence, iteration and choice. In addition, the use of inputs, outputs and files show how data flows throughout the procedures.

SUB-SYSTEM FLOWCHART - Batch Example
Merging Processing with Data

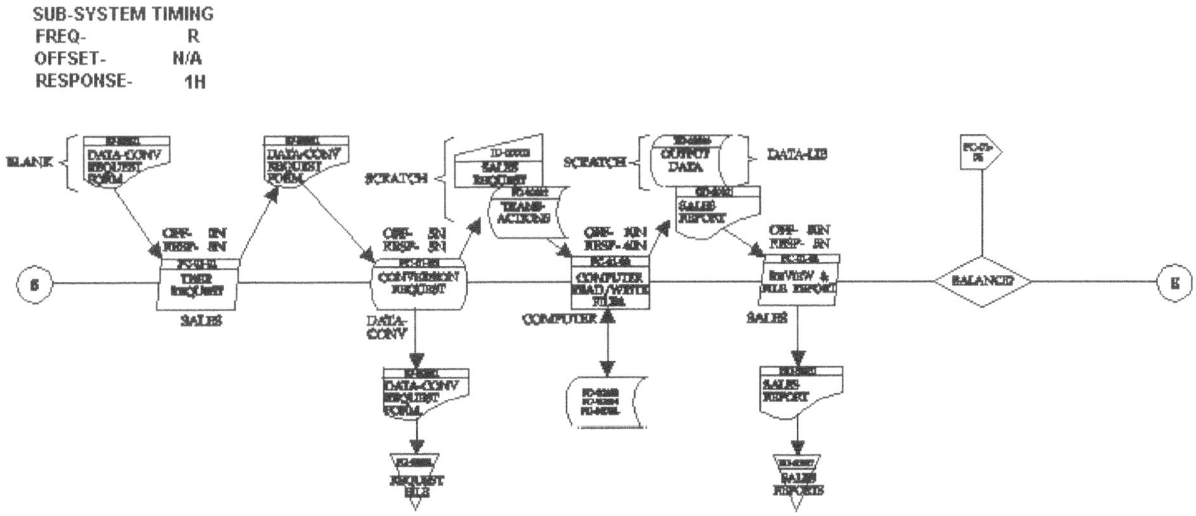

Flowchart can be drawn either horizontally or vertically.
Inputs, Outputs, and Files are related to each procedure in terms of C / U / R

SUB-SYSTEM FLOWCHART - Interactive Example

SUB-SYSTEM TIMING

FREQ -	**R**
OFFSET -	**N/A**
RESPONS -	**10S**

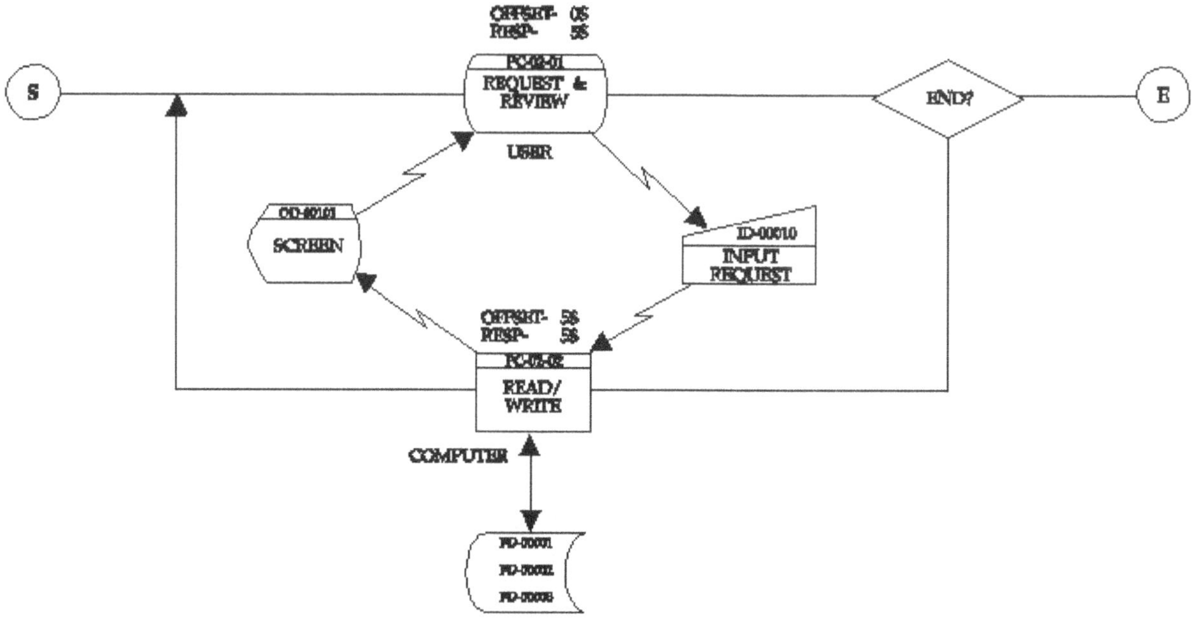

The **Computer Procedure Flowchart** shows the fourth level of detail in the system hierarchy: programs. It is drawn in a similar manner to the sub-system flowchart. Processing logic, inputs, outputs and files are all shown.

COMPUTER PROCEDURE FLOWCHART
Expresses Processing Dependencies

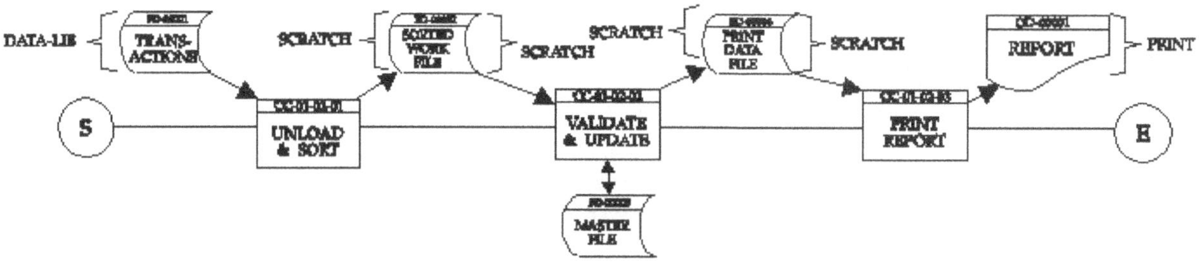

Uses ANSII Standard Symbols.
Can be drawn horizontally or vertically.

A **Software Structure Diagram** is used to graphically represent the processing logic either within a single program or a module. This type of diagram is based on the software engineering technique used which may vary from program to program.

SOFTWARE STRUCTURE DIAGRAM

Expresses software logic in acordance with the selected method of implementation;
Some examples:

BLOCK DIAGRAM

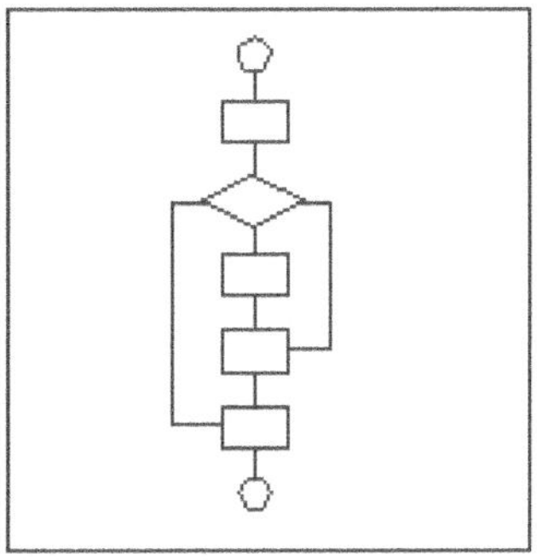

DECISION TABLE

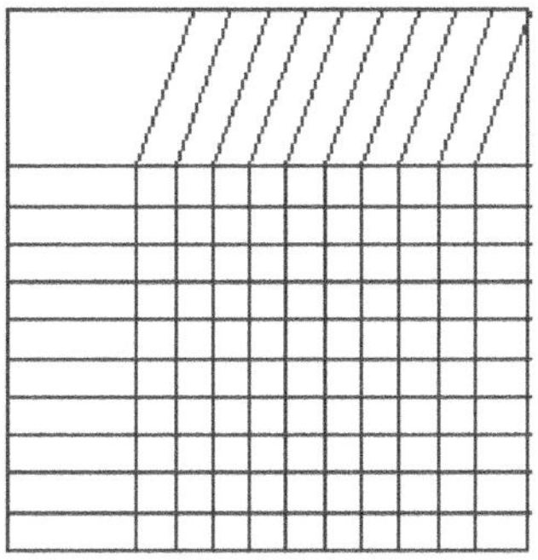

OTHER

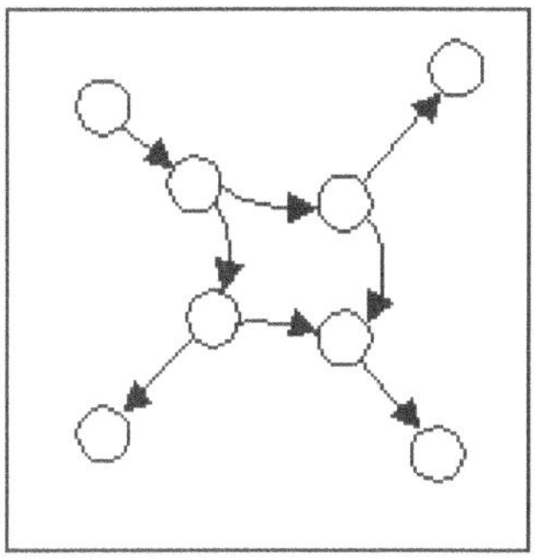

Each graphic is supported by narrative describing the processing logic for each level of detail.

This approach to documentation highlights the fact that system documentation can and should be completed prior to programming, not afterward. Graphics, processing logic, testing criteria, user instructions, input/output/file layouts, messages, etc. can all be prepared prior to the actual coding of the program. It all depends on whether design and documentation is viewed in the same vein as the approach used in architecture or engineering. Why should systems development be any different?

SOFTWARE ENGINEERING

From the perspective that computer procedures and programs represent only a portion of the information system architecture, software engineering is viewed as a subset of information systems engineering. Although programming ultimately represents nothing more than the translation of specifications into machine processable instructions, the intent of software engineering goes beyond this. It is concerned with developing computer software that is reliable, easy to maintain and modify, and performs in an efficient/cost effective manner.

Software exhibits some very "hard" characteristics. For example, it is typically not portable across computers. If software was truly portable, there would not be the many different "versions" of the same software product to suit different operating systems. Because of this, software engineering represents more physical design as opposed to the logical design of systems engineering.

The most important part of software engineering is in the architectural design and development of the computer procedure. The actual coding is merely a translation process, not unlike going from English to some other language. When the specifications for the procedure are developed, usually in a human understandable language, it is the programmer's responsibility to code or translate this into a language the computer understands. All the coder needs to know is the language, conventions, and the computer environment. This process is relatively simple and borders on being administrative which means that it can be automated.

Good systems engineering specifications will improve programmer productivity far better than any programming tool or technique. If good systems engineering specifications are prepared and software engineering practices are properly applied, then productivity will be maximized.

As we have seen, systems engineering is based on simple, straightforward, and commonly known principles. Some might say it is a discipline that stifles creativity. This is like saying that architecture, engineering, even music, lacks creativity, for these are all disciplines based on some form of science. There is science in every art, and an art to every science. Information systems engineering is a science; a teachable science.

METHODOLOGY CONSTRUCTION/NAVIGATION

The Information Systems Engineering Methodology (ISEM) consists of an assembly of nine phases detailing what is to be accomplished and by whom. Each phase consists of a defined set of activities (a total of 41); each activity consists of a series of operations or tasks to be performed. All phases, activities and tasks produce tangible deliverables that can be reviewed and checked. These deliverables substantiate adherence to the methodology and permits the measurement of progress. Both formal and informal review points are laced throughout the methodology which provides for an effective dialog between management, users, and systems developers.

The first phase is essentially used to plan the ISEM project. The remaining phases map the system structure as mentioned earlier. Phases 2, 3 and 4 are used to "explode" or decompose the system into its smallest pieces. The remaining phases are used to "implode" the structure through testing and implementation. The final phase (9) is used to evaluate the ISEM project.

METHODOLOGY CONSTRUCTION

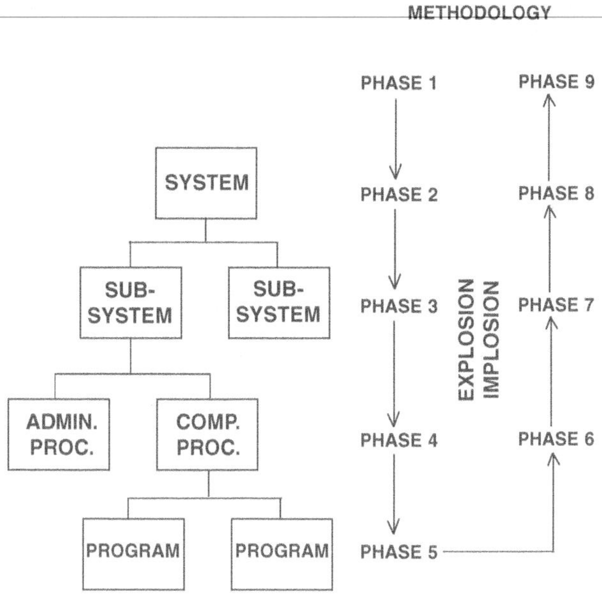

By carefully modularizing the work effort, phases can be conducted in parallel and concurrently, a much more effective way of utilizing resources than the linear approach found in the classic "five-step" project management oriented methodology (feasibility, design, program, test, review). This is the difference between an engineering approach versus an administrative one.

PHASE 1 - "System Study & Evaluation"

Phase 1 is used to identify the system(s) involved. It represents the feasibility study phase where current systems are analyzed, information requirements specified, and an overall system solution is proposed. The solution is represented by a complete rough design of the system, covering all levels, including programming. This becomes the basis for considering alternative solutions, including third party software available for purchase.

From the rough design, an estimate and schedule for the whole project can be developed along with a cost/benefit analysis. This is used by management to make project decisions, such as continuing, discontinuing or revising the project. Architects and engineers use a similar approach when proposing a new building or product.

The Information Flow Diagram and System Concept Diagram are prepared during this phase and included in the "System Study & Evaluation Report," the formal deliverable resulting from Phase 1.

Following this phase, the project proceeds to Phase 2. If more than one system has been identified in Phase 1, separate Phase 2's will be required. In this event, it is usually desirable to divide the project into separate projects to simplify design and management. This results in one project for each system identified.

PHASE 2 - "System Design"

This phase represents the formal definition of the sub-systems within the system. Using chronological decomposition, sub-systems are defined into unique time frames, along with their inputs, outputs and files.

Complete illustrative examples of the reports and screens are prepared for review by users. Prototyping tools can be used in the preparation of physical screens and reports. It should be emphasized that prototyping tools are not a substitute for determining information requirements, nor the logical design which must be performed prior to prototyping. Prototyping can only create a working model of the system as logically designed. It is used to "fine tune" the physical implementation.

The System Flowchart and Sub-System Concept Diagrams are prepared during this phase and is included in the "Systems Design Manual," the formal phase deliverable.

Based on the system design, the project scope is updated along with the project estimates and schedules. This is then reviewed with management to determine whether the project should be continued, discontinued, or revised.

Following Phase 2, a Phase 3 will be executed for each sub-system identified in the System Design. There is a one-to-one relationship between Phase 3 and a sub-system. If there are multiple sub-systems, there will be multiple Phase 3's which may be conducted in parallel or executed sequentially. This, of course, is a matter of available resources and schedule commitments. At this point, some organizations assign project leaders to each sub-system since they can be implemented independent of each other.

PHASE 3 - "Sub-System Design"

The logical sub-system is decomposed into physical procedures to process the data, both administrative and, where pertinent, computer. Design is based on efficiency and cost effectiveness. From this perspective, it represents the Industrial Engineering phase of the methodology.

The sub-system flowchart is prepared and included in the "Sub-System Design Manual," the formal deliverable from this phase.

Project estimates and schedules are revised as required, but only as they pertain to the delivery of this sub-system.

Phase 3 represents the last major review for management. It will be followed by Phase 4-I for Administrative Procedure Design and Phase 4-II for Computer Procedure Design, both of which can be conducted in parallel.

PHASE 4-I - "Administrative Procedure Design"

During this phase, the operational steps for each

administrative procedure in the sub-system are defined. The "playscript" procedure language is used to write the steps.

The written procedures are then packaged with samples of the inputs and outputs into an "Administrative Procedure Manual" for review by the end user. This manual represents the formal deliverable from the phase and will be used by the users of the system when installed.

PHASE 4-II - "Software Engineering"

This phase represents the point in the methodology where software design formally begins in the methodology. Although programming is consulted during the earlier phases, Phases 4-II, 5 and 6 represents their primary area of work. Whereas system engineering is preparing the administrative procedures for the sub-systems, the programmers now begin to concentrate on the design and development of the computer procedure and programs.

During Phase 4-II, a software implementation strategy is formed and programs defined. This includes, what inputs, outputs and files they will use, the processing logic for each program, including the processing flow to other programs, the control language statements (e.g., JCL, DCL, ECL, etc.) to be used, and the selected programming language, whether procedural or non-procedural (e.g., 4GL).

The logical files defined in the earlier phases are converted into physical files for use in the application by Data Base Administration. File layouts for use in programming are provided at this point, regardless of whether a DBMS is used.

A Computer Procedure Flowchart is prepared, along with pertinent Software Structure Diagrams, and included in the "Computer Run Book," the formal deliverable from the phase. This manual will be used by computer operations later and by programming in Phase 5.

A technical review is performed at the end of the phase, with computer operations assuming the role of the user.

Following Phase 4-II, a Phase 5 will be executed for each program identified in the Computer Procedure. This means that programs can be designed in parallel.

The remaining phases are used to "implode" or manufacture the system structure during implementation:

PHASE 5 - "Software Manufacturing"

Before Phase 5 begins, the program has been fully specified. So much so, that a program generator, report writer or 4GL should be able to interpret these specifications and produce the program automatically. In the absence of these devices, the program must be written manually.

If a procedural language is used, such as COBOL, the programmer should only have to produce the procedure division of the program. All other aspects should have been provided.

The intent of the phase is to produce a program, either manually or automatically, and perform a unit test of it. Test data is prepared and results collected. Testing should be performed in accordance with the test criteria as established in the Computer Run Book. Obviously, test data generators are useful for testing.

PHASE 6 - "Software Testing"

This phase is concerned with performing a "string" test of all of the programs within a specific computer procedure.

Although each program was individually tested in Phase 5, they are now brought together and tested as a whole. Again, testing should be performed in accordance with the specifications contained in the Computer Run Book.

Phase 6 represents the end of the formal software development phases of the methodology.

PHASE 7 - "Sub-System Test"

This phase represents parallel testing of all procedures in a sub-system.

PHASE 8 - "System Operation"

This phase represents a parallel test of the whole system (or selected sub-systems) and the implementation of the system. Users and computer operations personnel are brought together and trained in the use of the system and walked through the entire process. Following this, the system is ready for routine operation.

PHASE 9 - "ISEM Evaluation"

Phase 9 is used to conduct an operational review of the entire system. Based on the earlier design manuals, the system is reviewed to determine how well it performs according to specifications. Also, actual project figures are reviewed to determine how well the original estimates and schedules were met and to calculate the total cost of the project.

The formal deliverable resulting from this phase is a "System Evaluation Report" which contains the findings as mentioned above.

If a design change is introduced during a phase, it may be necessary to back-up a few phases to implement the change before moving forward again. In this situation, it is similar to a mini-modification/improvement project. Because design decisions are being captured and controlled, it is a relatively simple matter of implementing changes during the execution of the project. Further, the review points should not necessarily be considered an impediment to progress. Unless the Project Manager absolutely forbids continuation without approvals, projects can proceed to the next succeeding phases while the formal reviews are being conducted.

With ISEM, sub-systems can be designed, developed and installed individually as opposed to all at once. In fact, parts of a system can be installed and made operational while other parts are being designed, programmed and tested. This allows the maximum flexibility for controlling and delivering applications. Because systems exchange data, the Data Resource Management function oversees and coordinates the use of data throughout the development process, thus preventing data redundancy and promoting integration.

This system modularity also influences project management. Sub-systems can be packaged and priced for users in terms of the estimated time, cost, and delivery schedule for each. This provides the user with the flexibility to determine which portions of the system they wish to pay for and when to expect delivery.

ISEM PROJECT NETWORK
The Network Maps the System Structure

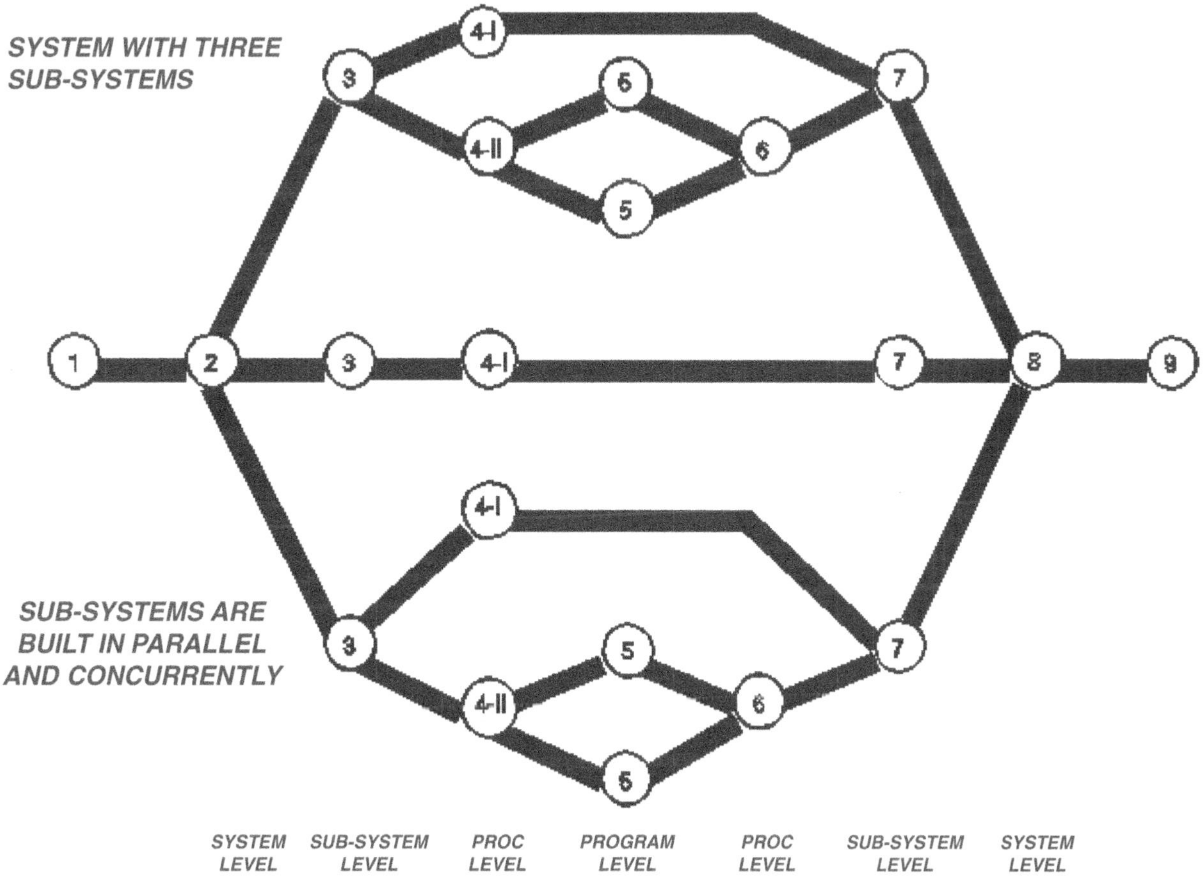

PROJECT PLANNING & IMPACT ANALYSIS

There are basically three types of work effort associated with information systems engineering: new systems development, maintenance, and modification/ improvements. New systems development represents totally new applications for an organization to build. Maintenance is the correction of errors or defects ("bugs"); it is making the system perform according to specifications. However, the majority of work for most systems development organizations is modification/improvements to existing systems.

ISEM recognizes that systems are built by evolution, not by revolution. The day a system is installed is the day it begins to undergo change. This is because information requirements are constantly undergoing change. Because of this, ISEM recognizes that not all projects will require the execution of every phase and activity in the methodology. After all, if a pane of glass is broken, should we redesign the whole building? Obviously not. We should only perform those steps necessary to repair the pane of glass. The same is true in systems; we should only perform those phases and activities needed to implement the objective.

All major projects featuring new development will require the execution of all ISEM phases. However, modifications and maintenance to existing systems will probably not. To determine which phases and activities need to be performed, we must first evaluate the existing system in terms of the resources used and the degree of definition. This is where an "impact analysis" is performed to study the possible effects of a proposed change that an information resource may have on other resources. For example, "impact analysis" may be used to study the effect of a potential change to a data element, output, file, information requirement, module, sub-system, etc. Based on resource relationships, a list of related resources can be assembled and analyzed.

Based on this "impact analysis," a decision is made as to the phases and activities required to implement the request. Only the required phases and activities are selected to implement the change. For example, if a modification to a single program is all that is required, then the project may simply consist of activities from Phases 4-II, 5 and 6 (the software engineering phases).

"Impact Analysis" and project planning is performed in Phase 1, Activity A of ISEM. From here, the project can either execute progressively through the methodology, or proceed to the required phases and activities.

This approach eliminates the need for varying methods based on the size of a project, e.g., small, medium and large. Systems design consists of the same logical sequence of events, only in moderation. The only effort required is that specified by the change or modification.

This approach also highlights the fact that there is no such thing as "systems life cycles" as promoted by some people in the industry. A system does not have a life cycle. It may go on forever if kept viable with change. Double-entry bookkeeping is a good example of this. Projects, not systems, have a life cycle. They have a beginning for planning, a middle for execution, and an end for review. This problem of "System Life Cycle" demonstrates the problem that the industry has with its vocabulary and concepts.

JAD, RAD, BPR & OTHER TECHNIQUES

A variety of techniques may be used throughout ISEM, all of which represent "how" to perform a specific activity or activities of work. Such techniques do not circumvent the need for these activities, only how to perform them. Two techniques gaining attention are Joint Application Development (JAD) and Rapid Application Development (RAD), both of which are closely related to what is called Agile or Extreme Programming. Although both techniques represent separate approaches, they are frequently implemented together.

The JAD technique is concerned with working closely with end-users to specify their business problem and develop a cost effective solution. This normally involves intensive meetings between end-users and developers. The RAD technique involves the use of application development aids (e.g., program generators, report writers, 4GL's, etc.) to quickly develop a program solution.

When coupled together and managed correctly, JAD and RAD can produce positive results. However, there are legitimate dangers to JAD and RAD if improperly implemented. Such techniques can easily turn into QAD ("Quick And Dirty"), which may satisfy the problem for the moment but may cause long-term headaches later on.

JAD itself represents active project participation from

end-users; this is vital for any development effort. But if the JAD session is used to only specify outputs (forms and screens), and not clearly define true business problems and information requirements, then the wrong programs will inevitably be developed.

RAD also shows great promise and great danger. If application development aids are implemented without a fundamental system or data architecture in place, then they will inevitably create redundant data and software resources, and erroneous information.

Because of their nature to quickly produce software, techniques such as JAD and RAD are oriented more to smaller applications involving one or two programs, as opposed to major enterprise-wide systems.

To implement a JAD/RAD project under "PRIDE"-ISEM, the following phase activities represent a checklist of work steps to be performed:

- Phase 1/Act C - Survey Information Requirements
- Phase 1/Act F - Develop System Approach
- Phase 2/Act A - Define Sub-Systems
- Phase 2/Act B - Define Sub-System Logic
- Phase 2/Act C - Illustrative OD/ID Examples
- Phase 3/Act A - Define Procedures
- Phase 3/Act B - Define Procedure Logic
- Phase 4-I/Act A - Define Operational Steps
- Phase 4-II/Act A - Design Software
- Phase 4-II/Act B - Define Software Logic
- Phase 4-II/Act C - Prepare File Layouts
- Phase 5/Act B - Produce Executable Program

Testing, of course, will need to be performed, but this will be based on the size of the application (e.g., number of procedures and programs). Also, consideration should be given to Phase 8/Activity B - "Educate/Train Users," execution of pertinent DBEM phase activities, and the Phase 9 "ISEM Evaluation."

Without a clear roadmap, as provided by "PRIDE"-ISEM, JAD/RAD projects can turn into chaotic and frustrating endeavors. As with any development effort, planning and organization greatly enhances your chance for success.

Business Process Re-Engineering (BPR) represents those activities within the "PRIDE" methodologies concerned with re-evaluating and, where necessary, re-designing the processes (sub-systems and procedures) needed to support the actions and/or decisions of a business function. Re-designing business processes is actually not a new concept. In fact, it is rather old and has been an inherent part of the "PRIDE" methodologies for over thirty years. Only recently have companies re-discovered the need for this type of work.

Unlike most application development efforts focused on programming, BPR is concerned with total business processes spanning organizational boundaries and implemented by either the computer, the human being, or other forms of office equipment. Because of its up-front orientation to planning and designing total systems, BPR should be considered a precursor to software engineering. In fact, BPR simplifies software engineering by taking the guesswork out of computer processing requirements. By concentrating on systems design, programming is greatly expedited.

BPR primarily affects Phase 2 of "PRIDE"-EEM, and Phases 2 and 3 of "PRIDE"-ISEM. Although a full "PRIDE"-EEM project would certainly be useful for understanding the enterprise, Phase 2 "Logical Enterprise Analysis" focuses on the business functions (the logical model of the enterprise) and the information requirements needed to support them. In Phase 2 of "PRIDE"-ISEM, sub-systems (business processes) are defined in terms of "what" and "when" information must be produced. In Phase 3 of ISEM, emphasis shifts to physically "who" and "how" the sub-system must be performed, using administrative and computer procedures. At this time, processing "work flows" are defined (from start to end), as well as "data flows" between the various processing components. For more information on BPR, see:

- "PRIDE"-EEM/Phase 2 - Logical Enter. Analysis
- "PRIDE"-ISEM/Phase 2 - System Design
- "PRIDE"-ISEM/Phase 3 - Sub-System Design

There are other techniques described throughout "PRIDE"-ISEM. For example:

- Client/server computing is described in Phase 1.
- Prototyping in Phase 2.
- Structured programming and object oriented programming (OOP) in Phase 4-II.

EEM/DBEM RELATIONSHIPS

The "PRIDE" Enterprise Engineering Methodology (EEM) precedes both the Information Systems Engineering Methodology (ISEM) and the Data Base Engineering Methodology (DBEM). It is used to formulate an Enterprise Information Strategy that includes development projects for ISEM and DBEM to implement.

Because of the close relationship between systems and data, ISEM has a close working relationship with DBEM. In Phase 2 of ISEM, sub-systems and logical files are defined. This will then normally initiate a DBEM project to support the data requirements. In Phase 2 of DBEM the Application Logical Data Base Model (ALDBM) is prepared by Systems Engineers under the supervision of Data Engineering. Data Engineering will then incorporate the ALDBM into the Enterprise Logical Data Base Model (ELDBM). Data Base Administration will then design/update the Enterprise Physical Data Base Model (EPDBM) and Application Physical Data Base Model (APDBM). The result of the APDBM is the physical file structures for programming. In other words, the DBA must deliver the file layouts to the Software Engineer in Phase 4-II of ISEM.

Data base design is separated from system design so that a neutral third party can accommodate the data requirements for the entire enterprise, not just a specific application.

During Phases 3 and 4-II of ISEM will define the need for working files, either manual or computer, for use in processing. This is communicated to DBEM for incorporation in the data base.

ISEM/DBEM PHASE RELATIONSHIPS

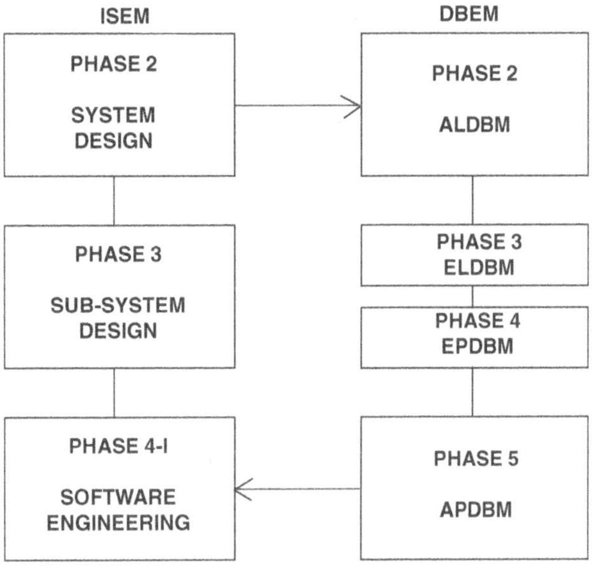

DBEM Phases can be added to an ISEM Project or managed as a separate project.

WHO SHOULD PERFORM ISEM?

There are significant differences between Systems Engineering and Software Engineering. The Systems Engineer tends to be more people and business oriented than the Software Engineer, who is more concerned with the effective implementation of technology.

The Systems Engineer fulfills the roles of chief architect, industrial engineer, industrial psychologist, and user liaison. Essentially, they must be proficient in interpersonal relations/communications, work measurement/simplification, possess good analytical/organizational skills, and knowledgeable about how the enterprise functions. They must also be able to interface with Software Engineering.

The Software Engineer is more in tune with computer hardware/software than the business. They typically lack the business and people skills as possessed by Systems Engineering. This is not to demean their role, only to clarify it. Although Phases 4-II, 5 and 6 are the principal phases where the Software Engineering works, this does not limit their involvement in the methodology. In fact, Software Engineers actively participate in the early phases of ISEM in a consulting capacity to Systems Engineers. During these phases, the Software Engineer analyzes the viability of designs as prepared by Systems Engineering, as well as participate in project estimating/scheduling activities. This participation in the early phases promotes cooperation and communications between the two functions.

OTHER FUNCTIONS PARTICIPATING IN ISEM

As in any other work effort, Project Managers are required to plan, estimate, schedule, and control an ISEM project. Further, Quality Assurance personnel review and verify deliverables resulting from the various phases of the methodology.

End-users are interviewed and consulted during the course of an ISEM project. As a result, their participation is vital. In fact, some of the most successful applications of ISEM has been where the user serves in the capacity of Project Manager.

What is absolutely critical to the success of any systems project is Data Resource Management. As such, Data Engineers and Data Base Administrators actively participate in a consulting capacity throughout ISEM. In the absence of a Data Resource Management organization, the Systems/Software Engineers must assume the responsibilities.

BENEFIT$

The benefits of Information Systems Engineering are numerous:

- It assures that systems development personnel are "doing the right things," building systems that satisfy the information needs of the business.
- Provides for improved control over the systems development process by simplifying the work effort.
- It demystifies the systems development process.
- It improves the performance of the development staff through a standard and consistent approach.
- Improved communications between end users and the development staff, as well as the development staff internally.
- Improved consistency and quality of systems.
- Systems behave reliably and are easier to maintain and modify.
- Improved software engineering productivity by promoting total systems engineering.
- Improved control over project planning and execution.
- Improved corporate control over information resources.
- Competitive edge through accurate and timely information under changing conditions.

END SECTION

Section 9

"PRIDE"-ISEM PHASE 1 SYSTEM STUDY & EVALUATION

If an information requirement is stated incorrectly at the beginning, then everything that follows will be incorrect.
- Bryce's Law

The purpose of this phase is to specify and analyze an information systems related problem and/or opportunity, and to propose to management a viable solution. It is the most critical phase of the ISEM methodology. Several events occur during this phase:

- The scope of the ISEM project is defined.
- An "impact analysis" is performed to determine the phases and activities required to implement the project.
- An analysis of the current system is performed.
- User information requirements are defined and reviewed.
- A system approach is defined for satisfying the requirements. The approach may include new development, modifications/improvements to existing systems, a packaged solution, or combinations of all three.
- An evaluation of alternatives is prepared, including estimates, costs, and schedules.
- A formal review with management is conducted to verify the phase findings.

METHODOLOGY NAVIGATION

An ISEM Phase 1 is normally initiated by a Work Request/Objective as defined by the Enterprise Information Strategy (EIS). This strategy was developed during Phase 4 of the Enterprise Engineering Methodology (EEM) and is issued to Systems Resource Management during EEM's Phase 5.

The Work Request/Objective defines the business problem/opportunity to be addressed by the project, which typically relates to new development, modification/improvements, or maintenance. During Phase 1, Activity A, the project scope is analyzed and a determination is made as to the phases and activities needed to satisfy the Work Request/Objective. For new development, the project proceeds normally through the activities of Phase 1. However, for mod/imps or maintenance an "impact analysis" is performed to study the information resources affected by the request. From this analysis, a determination is made as to the phases and activities required to complete the work; the project thereby proceeds to these steps and the remaining activities of Phase 1 may be by-passed. See the "General Discussion" section for more detail concerning Phase 1, Activity A.

Assuming a normal Phase 1, work proceeds sequentially through the remaining activities (B through H).

During Phase 1, Activity H a formal review of the phase findings is conducted with management. Consequently, management makes decisions regarding the project; they may elect to:

1. Approve the project for continuation as proposed.

2. Request revisions to the findings produced in the Phase.

3. Discontinue the project.

Assuming acceptance, the project will normally proceed to Phase 2 where the system will be decomposed into sub-systems. However, there may be exceptions:

• When a Phase 1 identifies the need for more than one system. In this situation, the Project Manager may elect to either separate the systems into different projects, or branch into multiple Phase 2's within the same project. This will make the work effort more manageable.

• Should the Phase 1 identify the need to simply modify an existing system or systems, it may not be necessary to perform a Phase 2. Instead, the project may proceed to other Phases and Activities in the methodology. In this situation, the decision to proceed is similar to the decision made in Phase 1, Activity A. This, of course, depends on what is required by the modification.

Although both of these scenarios are possible, they are highly unusual situations. However, these are the types of decisions which Project Management must make during Activity G.

GENERAL DISCUSSION

THE PHILOSOPHY OF PHASE 1

In its simplest form, Phase 1 represents a definition of the problem, an analysis of the current mode of operation, a definition of requirements, an evaluation of alternatives, and an agreed upon course of action. Phase 1 is based on generic principles of product evaluation. As such, the activities of Phase 1 can be applied to any type of project.

PHASE 1/ACTIVITY A

Activity A, "Preliminary Project Scope," is critical to the start-up of the ISEM project. A superficial analysis during this activity can result in severe project problems later. For new development, Activity A is used to specify the project scope and to produce a Detail estimate and schedule for the remaining activities of Phase 1.

The Project Scope is a concise definition of the business problem/opportunity to be addressed by the project and the areas of the enterprise affected both directly and indirectly (user departments). This textual definition is important. It establishes the project boundaries which will determine estimates, schedules and work effort. Projects tend to lose direction without clear boundaries and tend to wander into areas unintentionally. This will result in performing more work (or less) than what is necessary to satisfy project objectives.

Work Requests/Objectives are documented in the IRM Repository using the MI resource (Modification/Improvement). There is not necessarily a one-to-one relationship between an MI and a Project Description (PD). One MI may be implemented by many projects; one PD may implement many objectives. The Project Manager must have a clear understanding of these relationships.

The expression "Modification/Improvement" is based on the fact that most system development work is of this nature. The "PRIDE" methodologies recognize that change is constant; that systems are built by evolution, not by revolution. As soon as a system is installed, users will probably request changes to it.

ISEM was designed to accommodate change. Corporate information needs will fluctuate based on economic conditions, government regulation, competition, product changes, acquisitions, etc. The operating objective of the Information Resource Management organization, therefore, is to be sensitive and responsive to this changing environment.

System revisions or modifications will vary from minor program changes to complete system replacements. These modifications and improvements are not to be confused with Systems Maintenance. Systems Maintenance is an unscheduled activity and occurs primarily as a result of operating problems; that is, the system is not performing as intended or specified. Examples of this are program errors or incorrect procedures. Documentation changes are usually not involved and the action to be taken is corrective in nature. Modification/Improvements can be planned and scheduled, and should be in response to new or changing information needs. In fact, most new systems development activity should occur through this process if the basic systems have been designed correctly.

ISEM is a "closed loop" system. All projects, whether large or small, will begin in this phase. Based on an initial investigation, a determination is made as to what phases and activities must be used to implement the request. If the magnitude is determined to be major, the project should progress to a full Phase 1. If not, Project Management selects the appropriate phases and activities to execute. This means that a single consistent methodology is being used to implement all projects regardless of size. It does not have varying methods based on project size, e.g., small, medium and large projects.

IMPACT ANALYSIS

In order to accurately evaluate change, an "Impact Analysis" is performed to study the relationships between information resources and appraise what resources need to be modified. Some examples:

• If the length of a data element is to be modified, an analysis must be performed to identify the panels, maps, records, files, inputs, outputs, modules, programs, etc. affected.

• If a single report or screen needs to be changed, an analysis is required to identify the programs and procedures affected.

• If a software module is revised, the related programs and data structures must be identified.

This type of analysis reflects a "bill-of-material" philosophy. In other words, if one part is changed, what other parts will feel its effect? "Impact Analysis" is a natural part of an IRM Repository and should be used to perform this type of analysis. Based on the "Impact Analysis" Project Management can then determine what pertinent ISEM phases and activities are required to implement the change. When this has been done, an Order-of-Magnitude estimate and schedule can then be calculated for review.

Before proceeding with the project, the Project Scope and estimates/schedules are reviewed with the user sponsor(s). It is important that the Project Manager and the users have consensus in terms of the work to be performed, the approach, and anticipated estimated time/costs/schedule. This is true regardless of the type of work effort. If it is new development, the estimate may only reflect a Detail estimate/schedule for the remaining activities of Phase 1. If it is mod/imp related, the estimate may reflect an Order-of-Magnitude estimate/schedule for the entire project.

CURRENT SYSTEM ANALYSIS

The subject of current systems analysis is usually greeted with dismay or disdain by systems departments. There are many reasons for this. In many installations, the support of current systems takes more than 85% of the systems department's time, and the departments are more than ready to get on with new systems development and bury the old, non-working systems as quickly as possible. In cases where systems do not require a lot of maintenance, the systems department may find that the current systems are not giving management the kind of information it needs for effective decision making; these current systems become likely candidates for replacement.

However, there are some very legitimate reasons for documenting existing systems:

1. The documentation process will make development personnel more knowledgeable about the enterprise's business. Users often complain that the systems people are ignorant or naive about the business, and usually there is some justification in this complaint. The process of capturing current systems in ISEM will educate Systems Engineers in the business of the company.

2. Documenting current systems will clearly show:

• What information is being produced.

• Who is using the information and how (functions, jobs, resources).

• What data resources are being used (data, records, files, inputs, outputs).

• What processing is being executed, and what is not.

Quite often, user personnel do not use existing systems simply because they do not understand the purpose of the system or how it is to be used. If a system has been properly documented, perhaps only a few modifications are required as opposed to a major new application development effort. Unless we understand clearly what the old system is doing, how can we design a new system to replace the old? How can we plan the conversion from the old to the new? Most of the current system contains resources that will be re-usable for the new system. Data, records, files, inputs, outputs - most, if not all of these can be re-used in systems development.

Weaknesses and misusage of current systems will also be uncovered. Because of inadequate documentation, it is not unusual to see a good system be misinterpreted by users, including DP Operations. As a consequence, the system is only partially utilized and the benefits not fully realized.

3. Programs and files which only serve the purpose of unnecessarily occupying computer storage space can be identified and removed.

Degree of Definition

Capturing current systems can be a huge undertaking. A single company can have hundreds of systems and sub-systems, and thousands of procedures and programs, some of which are very old. It is important to remember that the intent here is to **ONLY IDENTIFY AND DESCRIBE THE CURRENT SYSTEM, NOT TO CORRECT DEFICIENCIES.** Quite often, there is a temptation to try to correct an obvious problem at this stage. As a consequence, the task of capturing current systems takes longer and longer. When errors or problems are spotted they should be documented as future Modification/Improvements and not corrected as part of this effort.

How much definition is enough? Ultimately it is based on the organization's needs. No two companies will approach the problem in the same way. Their requirements will vary. However, the vendor recommends simplicity when describing the various resources. The following guidelines should help:

System Resources:

Normally, system resources require a minimal amount of description. What is essential is that a clear and concise statement of the processing be defined. The vendor recommends a Phase 3 level of detail (Sub-System Design) in most instances. The descriptions should be accompanied by a System Flowchart (showing the sub-systems), and a Sub-System Flowchart (showing the procedures).

SYSTEM DESCRIPTION - Normally, the system narrative describes who the system will serve, what information requirements it implements, and what processing is included. When documenting current systems, it should suffice to explain who the system serves and a concise description of the overall processing.

SUB-SYSTEM DESCRIPTION - Timing is a key element here. The definition of Frequency (how often it occurs), Offset (when it must start in the cycle) and Response Time (how fast it must occur) are important aspects to the definition of a sub-system. Who is responsible for executing the sub-system, what Inputs, Outputs and Files are used are also important. For text, a simple description of the "Business Purpose" of the sub-system should suffice in most instances.

PROCEDURE DESCRIPTION - Procedures are primarily concerned with determining the work flow in an organization. Each procedure should specify who is to perform the work and when. Offset and Response Time should be uniquely defined for each procedure. As with sub-systems, each procedure should have prescribed Inputs, Outputs and Files. Again, for text, a simple description of the "Business Purpose" of the procedure should suffice in most instances. For Administrative Procedures, it may be necessary to detail the Operational Steps (Tasks) in "playscript" format.

PROGRAM DESCRIPTION - If program descriptions are important, then they should describe the functionality of the program, what it is to perform and how. The programming language, memory should also be defined, along with the Inputs/Outputs and Files.

Data Resources:

These resources contain the most essential intelligence in current systems analysis. They are the basic building blocks for all systems design. The definition of these resources must be taken seriously. Superficial definitions will cause problems later. Data resources have both logical and physical definitions. Both are important, but the physical characteristics will be the most visible during current systems analysis.

DATA DEFINITIONS - Each data element will have only one logical business definition, but can have more than one physical implementation. The logical definition is critical. The physical characteristics (length, precision, scale, picture, etc.) may vary from application to application. Select the most popular physical description when defining the data element. Express exceptions on the record (RD) where it resides.

RECORD DEFINITIONS - When capturing current systems, you will most likely be identifying physical records as opposed to logical. These can be easily identified from file layouts.

FILE DEFINITIONS - As with records, most files identified in current systems analysis will be physical and found in file layouts. The definition of the file must identify what records it contains. It should also describe whether it is a computer or manual file, and whether it is a primary or working file.

INPUT DEFINITIONS - Inputs help to identify the ultimate source of the data required to produce the information. They should each have a prescribed "Business Purpose," which includes who uses them and when, and what forms or screens are used to implement them. They should also describe what transactions they are used for (e.g., for data collection, or display/print requests).

OUTPUT DEFINITIONS - Outputs are used to identify the receivers of the information. Like inputs, they should specify the timing of the outputs (Frequency, Offset and Response Time), and what forms or screens are used to implement them. For text, each output should have a prescribed "Business Purpose." For customers requiring more detail, pertinent processing messages (Error/Warning/Information) should be documented as RD resources and related to the pertinent outputs (and inputs).

Corporate Entities:

To many people, the definition of corporate entities appears to be an overhead. In reality, the definition of these resources holds vital corporate intelligence, particularly in the area of Functional Entities. If carefully defined, they can be extremely helpful in understanding the business of the company and how it is implemented.

FUNCTIONAL ENTITIES - These are logical descriptions of either an enterprise or a business function. Defining business functions is particularly useful when interviewing Users for information requirements. What is basically required is a description of the "Functional Scope," along with a description of the "Duties and Responsibilities." These functions should also describe the specific Organizational Entities which implement them.

ORGANIZATIONAL ENTITIES - These resources represent the physical jobs which are used to implement the logical functions and are described by title, e.g., Manager, Clerk, Supervisor, Vice President, etc. It is important to note that the full job descriptions should be written by management and not necessarily by Systems Engineering.

HUMAN/MACHINE RESOURCES - Representing personnel and machine resources. The attributes of these resources should be defined by management and/or personnel.

NOTE: FE, OE and RE resources are normally defined under the Enterprise Engineering Methodology (EEM).

INFORMATION REQUIREMENTS - Current information requirements should be defined whenever possible. These can be defined by analyzing existing outputs and who they serve. When defined, it is extremely useful to compare and contrast existing requirements versus new. These requirements should be completed in full.

Tactics

Capturing current systems is an exercise in reverse engineering. Whereas, system design is a top-down effort, current systems analysis begins with an examination of the procedural flow and works up and down the system hierarchy.

According to the "PRIDE" definition of system structure, a sub-system consists of one or more administrative procedures and one or no computer procedures. With this in mind, the assumption can be made that a computer procedure represents a sub-system. Although a sub-system is limited to one computer procedure at most, a computer procedure is NOT limited to one computer program. A review of the computer "job streams" will reveal the computer procedures, their programs, outputs, and files.

A review of the computer procedure timings will show when jobs are being run, such as daily, weekly, monthly, year-end, etc. This can then be used to determine the timing specifications for the sub-systems. File back-up procedures can also be used to indicate timing considerations.

Another approach would be to analyze and sort inputs and outputs by common timing and receivers (such as those reports or screens used by Sales, by Customer Services, by Manufacturing, etc.). These groupings would ultimately represent sub-systems.

When sub-systems have been defined, they should be grouped into systems based on commonality. With this complete, you now have a skeleton of the system hierarchy.

What systems or sub-systems do you begin with? Start with those that provide operational information - those that are at the heart of the business operations. Those parts of the systems that provide policy and control information can be delayed until the operational systems have been captured, since they will require the same data created in the operational systems.

DEFINING REQUIREMENTS

Specifying information requirements is perhaps the most important task in Phase 1. All designs that follow must ultimately satisfy the requirements. Therefore, they must be defined with a high degree of precision and accuracy. This requires a highly skilled Systems Engineer; someone who is adept at analyzing business problems.

During Phase 1, end-users are first interviewed in terms of the information they need to support their area of the business. The Systems Engineer then interprets the user's description and prepares a formal information requirement definition which is reviewed with the user for accuracy.

Historically, analysts have been at a loss as to how to properly define information requirements. To a large extent, this was due to inadequate training. They simply did not know how to ask the right questions. Consequently, users would be interviewed in terms of the type and format of the outputs they wanted (reports and screens), as opposed to WHY they wanted it (the business rationale). Because of this, incomplete systems would be developed that would only satisfy some of the needs of the users (not all). Fortunately, the Information Requirement (IR) specification as provided in "PRIDE" helps to overcome this problem. Instead of physical media requirements, "PRIDE" focuses on the types of business actions and decisions that need to be supported. The IR is used to specify:

• The **BUSINESS PURPOSE** of the requirement.

• The **ACTIONS AND/OR BUSINESS DECISIONS** to be made, including WHO must make them and WHEN.

• The **BENEFITS** of having this information.

• The **OUTPUT(S)** and **SYSTEM(S)** used to satisfy the requirement.

• The **OBJECT(S)** associated with the requirement.

• The required **DATA ELEMENTS** to implement the requirement; including both primary and generated values.

The Systems Engineer must be wary of trying to specify too much (or too little) in a single requirement definition. As a general guideline, requirements should be separated and sorted according to:

• Unique time frame (Frequency/Offset/Response Time).

• Common "receivers" of the information.

• Common purpose - Policy/Control/Operational requirements should be separated.

In addition to a formal printed definition of each Information Requirement, an Information Flow Diagram (IFD) provides a graphical illustration of how information moves throughout a business. The IFD, as provided in ISEM, shows each information requirement, along with its timing and characteristics (P/C/O). Further, the graphic expresses users in the information flow, such as "originators," "participators," and "receivers." In this way, the graphic represents the boundaries of the application and the project scope.

The IR resource in the IRM can also be used to represent non-information related requirements, such as machine specifications.

PHASE 1/ACTIVITY E

Activity E represents the first major review with the end user. This is where the Information Requirements and Project Scope are reviewed and confirmed prior to seeking a systems solution. Each requirement must be evaluated by the user in terms of accuracy and clarity. The net result, the users and Systems Engineer reach consensus in terms of WHAT is required, WHO requires it, WHEN and WHY.

Glossary of Terms

During requirements definition the Systems Engineer defines the data elements needed to support the Information Requirements. As a by-product of this effort the Systems Engineer assembles a "Glossary of Terms." This glossary is an assemblage of all of the data elements in the applications and presented in a Webster/Oxford style format, complete with proper name and definition. Such a glossary is invaluable for clarifying the data items needed to support the requirements.

ROUGH DESIGN

Following confirmation of the information requirements, the Systems Engineer is then charged with the task of devising a system solution. To do so, the engineer prepares a complete rough design of a system. This involves the decomposition of the system into sub-systems, into procedures (both computer and administrative), into programs. In this respect, the design process takes on the appearance of a mini-Phase 2, 3, and 4-II.

Participating in the rough design are representatives from Software Engineering, Data Engineering, Enterprise Engineering, Data Base Administration, Data Communications, User Management, DP Operations; all of which consult with Systems Engineering during design. The intent is to arrive at a viable solution for the company to implement. In terms of programming, Software Engineering considers the method of implementation (e.g., languages, techniques and tools). Data Base personnel consider data storage/communications implications.

The system design is considered "rough" in the sense that it represents a tentative design for evaluation purposes. Regardless, there is considerable attention to detail in terms of identifying the system and data resources needed to satisfy the information requirements. During this activity, the Systems Engineer will consider alternatives. For example, the Systems Engineer will identify...

- **Re-usable resources** - system and data resources that have already been developed and are suitable for inclusion in the application.

- **Existing resources** requiring modification to accommodate the system.

- **New resources** that will have to be designed and developed.

- Resources that can be implemented by **commercial software packages.**

- **Data interface requirements** between all of the above.

The point is, rough designs are used to propose practical cost-effective solutions for management to consider. However, there are additional benefits. The rough design will become the basis for formulating a project plan, estimate and schedule. This can then be used for "make versus buy" decisions. It will also improve communications and cooperation between the development staff due to early participation in the planning phase.

One final word of advice on rough designs: Do not attempt to perform the whole project in Phase 1. It must be remembered that the intent of Phase 1 is to propose a viable solution for management to consider. Although the rough design may be rigorous, it must not be considered the final design; this is what the succeeding phases of ISEM are for. In other words, do not try to build the perfect system in Phase 1; simply take it to the level where the project team has confidence in the design and that a realistic project plan/estimate/schedule can be prepared.

System Concept Diagram

To graphically represent the system solution, the Systems Engineer prepares a System Concept Diagram. This is a free-form schematic that illustrates to the user how the envisioned system will operate upon completion. This is similar in intent to an artist's rendering as used in architecture.

Although there are no formal standards for this type of graphic, the System Concept Diagram normally shows the user departments involved, the business processes, and interfaces to other systems (if any).

EVALUATING A PACKAGED SOLUTION

A commercial software package provides one alternative for implementing an information system. Normally, it is not a complete information system but only tested computer programs that is sometimes accompanied by user procedures for input preparation. Most packages tell the purchaser what input is required to produce a specified output without telling the purchaser who does what, when and how. Because of this, the overall system, sub-system and administrative procedures may have to be documented to convert a package into an information system before installation can be accomplished. In other words, a package should be documented in the IRM like any other resource.

User Management typically buys more packages than system developers. This is primarily due to the proliferation of micro-computers throughout organizations in an attempt to solve information problems. Because of the traditional backlog or user requests, packaged software is an attractive alternative for the user.

There are two reasons why purchased packages can fail. One is because the User, having a definite information need or wishing to solve a business problem, purchases a package because it appears to solve the problem. An example of this is where the Order Processing Department purchases an Order Processing System because it seems to solve the problem. After the package is purchased, it is turned over to systems development for implementation. Shortly thereafter systems development discovers that: extensive changes are required to make the package interface with other systems; the package has poor documentation; the package does not completely meet the user's requirements; it will require considerable time to modify and implement.

A second reason is that systems development, behaving in a manner similar to the user, also conducts a superficial package evaluation. Little investigation is made as to how complete the documentation is, what modifications might be required to suit in-house needs, how it will interface with other systems, how long it may take to implement, etc.

In both examples, the solution to the problem was a package purchased before the real needs of the company were defined. Therefore, **AN ELEGANT SOLUTION TO THE WRONG PROBLEM SOLVES NOTHING.**

Even with this in mind, experience with many installations has been that the purchase decision is a viable alternative to 're-inventing the wheel'. This is especially true in those situations where resource and/or time constraints have necessitated the purchase consideration. When handled intelligently and adequately managed, packages can be successfully installed to the mutual satisfaction of both systems development and users to solve pressing business problems.

How then is it possible to avoid the common pitfalls and evaluate packages intelligently? Fortunately, the answer is simple: Phase 1. It provides all of the background information necessary to intelligently investigate packages. If performed correctly, the Phase 1 will have produced:

• Information Requirements.

• Rough Designs - containing tentative specifications about data and processing specifications.

This becomes the focal point for preparing a comparative analysis:

• Information Requirements/Outputs Analysis - by analyzing the package's outputs, the Systems Engineer should be able to determine which outputs satisfy the various Information Requirements, either in part or in full. A matrix is very useful for illustrating these relationships.

It must be remembered that packages are developed by vendors to solve general problems; not your specific information requirements. However, the vendor may have more insight into the business problem than the customer. As such, they may provide you with additional considerations in terms of specifying information requirements.

• Rough Design Analysis - because of the completed rough design, Systems Engineering can evaluate how well the package satisfies processing and data requirements. The Systems Engineer should be able to illustrate:

- Which sub-systems can be implemented by the package, either in part or in full. Timing requirements, in particular, should be considered.

- The types of file structures that accompany the package and if the files are accessible for additional customer specific processing requirements.

- How well the package's physical data characteristics conform to in-house requirements. For example, does the package express dates, times, names, monetary values, etc. in the same manner as is commonly used in the company? In other words, how does the vendor express data length, precision, scale, etc? Severe conflicts may be incompatible for in-house use.

For each packaged solution, Systems Engineering prepares a system evaluation. This allows all possible alternatives to be considered and clearly identifies the time and money saved by the purchase, if any. When preparing this evaluation, the engineer must consider additional time to customize and install the product. In addition, consideration must be given to interfacing the package to other parts of the customer's systems.

Installation Considerations

Whenever possible, package documentation should be incorporated into the IRM Repository for management and control. This means that narrative and specifications should be entered in the IRM Repository regarding:

• Systems, Sub-Systems, Administrative Procedures, Computer Procedures, Programs, Modules.

• Inputs, Outputs, Files, Records, Data Elements.

This does not necessarily mean entering the product's documentation in its entirety into the IRM, although this would be ideal. It may be more desirable to simply establish cross-references to the package's own documentation. However, because the package will, ultimately become a natural part of the company's systems, sufficient intelligence should be put into the IRM Repository so that the customer can merge the package's information resources with their own. This provides invaluable control to the customer, particularly when performing an "impact analysis." It also provides the customer with the ability to evaluate a product release prior to putting it into production.

The customer may elect to execute ensuing ISEM phases in order to properly document the package. This may be accomplished by either the 'in-house' development staff, by vendor personnel, or both working together.

Package Summary

The decision to purchase a package to reduce development time and costs can produce satisfying results for both the user and systems development staff. This does, however, require that the organization first develop the information requirements, evaluate and install the package that best meets the requirements, and finally, perform an audit of the project. The ISEM approach is a means for good communications among the participants and earns their commitment, thereby increasing the chances for a successful implementation.

HARDWARE CONSIDERATIONS

During the rough design consideration must be given to the physical implementation of the system. In many instances, the existing operating environment is assumed. However, the rough design may suggest the need for additional equipment and/or enhancements to the current operating environment. This is one reason why DP Operations and Data Base personnel participate in the preparation of the rough design; to appraise the impact of the proposed system on the current operating environment.

In studying the rough design, consideration is given to:

• The areas of the business served.

• Processing requirements: such as, Interactive, Batch, etc. Also, distributed versus central processing.

• Types of media (Inputs/Outputs): Graphical, Textual, Audio, Optical, etc.

• Data transmission/conversion requirements.

• Required processing speed and volume of transactions.

• Storage requirements (files).

• Back-up/file retention requirements.

From this, assumptions can be made regarding:

- Computer equipment and operating systems.

- Input/Output peripherals; e.g., terminals, printers, etc.

- Communication hardware/software, including networking.

- Data Base hardware and software.

- Other equipment and files (non-computer related).

From this evaluation, an estimate of anticipated expenditures can be prepared for inclusion in the overall project estimate. At this point, hardware decisions are only tentative. The final decision on these matters should be reserved for subsequent ISEM Phases as the system design is finalized.

Client/Server Computing

The expression "client/server" is used to describe a physical method of implementation for an information system. It suggests a network of computers where some are dedicated to managing key files ("file servers") and others that access the file servers remotely ("clients"). Obviously, this is but one approach to computing. Mainframe and mid-range solutions with "dumb" terminals will remain a viable solution for a long time to come. The larger machines are still a cost-effective solution for most of the "number crunching" required by major corporations. However, as the smaller machines gain in power, a truly distributed computing environment is becoming more and more attractive to companies.

Before a company jumps into "client/server" (a physical processing decision) they obviously should perform the logical system design and determine their processing requirements. As mentioned, consideration should be given to anticipated transaction volume, processing speed, security, file access, etc. Only after these areas have been specified should consideration be given to the required physical platform. The point is, "client/server" may or may not be right for you. Just because it is technically fashionable doesn't make it the right solution for your company. Again, emphasis should be placed on logical system design in order to properly specify physical requirements.

In terms of system design, aside from some special programming requirements, there is actually little difference between a normal "mainframe" based development project and a "client/server" project. Both types of projects will require the same phases and activities as prescribed by "PRIDE"-ISEM.

PROJECT PLANNING

Before a project estimate and schedule can be calculated, the path of the project must first be defined. Under ISEM, the project path is ultimately based on the system structure (as defined by the rough design). Following Phase 1, the remaining phases relate to the Standard System Structure as defined by "PRIDE":

PHASE	NAME	RELATED SYSTEM STRUCTURE RESOURCE
2	System Design	System
3	Sub-System Design	Sub-System
4-I	Admin. Proc. Design	Sub-System (all Adm-Procs incl.)
4-II	Software Engineering	Computer Procedure
5	Software Manufacturing	Program
6	Software Testing	Computer Procedure
7	Sub-System Test	Sub-System
8	System Operation	System

NOTE: A Phase 9 review is performed to conclude the project.

This means there is an explicit one-to-one relationship between the phase and the system resource; for example:

- For each system there will be a Phase 2 and 8.

- For each sub-system there will be a Phase 3, 4-I, and 7.

- For each computer procedure there will be a Phase 4-II and 6.

- For each program there will be a Phase 5.

This means that the remaining phases of the project can be deduced from the system structure. It also means that the project can branch into parallel phases. As an aside, this approach reflects the "explosion/implosion" design and development process inherent in the methodology.

Although the above phase/system-resource relationship is the normal way of conducting an ISEM project, there may be exceptions to the rule. For example:

• If multiple Phase 2's have been identified, it may be more practical to split them into separate projects.

• Although Phase 4-I is normally used to write all of the administrative procedures in a sub-system, there may be occasions where a complex procedure may require the execution of a separate Phase 4-I.

• Phase 5 is normally used to write and test a single program. However, if the program is extremely large and complex, it may be more practical to divide it into sub-Phase 5's where the modules of the program can be written and tested.

• Phase 8 marks the start-up of the system. If a large system, it may be more practical to execute a Phase 8 for a group of sub-systems. In other words, deliver the system in stages.

• Depending on the scope of the project and/or organizational considerations, the project may wish to include DBEM phases for Data Base Engineering. Normally this is treated as a separate project.

The point is, ISEM provides a good framework for executing a systems development project. However, Project Management must consider the practical implications for controlling the project and, as such, must make the final decision on project execution.

After the path of the project has been determined an Order-of-Magnitude (OOM) estimate and schedule can be calculated. This is an estimate and schedule of the amount of effort required to perform the remaining phases in the project. The estimate, thereby, is an expression of labor charges.

HUMAN RESOURCE REQUIREMENTS

Following the rough design and the project plan, consideration must be given to the types of human resources required to implement the project. Based on the type of application being developed, Project Management considers the types of Systems Engineers and Software Engineers required to work on the project, including the skills and proficiencies needed to perform the work. Based on this analysis, Project Management has four options:

• Use in-house personnel.

• Recruit additional personnel.

• Use outside personnel (contractors).

• Combinations of the above.

In all instances, the use of human resources is based on their qualifications, their availability, and their cost. Project Management, therefore, must balance these variables when selecting personnel. There may be trade-offs to consider; for example:

• One person may be more expensive than others but can deliver the work faster.

• A senior analyst could perform the work in fewer hours than a junior analyst, but is committed to other project assignments. Consequently, the analyst cannot devote sufficient time to the project. Whereas the junior analyst may have fewer conflicts and is more available for project work.

The use of known human resources has a direct impact on the Order-of-Magnitude estimate and schedule. However, in the situation where resources are unknown, the Order-of-Magnitude estimate and schedule can still be calculated. Here, the Project Manager must make some assumptions as to the type of skills and proficiencies required and the number of Systems and Software Engineers needed for the project. The estimate thereby becomes the basis for recruiting additional personnel or hiring contractors. It also becomes the basis for determining training requirements, the cost of which should be absorbed by the project.

COST/BENEFIT ANALYSIS

The purpose of a Cost/Benefit Analysis is to perform an economic impact analysis on the system and the project. Prior to performing this, Project Management has prepared an Order-of-Magnitude estimate to complete the project, along with any additional planned expenditures (such as for additional hardware/software acquisitions).

"PRIDE" provides a worksheet to assist in the preparation of the Cost/Benefit Analysis. However, in-house standards should be observed when performing the analysis.

During the analysis Project Management must consider how the proposed system solution will affect the company:

• Will it increase or reduce staffing?

• Will it increase or reduce equipment resources?

• What will be the annual savings and expenses of the system as it applies to the various user departments?

• What will be the one-time savings and expenses of the project as it applies to the various user departments?

• What will be the intangible benefits of the system?

• What is the investment evaluation? For example, what will be the Break Even Point of the project? What will be the Return On Investment (ROI)?

The Cost/Benefit Analysis becomes the basis for the preparation of the Cost and Evaluation Summary, a textual justification for the project.

PHASE 1 REVIEW

The formal deliverable resulting from Phase 1 is the "System Study & Evaluation Report." This represents a packaging of the various elements produced during the phase (e.g., Project Scope, Information Requirements, System Concept Diagram, System Logic Narrative, Project Plan, Order-of-Magnitude, Cost and Evaluation Summary, etc.). A formal meeting with management is conducted to review the deliverable. At this time, management may elect to approve the project as proposed in the deliverable, request revisions before proceeding, or cancel the project completely.

There may be many reasons for discontinuing a project at this time, perhaps the project team cannot produce an adequate solution for satisfying the business problem, or perhaps the economics of the project are not justifiable (the Return on Investment may be too low for example). Common sense would dictate never to expend human and financial resources on an activity which will never prove to be worthwhile.

Assuming cancellation of the project, the company retains specifications regarding:

• The current system.

• Information Requirements.

• A rough design of a proposed new system.

These specifications are maintained in the IRM Repository for use either at a later date, or by other applications. The point is the knowledge and design decisions formulated during Phase 1 have been captured and will not be lost even with personnel turnover. In fact, this intelligence can be used by others at a later date. It is entirely conceivable for a project to be discontinued after Phase 1 and be resumed at a later date with minor changes to the specifications.

DESCRIPTION OF PHASE ACTIVITIES

Activity A - Preliminary Project Scope

Information Resource Management submits a Work Request/Objective to Systems Resource Management who initiates an ISEM project. The request is normally prepared as part of the Enterprise Information Strategy (EIS) as prepared under the Enterprise Engineering Methodology (EEM).

Systems Resource Management initiates an ISEM project which includes the assignment of appropriate Project Management and Systems Engineering personnel.

Project Management studies the request and makes a determination of the ISEM phases and activities needed to satisfy the request. The Project Manager may decide to either:

1. Only complete a Phase 1.

2. Execute only selected phases and activities in the methodology, and possibly circumvent Phase 1 completely. This decision normally relates to a modification/improvement or maintenance request, and is based on an "impact analysis."

Project Management prepares/revises a Project Scope for the project. Based on the scope, Project Management prepares an estimate/schedule for review and approval by Systems Resource Management and User Management.

1. If the project is to proceed only through Phase 1, Project Management and Systems Engineering will prepare a Detail estimate and schedule for the PHASE.

2. If the project is to proceed to other phases and activities, then an estimate is prepared to complete the whole project.

If the project appears to be relatively small and easy to manage, then a Detail estimate and schedule is required.

If the project appears to be large and complex, then an Order-of-Magnitude estimate and schedule is required.

Activity B - Current Systems Analysis

Systems Engineering studies the current system. The engineer interviews users and DP Operations and samples work from this area. Data Base personnel also advises the engineer on existing file structures. Consequently, current system documentation is updated as required and a Current Systems Analysis is prepared highlighting the strengths and weaknesses of the current system. This analysis is reviewed with User Management for accuracy.

Activity C - Survey Information Requirements

Systems Engineering prepares an interview outline and schedule for determining information requirements and interviews User Management accordingly.

Activity D - Prepare Requirements & Scope

Systems Engineering interprets user information requirements and prepares a formal definition of each. An Information Flow Diagram is also prepared which shows how information within this application flows between users. As required, the Project Scope is updated at this time.

Activity E - Review Requirements & Scope

Systems Engineering reviews the information requirements, Project Scope and Information Flow Diagram with User Management for clarity and accuracy. This is the first major review point in the methodology.

Activity F - Develop System Approach

Systems Engineering prepares a system solution that will satisfy the User's information requirements. A complete rough design of the system is prepared at this time. This becomes the basis for considering alternative solutions, such as a "make versus buy" solution. Other functions participate in the rough design including Software Engineering, Data Base personnel, DP Operations, etc.

Activity G - Prepare System Evaluation

Based on the rough system design prepared in the previous activity, Project Management prepares a Project Plan which details the remaining phases in the project. From the plan an Order-of-Magnitude estimate and schedule is prepared to complete the project. This estimate combined with any other anticipated financial expenditures (for such things as hardware/software) is used in the preparation of a Cost/Benefit Analysis. This becomes the basis for developing the Cost and Evaluation Summary, a textual justification for the project.

Project Management and Systems Engineering then assemble the deliverables produced in the phase and packages them into a single phase review document which is evaluated by Quality Assurance prior to conducting the formal phase review with management.

Activity H - Phase 1 Review

Project Management conducts a formal review of the Phase 1 deliverables with Executive Management, User Management, Information Resource Management, and Enterprise/Systems/ Data Resource Management. The review is used to evaluate the work performed thus far and revise as required. At this time, management will review the formal Phase 1 **"System Study & Evaluation Report"** consisting of:

- Phase Cover Page
- Project Scope
- Current System Analysis
- Information Flow Diagram (IFD)
- Project Information Flow
- Information Requirements
- System Concept Diagram
- System Logic Narrative
- Project Plan
- Project Estimate/Schedule Recap
- Cost and Evaluation Summary
- Glossary of Terms
- Phase Review Checklist
- Supporting Matrices:
    - Requirements/System Matrix by Project
    - Entity/System Matrix by Project
    - Information Requirements/Outputs Matrix

Based on this report and subsequent review meeting, management may elect to revise parts of the report, discontinue the project, or approve it for continuation.

The next few pages include sample deliverables resulting from this phase.

SAMPLE PHASE 1 DELIVERABLES

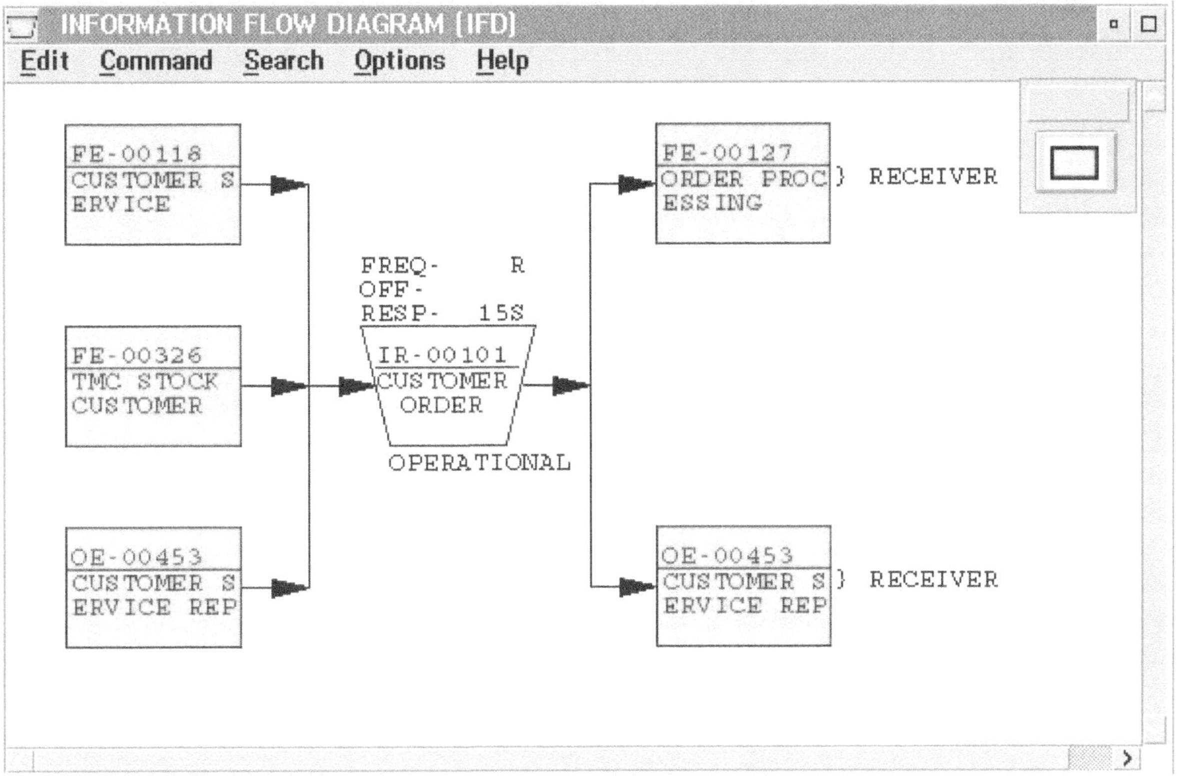

OD-90244 P R O J E C T I N F O M A T I O N F L O W

APR 30, 2004 RELEASE 1.2.0

FE / OE > INFO RQMTS V	FE-00118 CUSTOMER SERVICE	FE-00122 PRODUCT SHIPPING	FE-00127 ORDER PRO CESSING	FE-00152 SALES MANAGEMENT	FE-00154 CUSTOMER CREDIT	FE-00156 SALES REPRESENT ATION	FE-00326 TMC STOCK CUSTOMER	OE-00433 DIRECTOR, CUSTOMER SERVICE	OE-00440 CREDIT MANAGER
IR-00101 CUSTOMER ORDER	O		R				O		
IR-00102 CUSTOMER PROFI LE			R	O	O				O
IR-00103 CUSTOMER ORDER STATUS	O		R					R	
IR-00104 ORDER STATUS ANALYSIS	O	O	O	R					
IR-00105 CUSTOMER ORDERS TO PROCESS	R		O						
IR-00106 CUSTOMER CREDIT REQUEST			O		R				R
IR-00107 CUSTOMER SETUP/ MODIFICATIONS			O	R		O			

INFORMATION FLOW KEY: O = ORIGINATOR, P = PARTICIPATOR, R=RECEIVER

SAMPLE PHASE 1 DELIVERABLES

OD-90189	INFORMATION REQUIREMENTS	"PRIDE"-ISEM
SEP 01, 2004	RELEASE 1.2.0	PAGE 1

****
* IR-00101 CUSTOMER ORDER *
****

ENTITY(S) REQUIRING INFORMATION:	NUMBER	INFO. FLOW
1. CUSTOMER SERVICE	FE-00118	ORIGINATOR
2. ORDER PROCESSING	FE-00127	RECEIVER
3. TMC STOCK CUSTOMER	FE-00326	ORIGINATOR
4. CUSTOMER SERVICE REP	OE-00453	RECEIVER
5. CUSTOMER SERVICE REP	OE-00453	ORIGINATOR

TYPE REQUIREMENT: INFORMATION INFORMATION TYPE: OPERATIONAL

FREQUENCY:	OFFSET:	RESPONSE TIME:
UPON REQUEST	N/A	15 SECOND(S)

BUSINESS PURPOSE

The purpose of this requirement is to interact with a customer to place or revise an order for corporate products.

ACTIONS AND/OR BUSINESS DECISIONS

Order Processing will...

1. Check product availability and advise the customer of anticipated delivery dates.

2. Create, revise or delete an order.

NOTE: Deletion is only permitted if the order has not yet been processed by Customer Service.

3. Notify Customer Service to take necessary actions to fill the order.

4. Request changes, as required, to the customer's credit rating (i.e., increase, decrease) to Customer Credit.

BENEFITS

1. Improved communications/relations with the customer, thereby enhancing customer satisfaction.

2. Improved communications internally, particularly with Customer Service and Customer Credit. Such communications improves control over customer credit, thereby preventing unauthorized shipments.

OBJECTS/FILES AFFECTED:

* PRODUCT OBJECT	(FD-00100)
* CUSTOMER OBJECT	(FD-00101)
* ORDER OBJECT	(FD-00102)

REQUIRED DATA ELEMENTS:

* PRODUCT NUMBER	(DD-00003)
* NAME	(DD-00018)

(continued on next page)

SAMPLE PHASE 1 DELIVERABLES

(continued from previous page)

* UNIT PRICE	(DD-00111)
* ORDER NUMBER	(DD-00012)
* DATE	(DD-00016)
* QUANTITY ORDERED	(DD-00036)
* TERMS	(DD-00044)
* STATUS	(DD-00041)
* QUANTITY ON HAND	(DD-00508)
* CONTACT NUMBER	(DD-00042)
* DELIVERY PRIORITY	(DD-00050)
* CUSTOMER NUMBER	(DD-00001)

IMPLEMENTED BY INPUTS/OUTPUTS:

* PRODUCT AVAILABILITY (	OD-00010)
* CUSTOMER ORDER	(OD-00011)
* CUSTOMER PROFILE	(OD-00012)

SUPPORTED BY SYSTEM RESOURCES:

* TRUNDLE ORDER PROC. SYS (TOPS)	(OP)
* CUSTOMER ORDER INQUIRY	(OP-00)
* ORDER-PROC FILE MAINTENANCE	(OP-08)

*** END OF OUTPUT ***

OD-92028	SYSTEM CONCEPT DIAGRAM		"PRIDE"-ISEM
RESOURCE: OP	NAME: TOPS SYSTEM	PREPARED BY: BROWN	EFFECTIVE DATE: OCT 10, 2004
PAGE: 1	COPYRIGHT (C) TMC 2004	RELEASE 1.2.0	DATE PRINTED: OCT 10, 2004

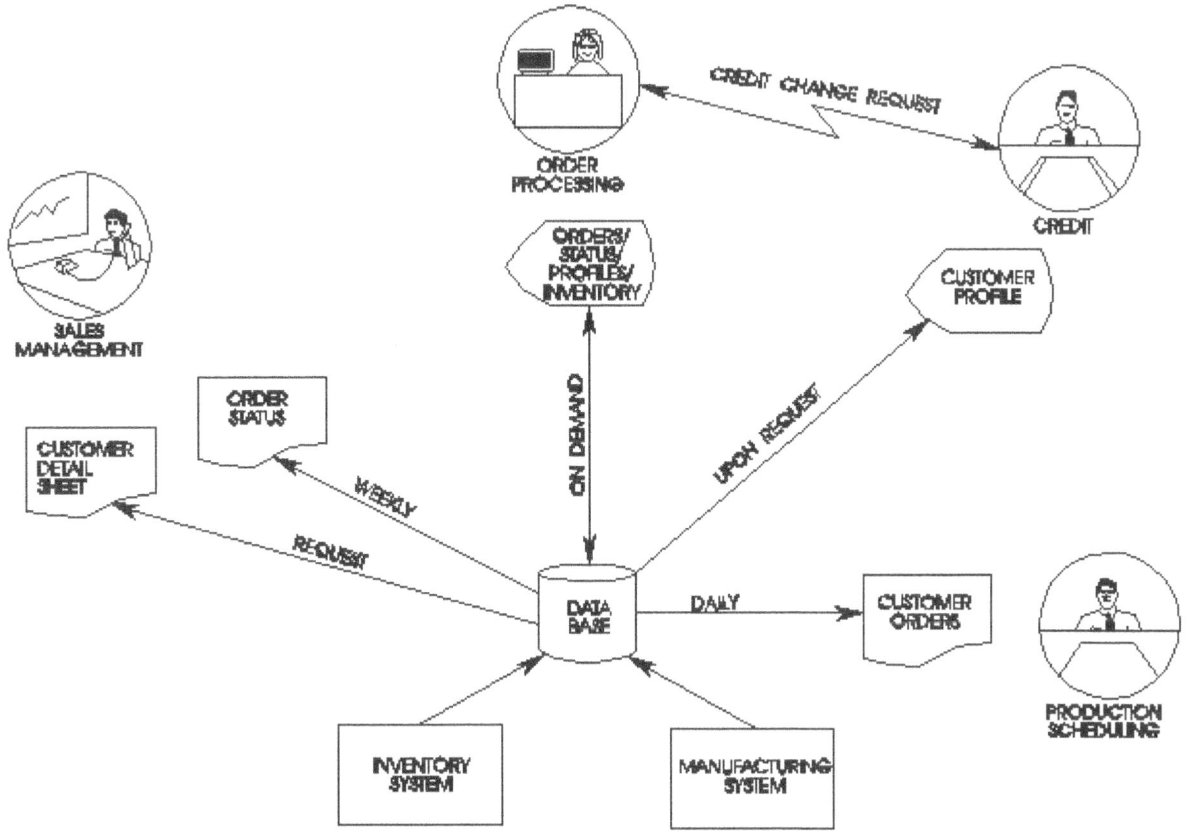

OD-90394	SYSTEM LOGIC NARRATIVE	"PRIDE"-ISEM
OCT 10, 2004	RELEASE 1.2.0	PAGE 1

* SYSTEM OP - TRUNDLE ORDER PROC. SYS (TOPS) *

BUSINESS PURPOSE

The purpose of the TMC Order Processing System (TOPS) is to process customer orders for corporate products. It is also used to monitor the status of customer orders in order to expedite problems and assure customer satisfaction with delivery.

This system affects business functions from the following areas: Marketing, Manufacturing, and Administration.

OVERVIEW

TOPS represents an integrated set of business processes that maintains a consistent flow of information between Order Processing, Credit, and Production. The system expedites customer orders thereby increasing customer satisfaction. It also implants effective controls over customer credit which reduces risk of loss of merchandise due to unauthorized processing of orders.

The system consists of six sub-systems; five to produce screens and reports in support of information requirements and a sixth to collect data for the data base.

CUSTOMER ORDER INQUIRY

This sub-system represents the formal interaction between the customer and order processing. A customer can call and place an order, query on the status of an order, or modify/ delete an order (within constraints). Through this facility, order processing can validate the customer's credit limit, check availability of product, and review general customer data (i.e., address).

WEEKLY ORDER STATUS

To monitor orders in progress, Sales Management is provided with a weekly report that tracks orders, back-orders, and shipments. In the event a problem arises, Sales Management contacts the necessary managers in Manufacturing to expedite and resolve the problem.

DAILY ORDER PRODUCTION

At the beginning of each work day, production is provided with an itemized list of orders to be processed for the day. Orders will either be picked and shipped accordingly or back-orders will be created and placed to manufacture the necessary goods.

CREDIT CHANGE REQUEST

In the event a customer wishes to increase (or decrease) their credit limit, Order Processing can submit an electronic request to the credit department who can quickly review and approve the request. This is implemented through electronic mail.

(continued on next page)

SAMPLE PHASE 1 DELIVERABLES

(continued from previous page)

CUSTOMER RECORD MANAGEMENT

Whenever a customer is initially established with the company, certain customer records must be established to allow the customer to request goods. In addition to establishing the customer in the computer files, a printed copy of the customer data is required for historical purposes. Further, whenever a change to the customer profile is incurred (i.e., change of address, contacts), the change is printed and filed for historical purposes, thereby creating an audit trail of changes for reference purposes. All changes to the customer profile must be approved by Sales Management.

ORDER PROCESSING FILE MAINTENANCE

This sub-system is collecting the necessary data to process customer orders.

* TEST CRITERIA

Parallel testing should begin with the "Customer Record Management" and "Order Processing Maintenance" sub-systems, followed by the remaining read-only sub-systems.

* * * END OF OUTPUT * * *

OD-90192	GLOSSARY OF TERMS	"PRIDE"-ISEM
SEP 01, 2004	RELEASE 1.2.0	PAGE 1

****
* REQUESTED FOR: PD-00132 ORDER PROCESSING SYSTEM *
****

DD-00008 ACCOUNT NUMBER

A unique number used to identify a financial account.

DD-00043 ADDRESS TYPE

A unique code used to identify a type of address.

DD-00104 AMOUNT PAID ON INVOICE

The total amount paid on a single invoice by a customer.

DD-00014 BACK ORDER NUMBER

A unique number used to identify a back-order.

DD-00055 BACK-ORDERS FOR WEEK

A count of the number of product back-orders for a week.

DD-00011 BILLING NUMBER

A unique number used to identify a billing (invoice).

DD-00030 BUSINESS TITLE

A proper name associated with a business position.

SAMPLE PHASE 1 DELIVERABLES

OD-90303 INFORMATION REQUIREMENTS/SYSTEMS MATRIX BY PROJECT

OCT 10, 2004 RELEASE 1.2.0

INFO RQMTS > / SYSTEMS V	IR-00101 CUSTOMER ORDER	IR-00102 CUSTOMER PROFILE	IR-00103 CUSTOMER ORDER STATUS	IR-00104 ORDER STATUS ANALYSIS	IR-00105 CUSTOMER ORDERS TO PROCESS	IR-00106 CUSTOMER CREDIT REQUEST	IR-00107 CUSTOMER SETUP/MODI FICATIONS		
CS CUSTOMER SERVICE SYSTEM		X							
OP TRUNDLE ORDER PROC SYS (TOPS)	X	X	X	X	X	X	X		
OP00 CUSTOMER ORDER INQUIRY	X	X	X						
OP04 WEEKLY ORDER STATUS				X					
OP05 DAILY ORDER PRODUCTION					X				
OP06 CREDIT CHANGE REQUEST						X			
OP07 CUSTOMER RECORD MANAGEMENT							X		
OP08 ORDER-PROC FILE	X	X	X	X	X	X	X		

OD-90304 ENTITY/SYSTEM MATRIX BY PROJECT

OCT 10, 2004 RELEASE 1.2.0

ENTITIES > / SYSTEMS V	FE-00118 CUSTOMER SERVICE	FE-00127 ORDER PROCESSING	FE-00152 SALES MANAGEMENT	FE-00154 CUSTOMER CREDIT	OE-00406 CHIEF INFO RMATION OFFICER	OE-00433 DIRECTOR, CUSTOMER SERVICE	OE-00440 CREDIT MANAGER	OE-00452 NAT'L SALES MANAGER	OE-00453 CUSTOMER SERVICE REP
OP TRUNDLE ORDER PROC SYS (TOPS)		EXECUTION			APPROVAL	APPROVAL		APPROVAL	REVIEW
OP00 CUSTOMER ORDER INQUIRY		EXECUTION			APPROVAL	APPROVAL			EXECUTION
OP04 WEEKLY ORDER STATUS	EXECUTION	EXECUTION	EXECUTION		APPROVAL	REVIEW			
OP05 DAILY ORDER PRODUCTION		EXECUTION	REVIEW		APPROVAL	REVIEW			
OP06 CREDIT CHANGE REQUEST	EXECUTION	EXECUTION	APPROVAL	APPROVAL	APPROVAL		APPROVAL		
OP07 CUSTOMER RECORD MANAGEMENT		EXECUTION	EXECUTION		APPROVAL	REVIEW			
OP08 ORDER-PROC FILE MAINTENANCE		EXECUTION			APPROVAL	EXECUTION			EXECUTION

SAMPLE PHASE 1 DELIVERABLES

OD-90299 I N F O R M A T I O N R E Q U I R E M E N T S / O U T P U T S M A T R I X

OCT 10, 2004 RELEASE 1.2.0

INFO RQMTS > OUTPUTS V	IR-00101 CUSTOMER ORDER	IR-00102 CUSTOMER PROFILE	IR-00103 CUSTOMER ORDER STATUS	IR-00104 ORDER STATUS ANALYSIS	IR-00105 CUSTOMER ORDERS TO PROCESS	IR-00106 CUSTOMER CREDIT REQUEST	IR-00107 CUSTOMER SETUP/MODI FICATIONS
OD-00010 PRODUCT AVAILAB ILITY	X						
OD-00011 CUSTOMER ORDER	X						
OD-00012 CUSTOMER PROFIL E	X	X				X	
OD-00013 CUSTOMER ORDER STATUS			X				
OD-00014 WEEKLY ORDER ST ATUS ANALYSIS				X			
OD-00015 DAILY CUSTOMER ORDERS					X		
OD-00016 CREDIT CHANGE R EQUEST						X	
OD-00017 CUSTOMER DETAIL SHEET							X

OD-90396	COST AND EVALUATION SUMMARY	"PRIDE"-ISEM
OCT 12, 2004	RELEASE 1.2.0	PAGE 10.1

AS DEFINED ON PD-00132 ORDER PROCESSING SYSTEM

The estimated cost for the project is $ 141,200.00 (consult the Project Estimate/Schedule Recap). Based on a Cost/Benefit Analysis performed in conjunction with the Sales Department, we conservatively estimate an annual savings of $80,000 in operating expenses. This is due to the fact that the new Order Processing System will require fewer employees to implement it (a 15% reduction of Order Processing clerks is projected) along with an improved cash flow of 5% due to improved control over production scheduling and inventory.

INVESTMENT EVALUATION

Break Even Point = $141,200 / $80,000 = 1.7 years

Return on Investment = 56.6%

BENEFIT$

1. More accurate and less redundant data due to systems integration of orders with credit, billing, inventory and shipping.

2. Improved customer service with fewer clerks.

3. Improved corporate image and market share.

- Ernest Bilko
Project Manager
October 12, 2004

* * * END OF OUTPUT * * *

END SECTION

Section 10

"PRIDE"-ISEM PHASE 2 SYSTEM DESIGN

Only when the Systems Engineer can walk in the moccasins of the user does the engineer have a right to design a system for the user.
- Bryce's Law

The purpose of this phase is to design sub-systems to implement information requirements. This is the logical system design phase of ISEM. Several events occur during this phase:

- Sub-Systems are designed with their appointed inputs, outputs and files.
- Actual examples of outputs (reports, screens, etc.) and pertinent inputs are prepared.
- The project plan and associated project estimates/schedules are updated.
- A formal review with user management is conducted to verify the phase findings.

METHODOLOGY NAVIGATION

A Phase 2 may be initiated either by following a normal Phase 1 or, if a modification/improvement, following Phase 1, Activity A. There is a one-to-one relationship between Phase 2 and an information system. In other words, a Phase 2 will be executed for each system identified in Phase 1 (normally there is just one).

The formal deliverable resulting from Phase 2 is a "System Design Manual" consisting of a description of the sub-systems, samples of outputs, along with an updated project plan. This is reviewed with management who must decide whether to:

- Approve the project for continuation.
- Request revisions to the phase findings.
- Discontinue the projects.

Following Phase 2, the ISEM project will branch to multiple Phase 3's "Sub-System Design." Because of the one-to-one relationship between sub-systems and Phase 3, the number of Phase 3's to be performed is based on the number of sub-systems identified in Phase 2.

Phase 2 may also trigger a supporting DBEM project. As part of the system design, Systems Engineering will have defined all of its major data requirements. Because of this, a Data Base Engineering project can then be initiated to begin the process of incorporating the system's data into the corporate data base.

GENERAL DISCUSSION

Before proceeding further, let's consider what is meant by the term "Information System." Sometimes the obvious is not always obvious. Is an "Information System" the same thing as Computer Software? Or the same as a Project? Or the same as a Computer? The answer is an unequivocal "NO."

So what is an "Information System"? Let's begin by examining its characteristics. In essence, it is no different than any other system; all of which have three fundamental attributes:

1. *A SYSTEM HAS A SPECIFIC PURPOSE* - Without a given purpose, the elements of a system are meaningless and unnecessary. The purpose of an "Information System" is to produce information (not data) in support of business needs.

2. *A SYSTEM IS A GROUPING OF ELEMENTS WHICH FORM A WHOLE* - The scope of a system is arbitrarily defined by those who must deal with it. No two systems will necessarily have the same number of elements. For example, one company may have a payroll system which simply calculates employee wages and issues checks. Another company may do this along with calculating corporate tax and personnel benefit figures. What this highlights is that the parts of an Information System, or any system for that matter, are grouped for convenience so that its dimensions are known and understood. A system can expand and contract based on what is necessary to achieve its goals or objectives. If there are parts of a system that do not contribute to the overall objectives of the system, then these parts are wasteful and create an unnecessary overhead. However, it is entirely conceivable, if not probable, that an organization has a single Information System; a system that includes Payroll processes, Accounting processes, Manufacturing, etc., etc.

3. *A SYSTEM OPERATES ROUTINELY* - It is the repetition of the parts of the system that causes it to behave in a predictable manner. Such is the case for "Information Systems" which must operate in a routine manner. There are certain processes which must occur on a daily basis, beginning of the week, at the end of the month, quarterly, year-end. Even query type processes (upon request) are used to satisfy common and routine objectives.

By definition, without any one of these basic characteristics, it is not a system.

We can, therefore, provide a standard definition:

INFORMATION SYSTEM - *a prescribed set of processes operating routinely to produce information for users to fulfill business actions and decisions.*

Information Systems are used to pay employees, manage finances, manufacture products, monitor and control production, forecast trends, process customer orders, etc. These are all excellent examples of how systems support information requirements.

But what is a Computer System? What does it produce? Information about computers? Or, are we merely trying to describe how a system was implemented? We beg the issue as to what is more important; the system or how it was implemented? To many in the field, implementation is more important than what is being implemented.

Are the terms "Information System" and "Computer Software" synonymous? Software is obviously the opposite of "Hardware." Software is machine processable instructions that permit the hardware to perform specific functions. Software has the same relationship to systems as robots have to a manufacturing process. Even if the robot performs properly or a program executes correctly, it does not necessarily mean you are producing anything worthwhile. What is most important in this situation is whether the logical system design is correct or not. The physical implementation can be less than perfect and you can still produce good results. The reverse is not true. "Systems" and "software" are as dissimilar as "information" and "data."

UNDERSTANDING SUB-SYSTEMS

A sub-system is a specific business process that operates routinely in a unique time frame. This time frame is predicated on the information produced by the sub-system. In other words, the timing of the sub-system is derived from the information requirements it supports.

The sub-system is also used to logically join data with processing. It simply defines what data is required and when. The succeeding phases of ISEM will be used to define how the data will be physically processed by the human being, office equipment, and by the computer.

To those people entrenched in programming, the sub-system concept is somewhat nebulous. Instead of thinking in terms of BUSINESS PROCESSES, they tend to think in terms of COMPUTER PROCESSES. This type of physical thinking does not apply in Phase 2 and should be reserved for Phases 3 and 4-II. During Phase 2, Systems Engineering must concentrate on enterprise-wide business processes.

There is no practical limitation to the number of subsystems in a system. A system can have as few as one, or as many as hundreds.

DESIGN CORRECTNESS

During Phase 1, approval was given for the Project Scope, the Information Requirements, and System Approach (the rough design). In Phase 2 each subsystem is finalized as to its Inputs, Outputs, Files, and processing specifications. The primary objective of Systems Design is to define the system in terms of:

- WHAT business processes make up the system.
- WHEN these processes need to occur.
- WHAT inputs and outputs will be produced during processing.
- WHAT data will be required for processing.

The emphasis in Phase 2 is on design correctness; designing a system that correctly satisfies information requirements. As such, ISEM's design philosophy during Phase 2 is to work backwards from the information requirements:

Information Requirements	>>>	Data Requirements
Receiver of Information	>>>	Originator of Data
Outputs	>>>	Inputs

Later, during Phases 3 and 4, the process is reversed. Here, the design expresses how data will physically be processed in order to produce information.

Source of Data	>>>	Destination of Info
Inputs	>>>	Outputs
Start	>>>	End

CHRONOLOGICAL DECOMPOSITION

"Chronological Decomposition" is a technique used in ISEM to decompose a system into its logical subsystems (business processes). Although several design concepts are implemented by Chronological Decomposition, "Information Driven Design" is the driving force behind the technique.

As the name indicates, timing is an essential part of Chronological Decomposition. This is because information is a perishable commodity. It only has value during a particular point in time. Users require information to support actions and decisions on a routine and timely basis, either instantaneously, daily, weekly, monthly, etc. All information systems operate routinely based on timing. Since this is true, why not make use of this timing consideration during system design as opposed to discovering it after the fact?

Timing will ultimately dictate how data will be collected and stored (availability requirements) and how data will be accessed to produce information. This approach implies that there are substantial differences between information and data, one of which is that data is the raw material that is used to produce information.

The supporting data must be defined in such a way that we can easily understand what primary data must be supplied by a User and what generated data must be calculated internal to the system. Data relationships can be extensive. For example, take NET-PAY which may be based on a complicated computation:

```
NET-PAY = GROSS-PAY - FICA - CITY-TAX -
UNION-DUES - (etc.)
```

The data elements used in the formula may also be calculated, such as:

```
GROSS-PAY = HOURS-WORKED X PAY-RATE
```

What this means is that in order to arrive at the correct value for NET-PAY, we must be able to reach all of the primary values, such as HOURS-WORKED and RATE, in a TIMELY manner. If we cannot do this, NET-PAY will be incorrect.

Defining these data dependencies has typically defaulted to the programmer who redefines the relationships with each application and buries it in the source code, making maintenance and change difficult.

The timing and data specifications resulting from the information requirements will ultimately dictate the type of system to be created. For example, if information is required upon request and within a matter of seconds, this will probably result in an "interactive" type of process. However, if the information is required upon request but within a few hours, this will probably result in "batch" type processing (it may even be processable manually). These specifications are the basic building blocks for all systems and software design.

Chronological Decomposition organizes all of the data required to support the application, into logical files (objects), or as DBEM refers to it, the Application Logical Data Base Model (ALDBM). As such, it synchronizes this data base with the application.

Perhaps the biggest benefits derived from Chronological Decomposition, as reported by users, has been that it forces the Systems Engineer to consider all of the required data and simplifies processing. It also emphasizes the need for sharing data. As a design develops, consideration is given to using data from other applications. After all, why create new files and processes if they already exist? To do this, Chronological Decomposition makes active use of the IRM.

With the logical design defined, consideration is then given to the most appropriate way to physically process the data, either manually or computer assisted. Here is where Functional Decomposition and Data Driven design excels. For software engineering, the characteristics of the data, its structures and what functions the computer must perform (e.g., create, update and reference) dictates the required programs. These specifications are the result of Chronological Decomposition. The physical characteristics of the data defines its validity. The data structures denote input, file and output relationships. The functional requirements determine how the data will be read and written in a program, whether sequentially, iteratively or selectively. In other words, Functional Decomposition and Data Driven Design will dictate physically "WHO" and "HOW" the data will be processed.

In no way should "Chronological Decomposition" be confused with "Functional Decomposition." Both have distinctly different purposes and missions. Whereas the former is concerned with logical system design based on the characteristics of information, the latter is concerned with how data is to be processed physically. They are not conflicting techniques. In fact, they are complementary as long as they are put into proper perspective. They simply have entirely different roles.

Steps In Design

The following is a brief explanation of the steps in Chronological Decomposition:

I. ANALYZE THE BUSINESS

This includes an analysis of the logical functions of the enterprise along with a study of the physical organization. The purpose of this is to evaluate how and when business actions and decisions are made. Included in this step are interviews with users.

II. SPECIFY INFORMATION REQUIREMENTS

Information Requirements are formally documented and reviewed with users for accuracy.

III. DEFINE DATA REQUIREMENTS

Define the data items needed to support each information requirement. Both primary and generated data elements are defined and related to the information requirements. Data definitions that were previously defined can be re-used. New elements are defined as required. What is important is that data dependencies be correctly defined (such as those primary data items needed to calculate generated items).

IV. DETERMINE OUTPUTS

Since, outputs are used as the vehicle for transmitting information, the Systems Engineer must determine what type of output(s) will be required to satisfy the information requirement(s). There is not necessarily a one-to-one relationship between information requirements and outputs. As part of this process, the engineer defines the timing of each output, along with the data that must appear on the output. This is derived from the information requirements.

V. BUILD NEW LOGICAL FILES

Based on the data requirements of the outputs, logical files are constructed using primary data (not generated). These files represent a logical view of the required data for this application only (not a global view).

VI. SUBSTITUTE EXISTING FILES FOR NEW FILES

Wherever applicable, existing files are substituted for the new logical files. This eliminates redundancy and promotes a shared data base environment. The criteria for substitution is:

• Uses the same "basic grouping" (logical key to the file).

• Uses the same data in the file.

• The file is maintained (updated) in a time frame that will accommodate the outputs.

VII. CREATE INPUTS TO COLLECT DATA FOR THE NEW LOGICAL FILES

Create inputs to collect data in a timely manner. This is only for the new files, not for the existing files that were substituted in the last step. After all, why create inputs for data that is already being collected elsewhere?

VIII. ORGANIZE INPUTS, FILES AND OUTPUTS INTO SUB-SYSTEMS

Inputs and outputs are sorted into separate business processes (sub-systems) based on compatible time frames and users (e.g., Sales, Customer Services, Manufacturing, etc.). Files are also attached to the sub-systems based on the data to be collected (through inputs) or retrieved (through outputs).

As Chronological Decomposition implies, there are basically two types of sub-systems:

• File Maintenance - to collect and store data in a timely manner.

• Display/Reporting - to access data in a timely manner to support outputs.

Sub-systems can be merged or separated as deemed practical by the Systems Engineer. However, the engineer must be highly sensitive to the timing issue since it will have a significant impact on the data base design.

In addition to the two types of sub-systems mentioned above, the Systems Engineer must determine the need for additional types of sub-systems; for example:

• File Conversion - a one time sub-system to move data from an old file format to a new file format.

• File Backup/Recovery - this is required in the absence of standard operating procedures that deal with recovering files in case of emergency.

• File Management - this is also required in the absence of standard operating procedures for file handling considerations (e.g., initialization, reorganization, auditing, repairing a damaged data base, etc.).

SYSTEM FLOWCHART

A system design is represented with the System Flowchart, one of the formal deliverables resulting from Phase 2. As with all ISEM flowcharts, the diagram graphically shows the next level of detail in the system structure: sub-systems. Each sub-system is represented on the flowchart with its appointed inputs, outputs, files and timing. A Table of Files accompanies the flowchart which shows how the various files are used in the sub-systems: either "C" Created, "U" Updated, "R" Referenced, or combinations of all three.

SUB-SYSTEM CONCEPT DIAGRAM

Following the system design, the Systems Engineer evaluates how each sub-system will be physically implemented. This is similar in intent as the rough design prepared in Phase 1, Activity F. From this rough design, the Systems Engineer creates a Sub-System Concept Diagram. Like the System Concept Diagram prepared in Phase 1, the Sub-System Concept Diagram is a free-form schematic used to graphically express how each sub-system will be implemented. As such, it will be used by User Management to visualize and evaluate the proposed sub-system. This diagram is prepared in Phase 2 and is confirmed in Phase 3 as the sub-system design is completed.

ILLUSTRATIVE EXAMPLES

After the decomposition of the system into sub-systems has been completed, Systems Engineering prepares illustrations of each Output and Input for review with users. These illustrative examples contain actual data values, so the users can visualize field alignment and contents to assure each Information Requirement is implemented in a usable form. Attention to detail on the part of Systems Engineering during the preparation and review of the illustrative examples dramatically reduces requests for changes later in the project. These actual Data Values also assist Systems Engineering in confirming the physical attributes of the Data Elements (such as class, length, precision, scale, etc.).

When ISEM refers to "inputs" and "outputs," it does not mean "files," although they may be related. Instead, an Input is a human intelligible medium that is used to collect data for storage in a file. An Output is a human intelligible medium used to convey information. Inputs can be represented by a keyboard, touch screen, audio input, optical input, printed forms, etc. Outputs can also take many forms: printed reports, screens, audio output, etc. The basic difference between the two is one collects data (inputs) and the other displays information (outputs). Sometimes, a printed form can represent both; it is used to collect data in a specific format that is ultimately passed on to a user for review as information. Such an input/output is commonly referred to as a "turnaround document" and should be represented as both an ID and an OD in the IRM. The common bond between the two is the FORM NUMBER. Also, the ID can be directly related to the OD in the IRM.

PROTOTYPING

With the advent of Fourth Generation Languages (4GL) and other application development aids, the concept of "Prototyping" has gained in prominence. To many, it represents a significant productivity improvement for their systems development efforts; others seem to be at a loss as to how to appropriately apply this technology. Some see it as a substitute for requirements definition; others see it as a way for evaluating the viability of a design. Which is correct? Well, let's begin by providing a standard definition for the word:

PROTOTYPING - *To develop an actual physical archetype of a product to some scale. The prototype is used for testing in order to visualize and evaluate the performance of the product and to make recommendations for improvement prior to the final design.*

Prototyping's "roots" are found in Engineering, Manufacturing and Construction. In the automotive field, working models of automobiles are prepared to scale prior to going into production. As an aside, prototypes are frequently used in car advertisements. If you were to look beyond the facade, you will find an automobile that is very poorly held together. One you definitely would not want to drive.

In engineering and construction, companies have whole departments dedicated to building physical models of buildings, plants, bridges, etc., that the company is planning to build. In fact, substantial money is invested into these models prior to construction.

In virtually all instances, the prototype is based on some form of logical design, usually in the form of a set of blueprints. Further, the logical design is based on a set of predetermined requirements. In other words, prototyping is applied in the following sequence:

REQUIREMENTS >>>> DESIGN >>>> PROTOTYPE >>>> BUILD

It is unimaginable to try and build an automobile, airplane, building, etc., in any other sequence. Companies do not allow modeling departments to develop prototypes of products before the logical design and requirements have been completed. It would be a senseless waste of time and money. So why should systems development be any different?

Until recently, it was not feasible to develop a prototype of an Information System. Systems departments could not develop a rudimentary model of their applications because it would have to be developed by hand and would take considerable time and money to develop. But these new application development aids quickly changed all of that. Systems developers suddenly discovered that they could build an application in a very short period of time. The productivity improvements seemed boundless.

People then began to ask how to apply prototyping tools in relation to their systems methodology. Some extremists felt that it eliminated the need for the programming phases completely. Some even feel the requirements and design phases can be eliminated, since systems can now be built on an "ad hoc" basis. What they did not realize was that the phases were not eliminated, but rather, some of the technical tasks were expedited. The same activities and tasks were required to build the application but at a much higher rate of speed.

There were also those who felt that prototyping was a good way for helping the user determine their information requirements. *"After all, users don't know what they want."* This is a lame excuse for not defining information requirements at the beginning. This "ad hoc" approach is simply a sign that the person was never trained in how to properly define requirements to begin with. The excuse is that it is fast and cheap to develop an application this way. *"If you don't like the application, simply throw it away."* Translation: *"Since I don't know what I'm doing, I'll keep hack-*

ing away at the problem until something worthwhile happens." If there is one place prototyping definitely does not belong, it is in the area of requirements definition. It only helps to determine physical requirements, not logical business requirements.

So where does prototyping apply in ISEM? Several places:

Prototyping and Phase 1 - "System Study & Evaluation"

During Activity F Systems Engineering prepares a rough design of the system. All sub-systems, procedures, programs, inputs, outputs and files are identified at this time. Following the rough design, a System Concept Diagram is prepared which is a rendering of the entire system. In most instances, this amount of detail is sufficient in order to make a practical Phase 1 decision to proceed with the project, modify it, postpone it or cancel it completely. However, there are instances where more detail is required.

There are basically two types of people that can assimilate ideas: The Conceptualist and the Detailist. For example, when building a house, the conceptualist can visualize what the structure will look like simply by looking at blueprints and/or an artist's rendering. However, the detailist often requires a more tangible idea as to what the house will look like. For the Detailist, it is necessary to be able to step through a model home to note the physical appearance of the house. The same is true in Phase 1. Some people will be able to visualize the entire system based on the System Concept Diagram and rough design. Others, will require something more tangible in order to assimilate the system. Prototyping can provide assistance in this respect. A "quick and dirty" application can be developed and reviewed in order for the user to visualize how the system will work and the types of screens and reports produced.

The only danger in using prototyping in Phase 1 is the possibility of devoting substantial time and money to a project that will only be cancelled and postponed after Phase 1. There is also a tendency by some to try to develop the entire system, including programming, in Phase 1. It must be remembered that the purpose of Phase 1 is to specify and analyze a business problem and/or opportunity and to propose to management a course of action. Regardless of how well the system design was thought out it is only a "rough design," not the final design. Changes will be incorporated into the design as it is detailed in the subsequent phases.

Prototyping and Phase 2 - "System Design"

During this phase, the system design begins to take on physical characteristics. Screens and Reports are designed and actual data values are used to give the user an idea of what the finished product will look like. Obviously, this is an area where prototyping excels.

Since the system has been logically defined, it should be relatively easy to mock-up the screens and reports using prototyping tools. Adjustments can then be made with a dialog between the analyst and the user.

The user should be forewarned; some prototyping tools are not as easy to use as they advertise. When a screen or report is designed, it can be quite cumbersome to modify. In a situation like this, it is recommended that the illustrative examples be prepared using simple text editors, screen layout masks or even a plain piece of paper. It will be much easier to accommodate changes (and there will be many) in a more cost effective manner. The manually prepared illustrations become the prototype's prototype.

In the final analysis, if the prototyping tool is easy to use and can readily accommodate change, then use it. If not, prepare the illustrative inputs and outputs by hand first.

Prototyping and Phase 3 - "Sub-System Design"

If the illustrative Output and Inputs were prepared manually in Phase 2, then Phase 3 represents the last opportunity to use prototyping prior to the programming phases. In this instance, because of the detail defined to this point, there should be few, if any, changes made to the screens or reports.

Summary

Prototyping is a valid concept and has an important role to play in systems development. The biggest problems that are associated with prototyping include:

1. Misapplication of prototyping as a concept and as a tool. It is like anything else; it should be used as intended. This specifically excludes using it as a tool for specifying information requirements or system design. This is the biggest danger with prototyping in

systems development.

2. When users see a working prototype, they tend to assume that the application is nearing completion and ready for operation. It must be clearly explained to the user the role that prototyping plays in systems development.

3. Users want to use the prototype as the finished product. This may suffice on an interim basis but the User must understand that it is far from being the final product. After all, who wants to drive a prototype of an automobile or fly in a prototype of an airplane?

4. Prototyping only emulates computer processes of a system. They are not helpful for other processes of a system, such as manual and office automation procedures.

5. Many prototyping tools are based on a specific DBMS. This may be fine in some cases, but it also has a tendency to force a company to use a single file management technique, one that may be inappropriate for the application.

Prototyping is not a substitute for requirements definition or system design, but rather a welcomed addition to assist in expediting the physical aspects of systems development.

UPDATING THE PROJECT PLAN

Before proceeding with the management review of Phase 2, Project Management must reappraise the Project Plan, along with the Order-of-Magnitude estimate and schedule. By the end of Phase 2 Project Management should have a clearer understanding of the system and should be able to determine if the project is proceeding as planned. If not, the plans, estimates, and schedules are updated accordingly.

PHASE 2 REVIEW

Although the Phase 2 is primarily aimed at User Management, other functions should participate in order to be kept current with project developments. This includes Data Resource Management (who may be asked to initiate a DBEM project), and Software Engineering who will consider the viability of the system design from a programming perspective.

DESCRIPTION OF PHASE ACTIVITIES

Activity A - Define Sub-Systems

Systems Engineering reviews the specifications resulting from Phase 1 and prepares a Detail estimate and schedule for the activities of Phase 2 (A - E), and obtains Project Management approval before proceeding.

Systems Engineering finalizes the system design using the Chronological Decomposition technique. A System Flowchart is used to represent the design.

Activity B - Define Sub-System Logic

Systems Engineering defines the processing logic for each sub-system. A Sub-System Concept Diagram is prepared for each sub-system which expresses the planned processing flow of the sub-system. In addition, relationships are defined between Inputs, Outputs, and Files. Further, Application Logical Files are explained.

Activity C - Illustrative OD/ID Examples

Systems Engineering prepares illustrative examples for each Output and Input in the system (inputs are optional). Because the examples use sample data values which are reviewed with User Management, Systems Engineering can update data definitions as required.

Activity D - Prepare Design Evaluation

Project Management reviews and revises as required the Project Plan along with the Order-of-Magnitude estimate and schedule for the remainder of the project.

Project Management and Systems Engineering then assemble the deliverables resulting from the phase and packages them into a single phase review document which is evaluated by Quality Assurance prior to conducting the formal phase review with management.

Activity E - Phase 2 Review

Project Management conducts a formal review of the Phase 2 deliverables with Information Resource Management, User Management, Systems Resource Management, and Data Resource Management. At this time, management will review the formal Phase

2 "System Design Manual" consisting of:

- Phase Cover Page
- System Flowchart
- System Logic Narrative
- Sub-System Concept Diagram
- Sub-System Logic Narrative
- Application Logical Files Discussion
- Input Discussion & Examples (optional)
- Output Discussion & Examples
- Project Plan
- Project Estimate/Schedule Recap
- Cost and Evaluation Summary
- Phase Review Checklist

Based on this report and subsequent review meeting, management may elect to revise parts of the report, discontinue the project, or approve it for continuation.

The next few pages include sample deliverables resulting from this phase.

SAMPLE PHASE 2 DELIVERABLES

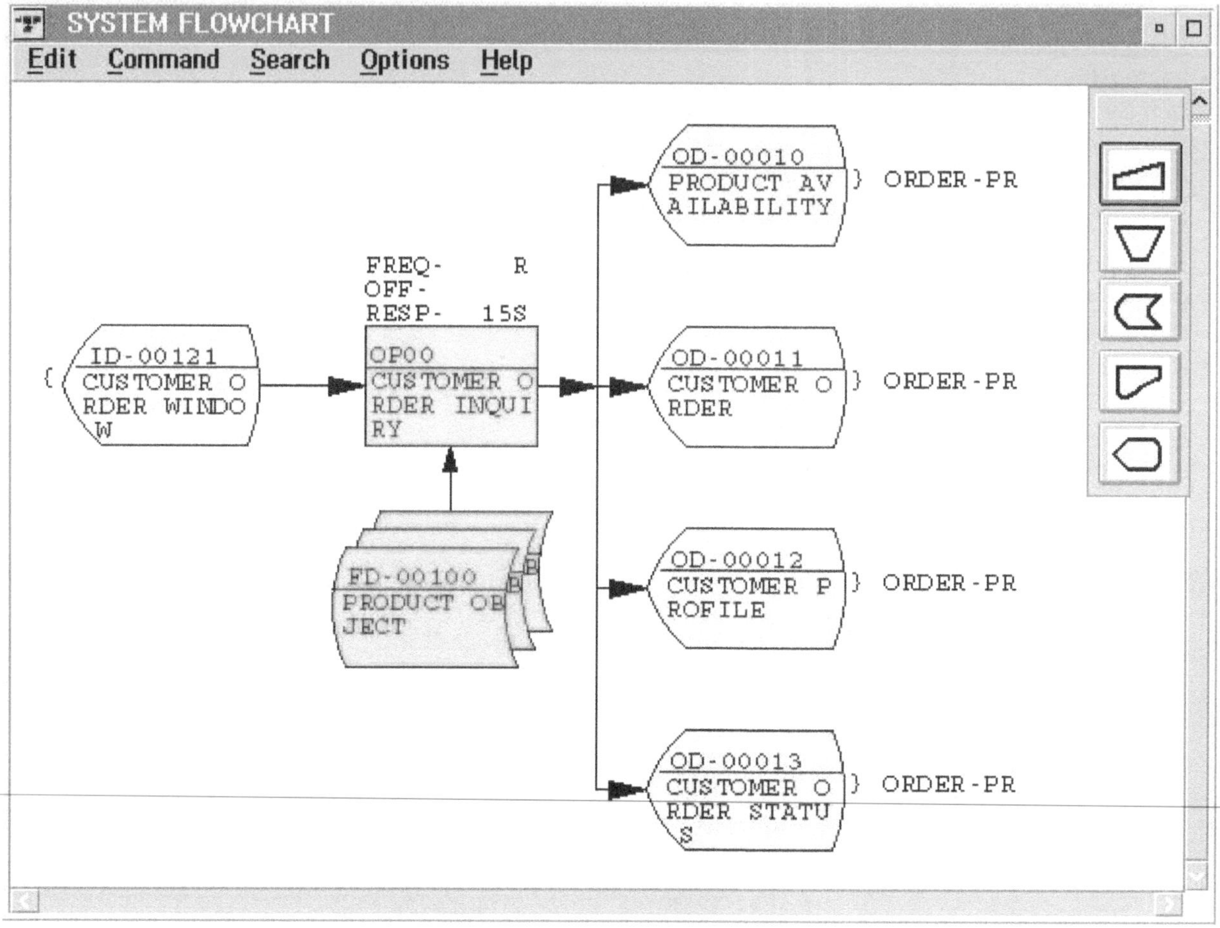

OD-90154	SYSTEM FLOWCHART	"PRIDE"-ISEM
OCT 19, 2004	VERSION 1.2.0	PAGE 7

* SYSTEM OP TRUNDLE ORDER PROC. SYS (TOPS)

* TABLE OF FILES FILES TO SUB-SYSTEMS	OP-00	OP-04	OP-05	OP-06	OP-07	OP-08
FD-00100 PRODUCT OBJECT APPLICATION LOGICAL	/ /R	/ /R	/ /R	/ /	/ /	C/U/R
FD-00101 CUSTOMER OBJECT APPLICATION LOGICAL	/ /R	/ /	/ /	/ /R	/ /R	C/U/R
FD-00102 ORDER OBJECT APPLICATION LOGICAL	C/U/R	/ /R	/ /R	/ /	/ /	C/U/R
FD-00103 TERRITORY OBJECT APPLICATION LOGICAL	/ /R	/ /	/ /	/ /	/ /R	C/U/R
FD-00104 ASSIGNMENT OBJECT APPLICATION LOGICAL	/ /R	/ /	/ /	/ /	/ /R	C/U/R
FD-00105 SHIPMENT OBJECT APPLICATION LOGICAL	/ /R	/ /R	/ /	/ /	/ /	C/U/R
FD-00106 BACK ORDER OBJECT APPLICATION LOGICAL	/ /R	/ /R	/ /	/ /	/ /	C/U/R
FD-00107 PAYMENT OBJECT APPLICATION LOGICAL	/ /	/ /	/ /	/ /R	/ /	C/U/R
FD-00108 GOVERNMENT AGENCY OBJECT APPLICATION LOGICAL	/ /	/ /	/ /	/ /R	/ /	C/U/R
FD-00109 ACCOUNT OBJECT APPLICATION LOGICAL	/ /	/ /	/ /	/ /R	/ /	C/U/R

OD-90395	SUB-SYSTEM LOGIC NARRATIVE	"PRIDE"-ISEM
OCT 25, 2004	RELEASE 1.2.0	PAGE 4.7

****
* SUB-SYSTEM OP-08 ORDER-PROC FILE MAINTENANCE *
****

FREQUENCY:	OFFSET:	RESPONSE TIME:
UPON REQUEST	N/A	15 SECOND(S)

BUSINESS PURPOSE

This sub-system is collecting the necessary data to process customer orders. Order Processing clerks will use this business process primarily to input customer orders and to query order status, customer credit limits, and product availability.

TEST CRITERIA

The screens for this sub-system should be rigorously tested in terms of:

Default entries
Tab sequencing.
Upshifting/Downshifting of characters.
F1 Field sensitive Help.
Correct computation of generated numeric values.
Acceptable characters & length of field entries.
Error/Warning/Information messages.
Data Base update.
Ease of use (its intuitiveness).

INPUTS/OUTPUTS USED	PREP/REC'D BY
ID-00100 ORDER PROCESSING WINDOW	VARIOUS
OD-00100 ORDER-PROC VALIDATION MESSAGES	VARIOUS

* * * END OF OUTPUT * * *

SAMPLE PHASE 2 DELIVERABLES

SAMPLE ILLUSTRATIVE OUTPUTS/INPUTS

00-00299 TOPS CONVERSION/RECONCILIATION REPORT DEC 21, 1992

VERSION 3.2 PAGE 1 OF 1

ORDER NUMB	CUST NUMB	CUSTOMER NAME	CUSTOMER PURCH ORD	ORDER DATE	ORDER BALANCE
A1001	T181941	RUDD, INC.	14-18121.1	11-27-92	$ 3,214.48
A1002	T101941	LEVINE BROTHERS	AB-12941	12-12-92	7,012.14
A1003	T182914	NATIONAL FURNITURE	4321-A	11-12-92	2,915.10
A1004	T118721	KINDLE'S	13-17-1431	12-10-92	245.15
A1005	T158691	MITCHELL MFG.	M-43791	10-10-92	4,247.12
A1006	T157172	MAYFAIR & MCDONNELL	M-91-41215	12-10-92	1,114.12
A1007	T198213	STANLEY, INC.	TS-437-198	10-19-92	3,804.16
A1008	T090148	CHAMBERLAIN MFG.	AL-143-1479	10-06-92	4,141.18
A1009	T116444	KINNEAR ROLLIN DOORS	324182	12-04-92	947.12
A1010	T180921	OVERHEAD DOOR	413-104115	12-18-92	3,014.12
A1011	T180922	KINDLE'S, INC.	13-17-1721	12-20-92	4,111.19
A1012	T180809	STRATFORD CO.	AB-143-14145	11-15-92	2,041.17
A1013	T199408	SERTA	SA-143-1951	11-15-92	3,041.41
A1014	T201712	SIMMONS, INC.	BD-4189-214	11-15-92	3,104.04

\- - - - - -

TOTAL DOLLARS CONVERTED $ 42,452.51

TOTAL NUMBER OF ORDERS 14

* * * END OF OUTPUT * * *

CUSTOMER ORDER

ABC COMPANY 1024-000

123 MAIN STREET MARCH 20, 1992
ANYTOWN, FL 34683
ATT: A.M. SMITH - PURCHASING AGENT

PRODUCT:	QTY ORD:	UNIT:	EXT. $
123-00	1	$10.00	$10.00
247-01	13	$20.00	$260.00

TOTAL $270.00

**DO NOT USE: X, Z, 9, 0 -
SHOW EXAMPLES OF HOW DATA WILL PHYSICALLY BE REPRESENTED**

CONSIDER PHYSICAL LIMITATIONS OF DISPLAY/PRINT DEVICE

KEEP LAYOUT SIMPLE, UNIFORM, EASY TO USE

OBSERVE STANDARDS WHENEVER POSSIBLE

SAMPLE PHASE 2 DELIVERABLES

OD-90163 INPUT DISCUSSION & EXAMPLES "PRIDE"-ISEM

OCT 25, 2004 RELEASE 1.2.0 PAGE 6.1

INPUT	NAME	PREP. BY	CHECKLIST
ID-00100	ORDER PROCESSING WINDOW	BROWN	(X)
ID-00121	CUSTOMER ORDER WINDOW	BROWN	(X)
ID-00125	WEEKLY ORDER STATUS REQUEST	BROWN	(X)
ID-00126	REQUEST FOR CUSTOMER ORDERS	BROWN	(X)
ID-00127	CREDIT CHANGE QUERY	BROWN	(X)
ID-00128	CUSTOMER DETAIL REQUEST	BROWN	(X)

* * * END OF OUTPUT * * *

OD-90163 INPUT DISCUSSION & EXAMPLES "PRIDE"-ISEM

OCT 25, 2004 RELEASE 1.2.0 PAGE 6.2

```
***********************************************************************
* ID-00100          ORDER PROCESSING WINDOW                           *
***********************************************************************
```

MEDIA- TV FORM NUMBER- KEYBOARD

DESCRIPTION-

BUSINESS PURPOSE

The purpose of this input is to collect data from the Order Processing department.

INPUT SEQUENCE (SORT/ACCESS KEY)-

NUMBER	NAME
DD-00001	CUSTOMER NUMBER
DD-00003	PRODUCT NUMBER
DD-00012	ORDER NUMBER
DD-00122	TERRITORY CODE
DD-00059	TERRITORY ASSIGNMENT NUMBER
DD-00010	SHIPMENT NUMBER
DD-00014	BACK ORDER NUMBER
DD-00192	PAYMENT NUMBER
DD-00009	GOVERNMENT AGENCY NUMBER
DD-00008	ACCOUNT NUMBER
DD-00011	BILLING NUMBER

RELATED INPUTS-

NUMBER NAME

(NONE FOUND)

RELATED OUTPUTS-

NUMBER	NAME
OD-00100	ORDER-PROC VALIDATION MESSAGES

(continued on next page)

SAMPLE PHASE 2 DELIVERABLES

RELATED FILES-

NUMBER	NAME
FD-00100	PRODUCT OBJECT
FD-00101	CUSTOMER OBJECT
FD-00102	ORDER OBJECT
FD-00103	TERRITORY OBJECT
FD-00104	ASSIGNMENT OBJECT
FD-00105	SHIPMENT OBJECT
FD-00106	BACK ORDER OBJECT
FD-00107	PAYMENT OBJECT
FD-00108	GOVERNMENT AGENCY OBJECT
FD-00109	ACCOUNT OBJECT
FD-00110	BILLING OBJECT

INPUT TRANSACTIONS-

NUMBER	NAME
RD-00451	CUSTOMER PROFILE
RD-00452	CUSTOMER CREDIT LIMIT
RD-00453	ORDER DETAILS
RD-00454	ORDER STATUS QUERY
RD-00455	PRODUCT AVAILABILITY QUERY

*** END OF OUTPUT ***

SAMPLE PHASE 2 DELIVERABLES

OD-90164 — OUTPUT DISCUSSION & EXAMPLES — "PRIDE"-ISEM

OCT 25, 2004 — RELEASE 1.2.0 — PAGE 7.1

OUTPUT	NAME	PREP. BY	CHECKLIST
OD-00010	PRODUCT AVAILABILITY	BROWN	(X)
OD-00011	CUSTOMER ORDER	BROWN	(X)
OD-00012	CUSTOMER PROFILE	BROWN	(X)
OD-00013	CUSTOMER ORDER STATUS	BROWN	(X)
OD-00014	WEEKLY ORDER STATUS ANALYSIS	BROWN	(X)
OD-00015	DAILY CUSTOMER ORDERS	BROWN	(X)
OD-00016	CREDIT CHANGE REQUEST	BROWN	(X)
OD-00017	CUSTOMER DETAIL SHEET	BROWN	(X)
OD-00100	ORDER-PROC VALIDATION MESSAGES	BROWN	(X)

* * * END OF OUTPUT * * *

OD-90164 — OUTPUT DISCUSSION & EXAMPLES — "PRIDE"-ISEM

OCT 25, 2004 — RELEASE 1.2.0 — PAGE 7.2

****
* OD-00100 ORDER-PROC VALIDATION MESSAGES *
****

MEDIA- TV FORM — NUMBER- SCREEN

DESCRIPTION-

BUSINESS PURPOSE

The purpose of this output is to validate data entries and respond to the user with pertinent processing messages.

OUTPUT SEQUENCE (SORT/ACCESS KEY)-

NUMBER	NAME
DD-00001	CUSTOMER NUMBER
DD-00003	PRODUCT NUMBER

RELATED INPUTS-

NUMBER	NAME
ID-00100	ORDER PROCESSING WINDOW

RELATED OUTPUTS-

NUMBER	NAME
(NONE FOUND)	

RELATED FILES-

NUMBER	NAME
FD-00100	PRODUCT OBJECT
FD-00101	CUSTOMER OBJECT
FD-00102	ORDER OBJECT

(continued on next page)

SAMPLE PHASE 2 DELIVERABLES

(continued from previous page)

FD-00103	TERRITORY OBJECT
FD-00104	ASSIGNMENT OBJECT
FD-00105	SHIPMENT OBJECT
FD-00106	BACK ORDER OBJECT

MESSAGES-

NUMBER	NAME	TYPE
RD-E0001	INVALID ENTRY	ERROR
RD-W0003	UNUSUAL VALUE (DEFAULT ASSUMED)	WARNING
RD-I0100	CUSTOMER CREDIT VIOLATION	INFORMATION
RD-I0101	PRODUCT OUT OF STOCK	INFORMATION

DATA DESCRIPTIONS-

NUMBER	NAME
DD-00003	PRODUCT NUMBER A unique number used to identify a product.
DD-00018	NAME A given name.
DD-00111	UNIT PRICE The list price for one unit of a particular product.
DD-00508	QUANTITY ON HAND The total quantity of a product that is in stock.

/\

* * * END OF OUTPUT * * *

END SECTION

Section 11

"PRIDE"-ISEM PHASE 3 SUB-SYSTEM DESIGN

Forgetting the human-being during design will cause the human-being to forget the system at time of startup; it will be DOA, Dead On Arrival.
- Bryce's Law

The purpose of this phase is to design procedures to implement the sub-systems in a cost effective manner. Several events occur during this phase:

- Procedures are designed with their appointed inputs, outputs and files.
- Inputs and outputs are finalized, complete with input transactions, messages, maps, and panels.
- Narrative is written to describe the processing logic and test criteria.
- The project estimates/schedules associated with the sub-system are reappraised.
- A formal review with user management is conducted to verify the phase findings.

METHODOLOGY NAVIGATION

A Phase 3 may be initiated either by following a normal Phase 2 or, if a modification/improvement, following Phase 1, Activity A. There is a one-to-one relationship between Phase 3 and a sub-system. In other words, a Phase 3 will be executed for each sub-system identified in Phase 2.

There can be several Phase 3's executing in parallel or, depending on resource allocations and/or project priorities, some Phase 3's may be postponed or delayed.

The formal deliverable resulting from Phase 3 is a "Sub-System Design Manual" consisting of a description of the procedures, input/output definitions, along with an updated project estimate/schedule for the sub-system. This is reviewed with management who must decide whether to:

- Approve the sub-system for continuation.
- Request revisions to the phase findings.
- Discontinue the sub-system.

Following Phase 3, this branch of the ISEM project will branch further to multiple Phase 4's; Phase 4-I for Administrative Procedures and Phase 4-II if a Computer Procedure has been identified.

Phase 3 may also affect DBEM Phase 5 where the Application Physical Data Base Model is designed. During ISEM Phase 3, physical working files are identified. Data Base Administration must then accommodate these new files as required.

GENERAL DISCUSSION

Based on the rough designs of the sub-system as prepared in Phases 1 and 2, Systems Engineering prepares the final sub-system design. Whereas in Phases 1 and 2 the emphasis was to design backwards, from Output to Input, in Phase 3 (and Phase 4) the emphasis is on designing forward, from beginning to end. Each procedure is detailed as to WHAT is to be processed, WHEN it will be processed, WHO will process it, and HOW. All of the Inputs, Files and Outputs identified for the sub-system in Phase 2 are detailed, as well as additional files, such as manual files and computer input/output files.

The principal objective of Phase 3 is to define and synchronize processing within the sub-system, particularly the human/machine interface. Sufficient detail should be expressed in Phase 3 so that the administrative procedures can then be written and developed separately from the computer procedure. To do this, the sub-system design must express two things:

1. **Processing Flow** - this is an expression of how procedures will be executed, from start to end. Consequently, dependencies between procedures are defined. Such procedure dependencies reflect sequential, iterative, and parallel processing.

2. **Data Flow** - how data moves within a sub-system is expressed by how inputs, outputs and files are used to process the data in the various procedures. Here, data is expressed in terms of how it is created, updated, and referenced.

Processing and data dependencies must be explicitly defined before the sub-system can branch to "Administrative Procedure Design" (Phase 4-I) and "Software Engineering" (Phase 4-II). Failure to define such detail will result in processing inconsistencies later.

The successful implementation of a system depends upon the clearness and completeness of its procedures. Many systems are ineffective or have failed in operation due to the lack of procedures. Procedures can minimize the need for trained specialists in each department in order to insure correct performance of the system. System modifications can be implemented readily through procedure revisions. Many times in the past, systems have been implemented by "word of mouth" under the guise of quicker system start-up. This increase in speed was an illusion and the end result was an increase in expense and system errors due to incorrectly prepared data and poor communications between functions. More importantly, a poor start-up of a system can create a credibility gap with the user that may never be overcome, even after all problems have been corrected. Careful delineation of procedures during sub-system design will minimize these problems.

TYPES OF PROCEDURES

Each sub-system consists of one or more Administrative Procedures and, at most, one Computer Procedure. It is quite possible for a sub-system to be implemented without a computer procedure; a manually implemented sub-system for example. However, a sub-system will always have at least one administrative procedure, even if it is nothing more than to initiate/schedule a computer procedure.

A separate procedure should be defined any time the "work flow" crosses an organizational boundary (such as between business functions). If the procedural "work flow" within any single part of the organization appears overly complex, the engineer may define multiple procedures without necessarily crossing an organizational boundary.

An Administrative Procedure deals specifically with the human being. The Computer Procedure deals with the internal processing of the computer.

Administrative Procedures

There are three types of Administrative Procedures:

1. **Manual Procedure** - where the human being must perform an operation without any machine assistance. This may consist of steps involving nothing more than the use of pencil and paper. Manual procedures include the preparation of data, balancing of reports, processing documents, distribution, review of information, etc.

2. **Office Automation Procedure** - is concerned with providing a set of human readable instructions to perform functions in conjunction with specific office equipment. Office Automation procedures include Word Processing, Data Entry, Facsimile Transfer, Electronic Mail, etc.

3. **Computer Terminal Session** - where work pertains to the direct interface between the human being and the computer. For example, Data Conversion/Entry (keying) would fall into this category. Also, an interactive session where the user responds to the computer to reference information via screens is another example. Another would be where the computer operator must submit a batch job.

Computer Procedures

These procedures consist of a series of computer programs, one or more, required for the computer to perform during the the sub-system. A Computer Procedure could be as simple as one program or could have several programs within it. It ultimately depends on the type of work required. The decision as to the number of programs required to implement the computer procedure is made during Phase 4-II "Software Engineering."

Standard Operating Procedures

ISEM does not cover standard operating procedures for computer operation (e.g., logon/logoff), data control, data transmission and conversion. These procedures can vary depending on equipment, systems, software and organization. However, ISEM does assume that these procedures exist. Therefore, during sub-system design, procedures are prepared only when the system requires an action that is non-standard. For example, batch balancing pertains specifically to a particular input and requires that a specific procedure be followed.

TIMING CONSIDERATIONS

As the sub-system is decomposed into procedures, timing is refined to express when each procedure is executed within the sub-system and how fast it must be performed. The timing for the overall sub-system is expressed in terms of:

- Frequency - how often the business cycle occurs.

- Offset - when it begins in the cycle.

- Response Time - the total amount of time allowed to deliver the information.

However, "Frequency" is not applicable for a single procedure, and "Offset" and "Response Time" take on new meaning:

- Offset - when the procedure is to begin within the time frame of the sub-system.

- Response Time - the speed by which the procedure must execute.

The objective here is to synchronize the timing of the procedures. Offset and Response Time, therefore, will not be the same as specified for the overall sub-system. As an example, assume that a sub-system represents a weekly process with an offset of 1D (the first day of the week) and a response time of three hours. Let's also assume that the sub-system can be executed with three sequential procedures. The timing of the three procedures could possibly be defined as:

PROC-1	PROC-2	PROC-3
Offset - 0 hour	Offset - 1 hour	Offset - 2 hour
Resp. - 1 hour	Resp. - 1 hour	Resp. - 1 hour

This is a simple example; sub-systems can obviously be much more complicated than this. Yet, it demonstrates the synchronization of the three procedures. Timing at this level also provides insight into the type of responsiveness which both the human being and computer must have during the sub-system.

One last note, the accumulated response time of the procedures must not exceed the response time specified for the overall sub-system.

SUB-SYSTEM FLOWCHART

The sub-system design is graphically represented by the Sub-System Flowchart, one of the formal deliverables resulting from Phase 3. As with all ISEM flowcharts, the diagram shows the next level of detail in the system structure: procedures. Each procedure is represented on the flowchart with its appointed inputs, outputs, files and timing. A Table of Files accompanies the flowchart which shows how the various files are used in the procedures: either "C" Created, "U" Updated, "R" Referenced, or combinations of all three.

The sub-system flowchart expresses both the processing flow and data flow of the sub-system. As a "Process Diagram," the flowchart shows the flow of work; what the human being is responsible for performing and what is expected from the computer. How the flowchart is drawn, horizontally or vertically, is irrelevant. All that is important is that the graphic express the dependencies between procedures, inputs, outputs, and files.

METHODS OF PROCESSING

During sub-system design the Systems Engineer must be reminded of the basic processing concepts as defined in ISEM:

1. There are three basic processing constructs:

- **Sequence** - representing consecutive steps in processing.
- **Iteration** - representing repetition until a condition is met.
- **Choice** - representing a selected path based on a prescribed criteria.

The Systems Engineer will use any one of these constructs, or combinations, in designing the processing flow.

It must be remembered that as a sub-system starts, it continues from procedure-to-procedure, until it reaches a conclusion within the specified sub-system time frame. This does not mean that a sub-system will have only one start or one end; in fact, it can have multiple ways of starting and multiple ways of ending. If sub-system processing does terminate, yet another procedure is waiting, the other procedure may represent a separate sub-system.

2. All processing involves the use of transactions, which may represent:

• How data is collected for storage.

• How a report, screen or other forms of output is requested.

3. Systems communicate through data.

In Phase 2, primary files were defined and related to the various sub-systems. In phase 3, the Systems Engineer must consider the use of additional physical file requirements. This specifically includes working files, such as:

- Computer input transaction request files.

- Computer output data files.

- Manual files for storing/retrieving documents.

Computer working files for use between programs are not considered during Phase 3; this is reserved for Phase 4-II "Software Engineering."

These additional file requirements are communicated to Data Resource Management for implementation.

For additional information on "Methods of Processing," consult the Methodology Introduction.

ERGONOMICS

In a nutshell, ergonomics is concerned with adapting our physical work environment to suit the needs of the human being, not the other way around. The theory here is if we, as human beings, can easily adapt to our physical surroundings, the better we can accept and implement our work environment. This is a valid point which should be considered in any systems development undertaking, large or small. In reality, the concept is an old one as it was used by "Systems & Procedures" departments which predated the introduction of commercial computers in the workplace. But it was in the 1980's where the concept was reborn and renamed "ergonomics." This resulted in a movement where we reconsidered the design of everything from computer screens (the advent of the Graphical User Interface or "GUI"), keyboards, office furniture, etc.

Ergonomics is applied when we move from logical design to the consideration of the most suitable physical implementation. In "PRIDE"-ISEM, this occurs here in Phase 3 which is where we design the physical implementation of the business process, including the human/machine interface. It is here where we must consider the human being. A poorly designed and programmed system that can be easily used is far better than an elaborate system that alienates the users. Obviously, the objective of system developers is to produce a superior system that is easy for the users to implement and use.

In order to devise a suitable physical implementation, consideration must be given to the intelligence level of the humans who will implement the business process. To illustrate, I know of a popular electronics retailer who devised a totally new system for their service shops across the country. The idea was to give the store clerks easy access to reference products, parts and warranties. Prior to this, the service shops relied on massive printed catalogs which, although they were accurate, required considerable time to lookup components and warranties. The company wanted to expedite this process thereby improving customer satisfaction. As a result, they devised an elaborate system involving robust screens with computer graphics, and a mouse. True, the system could rapidly access data, but it tended to intimidate the clerks in the service centers (most did not have the inclination or temperament to use a computer). So much so, they refused to use it, opting instead to use their old catalogs. The new system quickly died a painful and expensive death. In this instance, the company failed to recognize the intelligence level of the clerks and their adaptability to new technology. Many such snafus can be found in the corporate world, all because developers failed to recognize the intelligence level of their targeted users.

In addition to the user's intelligence level, developers should take into consideration the human senses: hearing, seeing, touch, taste, smell. For example, what is the point of devising an elaborate color scheme for screens or forms if there is a possibility that someone may be color blind? Or to add audio responses to computer prompts if the person is either hard of hearing or the device is placed in a noisy environment? Systems are for people, not for the computer. If people cannot assimilate new technology or find it awkward to use, they will resist it wholeheartedly.

There are two basic requirements for making something ergonomically correct: It must be easy to learn and it must be easy to operate. The features required for learning a new tool or technique may or may not be the most appropriate for routine operations. For example, "menus" and "windows" are very easy for users to learn and understand initially since they can guide users through a process while sitting in front of a screen. However, in normal operation, the need to step through a number of menus to achieve a goal can be extremely frustrating and deemed a considerable waste of time by the user

Ergonomics involves any equipment, furniture or supplies to be used in performing the work. Designing an office layout is just as important as the equipment used. Because systems affect the human senses, the Systems Engineer should consider such things as lighting, posture, sound, hand and eye movement, etc., all of which directly affects the workers.

Facade should not be confused with substance. Ergonomics is much more than being "on-line" to a computer with elaborate color graphics. It is a matter of determining and implementing the technology to suit the users. This requires a combination of skills in industrial engineering and industrial psychology. Industrial engineering provides for the synchronization of the work flow. Industrial psychology provides for the human element to be factored into the design while the system is being built, not added on later.

In terms of equipment, there are many tools for the user to operate such as monitors, keyboards, printers, scanners, microfilm/microfiche readers, multimedia equipment, computers, etc. There is often a tradeoff between a device's ease of use and its functionality. Quite often, the simpler the device is to use, the less capability it may have. The following is a partial list of devices commonly used:

1. **Keyboards** - System designers should be aware if the users are sufficiently proficient at using a keyboard for a particular use. Not all applications require massive typing, most just require short and simple keying. The keys themselves should be evaluated; do users prefer a flat touch panel, or the traditional keyboard with individual keys? Ultimately, it depends on the functions desired and who will operate it. Touch panels can be extremely efficient for simple operations. However, the traditional keyboard is more desirable for more complicated operations. Also, a key click response is desirable to acknowledge the key stroke.

2. **Mouse, Joystick, and Pen based entry** - these are effective devices for simple operations but are questionable for major production purposes (where they can actually become annoying). This is why programs are usually written to accommodate multiple input devices, such as for both a mouse and a keyboard. Perhaps the best application for these type of devices is in the area of graphics.

3. **Optical Scanners** - These are tools that are widely used in retail, inventory, shipping, and other routine tasks. However, there is little point to these devices if they are highly susceptible to error, requiring corrective keying. As easy as these tools are to use, it ultimately depends on the application to determine their

practicality.

4. **Voice Recognition** - This is a field still in transition. Although it holds great promise for the future, there are currently only a few practical applications. Voice recognition systems are heavily dependent on voice tone and inflection. Speech must usually be very specific and precise, which is often difficult for people to adapt to.

5. **Buttons and switches** - In general, buttons and switches are very easy to use if properly labeled. However, like touch panels they have limited functions, sometimes serving a single purpose.

6. **Manual Forms** - Paper documents are one of the easiest forms of media for users to understand. This is because offices have operated with paper systems for years. However, forms can alienate users as easily as any other media if poorly designed. The emphasis of forms design should be on simplicity, not complexity. Fields must be self explanatory or must include some form of instructions. Screens are merely electronic analogues of paper forms. They have some advantages and disadvantages to manual forms. Systems Engineering should know the differences. Use company standards for designing forms. In general, use the following list as suggested guidelines. Many of these guidelines relate to screens as well as they do to manual forms and printouts.

• Design the form to be easy to read and use. "Cluttered" forms that are difficult to read and use will alienate users and cause irritation.

• Do not design the form to the edge of the paper. Leave a margin for printing. A ½" margin is good; ¾" is better.

• "Zone" the form by the records and/or the people who will be using the form; e.g., shipping, accounting, inventory, sales, etc., and by the sequence by which it is completed.

• Place the form name and number in a standard position, one that would be easy to read when paging through copies. Several companies use the upper-left hand corner as their standard.

• For forms requiring typing or keying, determine the column alignment required for the machine to be used. Minimize as many keystrokes as possible when designing the form.

When using computer screens, there are several options available for design:

1. **Menus** - Because of their emphasis on simplicity, menus generally offer limited functions; e.g., maintain files and generate outputs. They can be designed several ways: by the output or input to be used; by the sub-system or process to be executed; by the person having to execute the process; or combinations of the above. But it should be noted that menus are tailored more for the user than they are for system functions. As a result, they may not be as efficient as other approaches.

2. **Command Language** - In contrast to menus is the command language approach which is aimed at greater functionality and speed. When complicated and diverse operations are required, a command language provides the greatest form of flexibility and efficiency. It allows the user to perform their duties in the fastest amount of time. As with any language though, it may be difficult for the user to learn. As such, there must be great care in the vocabulary selected. Cryptic commands may produce cryptic results. Data validation and "help" explanations are highly desirable features for a command language.

3. **Graphical User Interface (GUI)** - this is like having a combined form of Menus and Command Languages. This type of interface, which is normally provided on PC's, provides screen panels, windows and other conventions (action bars, pull-down-choices, checkboxes, push buttons, etc.) which the user will interact with to submit commands. The GUI provides the same type of structure and facilities as used with menus, yet provides the user with the freedom to perform different functions rapidly as under a command language.

INPUT/OUTPUT DEFINITIONS

When the procedural flow is established, Systems Engineering reviews and completes the definition of all inputs, outputs, files, records and data elements related to the sub-system. The illustrative input/output examples as prepared under Phase 2 are finalized for review. This includes the preparation of input transactions, panels, maps, messages, etc. associated with the inputs and outputs.

The technique of working backwards from outputs to inputs is a powerful concept. If an output is defined thoroughly, defining the required input is a simple task. When defining the output, the Systems Engineer must...

• Provide an example of how the output will physically appear. This will include live data values to show how the completed output will appear in practice. Ideally, the engineer should have multiple examples to express various conditions and uses of the output. Also, sample messages should be shown as required.

• Provide a textual description of the output in terms of:

-How headings and footers are created (including pagination).

- How the body of the output is produced, including where the data is derived from (files), and any special processing logic associated with the output.

• Define pertinent processing messages associated with the output.

The Output Description is ultimately the driving force behind the design of the Input. From these specifications the Systems Engineering should be able to define the input in terms of:

• A full example.

• A textual description of the input including:

Its business purpose (what outputs does it support).

The physical media of the input and how it works.

• The transactions supported by the input. Each transaction should be fully defined, in terms of field entries and default parameters. If an input relates to an input transaction file, then this relationship should be explicitly defined.

• Processing messages associated with the input.

TEST CRITERIA

The input/output logic provides the basis for defining the TEST CRITERIA of the sub-system, and, as such should be posted to the sub-system narrative (descriptive text). By Phase 3, input transactions should be fully defined, including the entries to be validated, along with default values. From this, Systems Engineering can begin to establish testing criteria for use in Phase 7. Further, Systems Engineering should anticipate the volume of transactions that the sub-system will process on the average, as well as maximum volumes. This will be used to establish processing benchmarks.

HOW THE RD IS USED

The Record Description serves a variety of purposes, the most obvious of which is to represent a storage record as used in a File (FD). However, there are other things it can represent:

1. Input Transactions

2. Screen Panels

3. Print Maps

4. Messages & Help Text

Messages and Help Text are actually conveyed through panels and maps and should be defined as such. Typical types of messages include Errors, Warnings, and Information Messages.

As RD's, transactions, panels and maps can be easily related to pertinent Inputs and Outputs. Further, RD-to-RD relationships can be established in the IRM Repository for control purposes.

This approach to decomposing the physical aspects of screens and reports is very conducive for translation considerations. Translation files can be defined and developed where the panels or maps can be physically stored. As required, the panels and maps can be extracted, translated, before being returned to the file.

UPDATING THE PROJECT PLAN

Before proceeding with the management review of Phase 3, Project Management must reappraise the Project Plan, along with the Order-of-Magnitude estimate and schedule. By the end of Phase 3 Project Management should have a clearer understanding of the sub-system and should be able to determine if the project is proceeding as planned. If not, the plans, estimates, and schedules are updated accordingly.

The principal difference between updating the project plan in Phase 3 and Phase 2, is that the Phase 3 update is used FOR THIS SUB-SYSTEM ONLY, not for others in the system. Updates to other sub-systems will occur as the various Phase 3's are executed.

PHASE 3 REVIEW

This represents the last major management review prior to branching out to Administrative Procedure Design (Phase 4-I) and Software Engineering (Phase 4-II). The final result of Phase 3 is a complete Sub-System Design Manual. Quality Assurance reviews this manual to assure adherence to standards. It is also reviewed with management as part of the last activity In Phase 3.

It is important to understand that in this review, User Management is not just approving the design of the sub-system. They are also committing that they understand the procedures and will be responsible for their implementation.

Although the Phase 3 is primarily aimed at User Management, other functions should participate in order to be kept current with project developments. This includes Software Engineering (who will evaluate computer processing requirements) and Data Base Administration (who will be responsible for implementing the physical work files identified during this phase).

DESCRIPTION OF PHASE ACTIVITIES

Activity A - Define Procedures

Systems Engineering reviews the specifications for the sub- system from the Phase 2, "System Design Manual." Based on these specifications, Systems Engineering prepares a Detail estimate and schedule to complete the Phase 3 activities for the sub-system which is reviewed with Project Management for approval. If acceptable, Systems Engineering designs the procedural "work flow" of the sub-system. A Sub-System Flowchart is used to represent the design.

This becomes the basis for confirming or revising the Sub-System Concept Diagram as prepared in Phase 2.

Activity B - Define Procedure Logic

During this activity Systems Engineering:

1. Completes the design of the Inputs/Outputs. This includes a complete description and example of each screen and report. The inputs and outputs are decomposed, as required, into input transactions, panels, maps, and messages.

2. Defines the processing logic for each procedure in the sub-system.

3. Defines the Testing Criteria for the sub-system. This is based on the input, output, file, and procedure specifications. The Testing Criteria is added to the sub-system narrative.

4. Reevaluates the rough design of the computer procedure.

Activity C - Prepare Design Evaluation

Project Management reviews and revises as required the Project Plan along with the Order-of-Magnitude estimate and schedule for the remaining phases in the project pertaining to this sub-system only.

Project Management and Systems Engineering then assemble the deliverables resulting from the phase and packages them into a single phase review document which is evaluated by Quality Assurance prior to conducting the formal phase review with management.

Activity D - Phase 3 Review

Project Management conducts a formal review of the "Sub-System Design Manual" with User Management and Systems Engineering. This is the last major design review meeting for User Management. As in Phases 1 and 2, the user must decide to either: approve the report, at which time the project progresses to Phase 4; disapprove of the report and cancel this part of the project, or; have the report modified before continuing. Other functions participating in the review include Software Engineering and Data Base Administration.

The formal Phase 3 **"Sub-System Design Manual"** consists of:

- Phase Cover Page
- Sub-System Concept Diagram
- Sub-System Logic Narrative
- Test Criteria
- Sub-System Flowchart
- Procedure Logic Narrative
- Input Discussion & Examples
- Output Discussion & Examples
- Project Estimate/Schedule Recap
- Phase Review Checklist

Based on this report and subsequent review meeting, management may elect to revise parts of the report, discontinue the project, or approve it for continuation.

The next few pages include sample deliverables resulting from this phase.

SAMPLE PHASE 3 DELIVERABLES

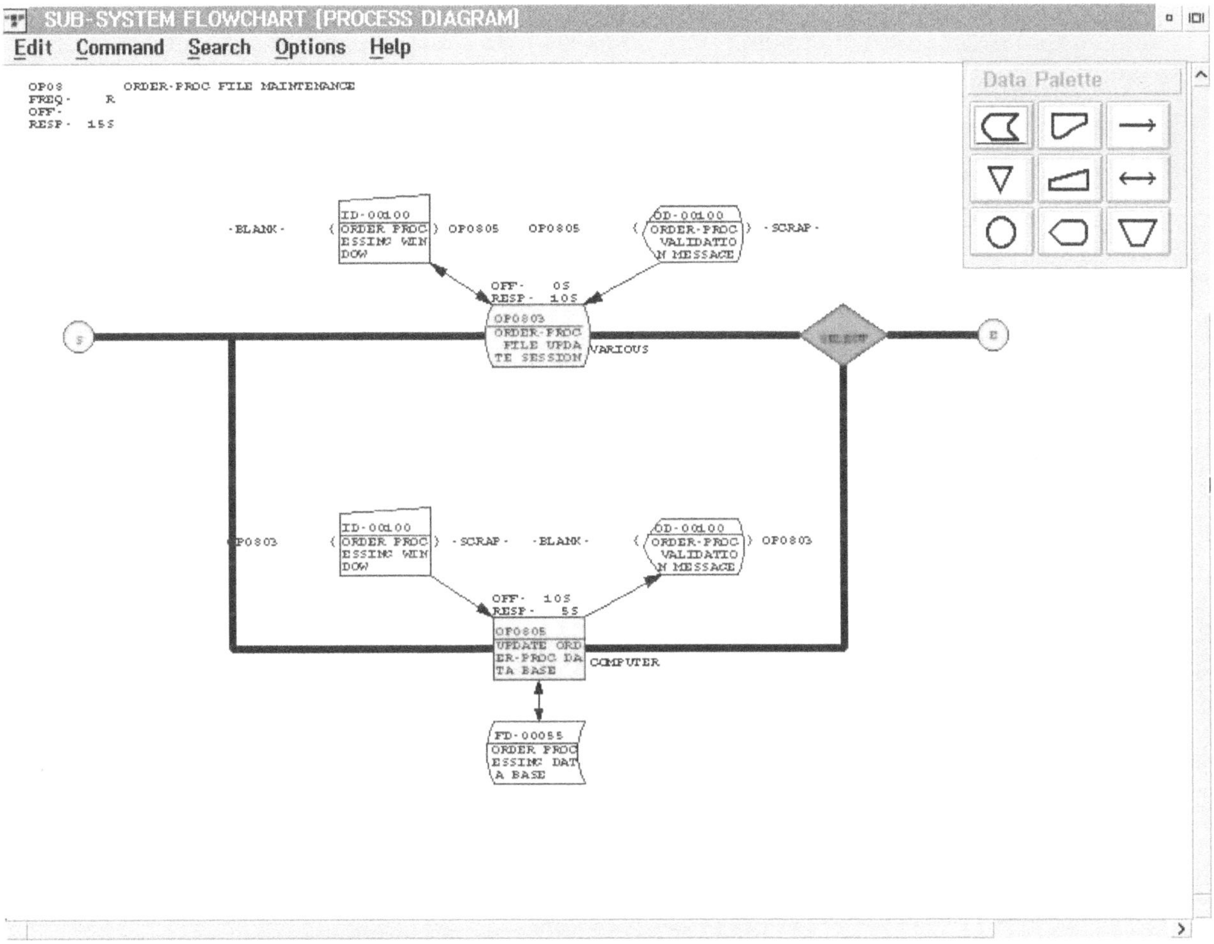

OD-90397	TEST CRITERIA	"PRIDE"-ISEM
NOV 07, 2004	RELEASE 1.2.0	PAGE 3.1

****
* SUB-SYSTEM OP-08 ORDER-PROC FILE MAINTENANCE *
****

The sub-system must, of course, conform to the testing requirements as outlined in standard ST-01-03 "TMC Sub-System Testing Criteria." In addition to this, the sub-system must validate all data entries. When an error, warning, or information message is displayed, suitable help text must be provided to guide the user through the proper processing.

* * * END OF OUTPUT * * *

OD-90398	PROCEDURE LOGIC NARRATIVE	"PRIDE"-ISEM
NOV 07, 2004	RELEASE 1.2.0	PAGE 5.1

****
* PROCEDURE OP-08-03 ORDER-PROC FILE UPDATE SESSION *
****

TYPE: KEYING FUNCTION: ORDER-PROC

OFFSET: 0 SECOND RESPONSE TIME: 10 SECOND(S)

BUSINESS PURPOSE

The purpose of this administrative procedure is to enter and validate data through a computer terminal using the Order Processing Window. Entries are made in accordance with the prompts and messages associated with the panel. HELP text is also available upon request.

INPUTS/OUTPUTS USED		FROM	TO
ID-00100	ORDER PROCESSING WINDOW	-BLANK-	OP-08-05
OD-00100	ORDER-PROC VALIDATION MESSAGES	OP-08-05	-SCRAP-

FILES USED	FROM	TO

(NONE FOUND)

* * * END OF OUTPUT * * *

END SECTION

Section 12

"PRIDE"-ISEM PHASE 4-I ADMINISTRATIVE PROCEDURE DESIGN

Information is for people, not for the computer.
- Bryce's Law

The purpose of Phase 4-I is to prepare clearly written Administrative Procedures for the User in accordance with the specifications contained in the Phase 3, "Sub-System Design Manual." Several events occur during this phase:

- The operational steps of each Administrative Procedure are defined.
- If applicable, on-line "help" text is written.
- A formal user manual is prepared, including procedures, and descriptions about inputs and outputs.
- The documentation and procedures are reviewed with users.

METHODOLOGY NAVIGATION

A Phase 4-I may be initiated either by following a normal Phase 3 or, if a modification/improvement, following Phase 1, Activity A. There is not necessarily a one-to-one relationship between Phase 4-I and an administrative procedure. In most instances, all administrative procedures within a sub-system are written during a single Phase 4-I. In other words, a Phase 4-I will normally be executed for each sub-system, not for each procedure.

If a large and complex procedure has been identified, then the Project Manager may elect to manage it separately in a different Phase 4-I. In this situation, the Procedure number can be put into the project/phase key instead of the sub-system identifier.

The formal deliverable resulting from Phase 4-I is an "Administrative Procedure Manual" consisting of the procedures, along with input/output explanations. This is reviewed with User Management for clarity.

Following Phase 4-I, this branch of the ISEM project will proceed to Phase 7 where it may rejoin the Computer Procedure in a parallel test. If there is no Computer Procedure in the sub-system, then Phase 7 is still performed in order to walk through and test all of the Administrative Procedures.

GENERAL DISCUSSION

Clearly written Administrative Procedures immeasurably improve the processes of system implementation and operation. Often, preparation of these procedures has been eliminated in order to expedite system start-up. Experience has shown that this approach can cause considerable expense, frustration and problems. It also causes personnel inflexibility in both the system and user areas. For example, since Systems Engineering is the primary group that understands the operation of the entire system, without procedures they would be constantly involved with routine operations. At the same time, user personnel tend to become experts in certain aspects of system operation and without written procedures they cannot be reassigned to other duties since a means for transferring knowledge (the procedure) is not available.

A new system, during start-up, can experience various problems due to lack of knowledge and experience. Written procedures provide an excellent basis for avoiding and correcting these problems. The error possibilities can be outlined in the procedures and serve as cautions for the employees executing the procedures. Procedural improvement or revisions can be incorporated by issuing a new procedure. The "debugging" of Administrative Procedures is just as essential as "debugging" computer programs. This is why ISEM specifies that all sub-system tests be accomplished with all procedures completed.

All procedures should answer these questions for the User:

- What is the purpose of the procedure?
- Who should perform the procedure? When?
- How should the procedure be accomplished?
- What is needed to accomplish the procedure?
- What are some examples?
- What should be done after the processing is accomplished?

PLAYSCRIPT TECHNIQUE

Administrative procedures are written using a manual procedure language called "Playscript" as developed by Leslie H. Matthies. A playscript procedure includes sections describing its Purpose, Set-Up (the inputs, outputs and files required to perform the procedure), and action instructions consisting of verbs and nouns. The technique, therefore, is patterned after a script to a play. These procedures direct the user through the execution of the administrative procedure. As an aside, "Playscript" was the technique which procedural programming languages (such as COBOL) were patterned after.

"HELP" TEXT

Help text is normally associated with interactive processing at the computer screen where the user requires instructions from the computer to guide them through processing. This is needed to answer both common and technical questions regarding processing. Development of help text is almost a prerequisite for all PC processing. It is in Phase 4-II that the help text is physically assembled for the sub-system, but it is written here in Phase 4-I.

There is basically three areas requiring HELP text:

1. For Window/Screen Processing - providing tutorial describing the purpose of the screen, who it serves, and the basic processing action of the screen. In the industry, this is often referred to as "general help" or "extended help." The narratives developed in Phases 2 and 3 regarding the overall sub-system and procedures are particularly useful for developing such text. Also, the narratives describing inputs (ID's), outputs (OD's), and screen panels (RD's), as well as the "Playscript" instructions, are excellent sources for assembling help text.

2. The user will also require help in making field entries. This includes acceptable values, what the values mean, and the physical characteristics of the entries (e.g., length). The logical/physical characteristics of the field entries (DD's) on the input transactions are an excellent source for assembling this type of help text.

3. Special function keys - like field entries, HELP text should be provided to explain special keys (function keys or combination keys (e.g., CTRL + C)).

Help indices are also very useful for the reference purposes, such as by subject, by field entry, by keys, etc. Fortunately, standards are emerging in the industry for writing help text.

Help text can take many physical forms e.g., MS

Windows HLP files, OS/2 INF files, Web files (HTML), etc. Selection of the suitable implementation is based on the physical design decisions made in Phase 3.

There is absolutely no incompatibility between "Help" text and "Playscript." In fact, "Help" text makes a convenient vehicle for accessing "Playscript" instructions.

The final result of Phase 4-I is a completed Administrative Procedure Manual for the Sub-System. All procedures are checked by Quality Assurance to assure conformance with standards. The procedures are reviewed with the users as part of the last activity of Phase 4-I. When approved and accepted, the procedures are used to execute Phase 7, "Sub-System Test," and for the routine operation of the system.

In summary, administrative procedures should receive the same attention as procedures written for computers. Without these procedures, which are the critical human interface to systems, a well designed and programmed system may be useless.

DESCRIPTION OF PHASE ACTIVITIES

Activity A - Define Operational Steps

The phase begins with Systems Engineering preparing a Detail estimate and schedule for the activities of Phase 4-I. This is followed by Systems Engineering defining the operational steps for each Administrative Procedure. The "Playscript" technique is normally used for this purpose.

Activity B - Prepare Administrative Procedure Manual

Systems Engineering prepares the Administrative Procedure Manual which is reviewed in detail with Quality Assurance for adherence to standards. Revisions are implemented as required.

Where applicable, "Help" text is written.

Activity C - Phase 4-I Review

Systems Engineering conducts a review of the **"Administrative Procedure Manual"** with User Management. The manual consists of:

- Phase Cover Page
- Table of Contents
- System Concept Diagram
- System Logic Narrative
- System Flowchart
- Sub-System Concept Diagram
- Sub-System Logic Narrative
- Sub-System Flowchart
- Procedure Logic Narrative
- Playscript Instructions
- Input Discussion & Examples
- Output Discussion & Examples
- Procedure Index by Function
- Procedure Index by Input/Output
- Phase Review Checklist

Based on this report and subsequent review meeting, management may request revisions to the manual.

The next few pages include sample deliverables resulting from this phase.

OD-90162	PLAYSCRIPT INSTRUCTIONS	"PRIDE"-ISEM
NOV 17, 2004	RELEASE 1.2.0	PAGE 2.5.1

```
*********************************************************************************
* PROCEDURE OP-08-03      ORDER-PROC FILE UPDATE SESSION   *
*********************************************************************************
```

PURPOSE-

The purpose of this administrative procedure is to enter and validate data through a computer terminal using the Order Processing Window. Entries are made in accordance with the prompts and messages associated with the panel. HELP text is also available upon request.

SETUP-

1. SUPPLY OF BLANK "ORDER PROCESSING WINDOW" (FORM = KEYBOARD) — ID-00100

2. OBTAIN THE "ORDER-PROC VALIDATION MESSAGES" FROM OP-08-05 "UPDATE ORDER-PROC DATA BASE" (FORM = SCREEN) — OD-00100

OPERATIONS-

ORDER-PR

1. Logon to computer using the standard logon procedure.

2. Access "Order Processing System" screens by double-clicking on the "Order Processing" Icon on the computer desktop.

3. Select the "Order Processing Maintenance" icon to invoke the "Order Processing Screen."

   A. Enter your PASSWORD when prompted by the computer.

4. Select the type of data entry you wish to perform from the window's action-bar:

   A. If you wish to enter a customer's order, select the "Orders" pull-down-choice from the "Customer" action-bar-choice. Goto step 6 for additional instructions.

   B. If you wish to query the status of a customer's order, select the "Status" action-bar-choice from the "Customer" action-bar-choice. Go to step 8 for additional instructions.

   C. For all other queries, proceed to the next step (5).

/\

*** END OF OUTPUT ***

END SECTION

Section 13

"PRIDE"-ISEM PHASE 4-II SOFTWARE ENGINEERING

Good Systems Design + Good Programming = Great Systems
Good Systems Design + Bad Programming = Good Systems
Bad Systems Design + Good Programming = Bad Systems
Bad Systems Design + Bad Programming = Chaos

- Bryce's Law

The purpose of this phase is to design software to satisfy the computer procedure as specified in the Phase 3, "Sub-System Design." Several events occur during this phase:

- Processing and data requirements are evaluated.
- A software strategy is defined to satisfy the requirements.
- Software is designed, complete with specifications to represent processing logic.
- The computer operating environment is defined and evaluated.
- A Computer Run Book is assembled containing the design specifications suitable for review.

Phase 4-II marks the beginning of the software development phases of ISEM. Phase 4-II is used to design the software; Phase 5 is used to produce and test individual programs, and Phase 6 is used to test the software ("string test"). These phases represent where supplemental application development tools can be used in the methodology. For example, CASE tools (Computer Aided Software Engineering), program generators, report writers, fourth generation languages, test data generators, etc.

METHODOLOGY NAVIGATION

A Phase 4-II may be initiated either by following a normal Phase 3 or, if a modification/improvement, following Phase 1, Activity A. Normally, there is a one-to-one relationship between Phase 4-II and a computer procedure.

The formal deliverable resulting from Phase 4-II is a "Computer Run Book" consisting of design and operating specifications for the software. This is reviewed with DP Operations, Systems Engineering, Data Resource Management, and Systems Resource Management for clarity and viability.

Following Phase 4-II, this branch of the ISEM project may branch into multiple Phase 5's ("Software Manufacturing"); one Phase 5 per program.

GENERAL DISCUSSION

Based on the rough designs of the computer procedure as prepared in Phases 1, 2 and 3, Software Engineering prepares the final computer procedure design. This requires the highest level of technical competence on the part of the Software Engineer.

During Phase 3, the sub-system was decomposed into procedures, both administrative and computer. The sub-system designer specified several things, including:

- Required machine responsiveness (timing).

- Input/Output descriptions, including transaction details, special processing rules to produce outputs, and messages.

- External computer files for input request transactions and output data.

Following Phase 3, control over the design and development of software is turned over to the Software Engineer. Because of this delineation of responsibility, the Software Engineer must have a clear understanding of the sub-system. This implies that a knowledgeable Software Engineer can exert influence over the Systems Engineer to prepare specifications correctly. This eliminates the problem of accepting vague or incomplete specifications from the sub-system designer.

The difference between Phase 4-II and 5 is the difference between engineering and manufacturing. Whereas, Phase 4-II is concerned with the design of computer software, Phase 5 is concerned with actually developing the program according to specifications. This can be performed by either manually writing software according to the rules of a particular language or the execution of an application development aid, such as a program generator, report writer or fourth generation language. As such, Phase 5 is dependent on the method of implementation as specified in Phase 4-II

The separation of the programming activities between Phases 4-II and 5 provides project flexibility and a more efficient utilization of programming resources. At the same time, this division of effort assures a complete design prior to program implementation where additional resources, primarily the computer, are utilized. This approach also lends itself readily to using senior programmers to specify the programs, and less experienced programmers to code and test the programs under supervision or implement using automated programming tools.

Because ISEM recognizes that there are many ways to implement software, it is programming language/tool/technique independent. This is why it can comfortably coexist with various approaches to software engineering. This is also why systems engineering should be considered a prerequisite to software engineering. Without adequate systems specifications, software can only be prepared based on unsubstantiated assumptions, speculation and conjecture. It simply will be counterproductive.

The principal objective of Phase 4-II is to define and synchronize processing within the computer procedure. To do this, the computer procedure design must express two things:

- **Processing Flow** - this is an expression of how programs will be executed, from start to end. Consequently, dependencies between programs are defined. Such program dependencies reflect sequential, iterative, and parallel processing.

- **Data Flow** - how data moves within a procedure is expressed by how inputs, outputs and files are used to process the data in the various programs. Here, data is expressed in terms of how it is created, updated, and referenced.

Failure to define such detail will result in processing inconsistencies later.

DBEM INTERFACE

During Phase 2 (System Design), the primary logical files of the system were defined. This triggered a DBEM project where the logical files were merged into the enterprise data base. Data Base Administration thereby devised a physical design to implement the logical files. The net result are physical file layouts for inclusion in programming.

During Phase 3 (Sub-System Design), external computer files, such as input transaction files and output data files, were identified for implementation by Data Base Administration. In other words, by Phase 4-II all of the major computer files should have been defined and implemented by the DBA. However, during Phase 4-II, the Software Engineer may define the need for working files to pass data between programs. This is also communicated to the DBA who must implement the physical files accordingly.

What this highlights is that in a pristine "PRIDE" environment, the Software Engineer is the recipient of the file layouts, not the creator. This approach to data base design promotes integration between applications. Instead of the System/Software Engineers designing a data base to suit the requirements of their application, the data base is implemented by Data Resource Management who is responsible for implementing the data requirements of the enterprise on a global basis.

In the absence of a viable Data Resource Management organization, implementation of the physical files defaults to the Software Engineer.

PHASE 4-II DELIVERABLES

The formal deliverable resulting from Phase 4-II is the "Computer Run Book." This document serves two purposes:

1. It contains the specifications and designs required to manufacture software in Phase 5.

2. It will be used as reference documentation by Operations when the software goes into production.

Within the Computer Run Book are two graphics:

Computer Procedure Flowchart

The computer procedure design is graphically represented by the Computer Procedure Flowchart. As with all ISEM flowcharts, the diagram shows the lowest level of detail in the system hierarchy: programs. Each program is represented on the flowchart with its appointed inputs, outputs, files and timing. A Table of Files can accompany the flowchart which shows how the various files are used in the programs: either "C" Created, "U" Updated, "R" Referenced, or combinations of all three.

The flowchart expresses both the processing flow and data flow of the procedure, which is similar to the Sub-System Flowchart. How the flowchart is drawn, horizontally or vertically, is irrelevant. All that is important is that the graphic express the dependencies between programs, inputs, outputs, and files. This is useful for expressing design issues as well as for production purposes.

Software Structure Diagram

This refers to a graphic used to express the internal processing logic of either a program or module. It is a generic graphic that can take many forms and use different symbols and notation. It is totally dependent on the selected method of implementation for software. If a particular programming technique or tool observes certain graphical conventions, then the diagram is prepared according to these conventions. However, there may be occasions where the method of implementation does not require a graphical explanation. Instead, software specifications need to be conveyed in some other manner (perhaps textual). In this event, a Software Structure Diagram may not be necessary. As such, the diagram is considered optional.

THE PHILOSOPHY OF SOFTWARE ENGINEERING

The design of software is a highly complex and technical issue. It cannot be performed well using unorganized means. The key to success is to treat software development as a science as opposed to an art-form. To be recognized as a legitimate science, there must be some governing principles. To begin with, there are three basic objectives to Software Engineering:

1. To produce reliable programs that will perform according to specifications.

2. To produce programs that are easy to maintain and modify.

3. To implement programs in the most practical, efficient and cost effective manner possible.

These objectives are essentially no different than any other engineering discipline. As such, Software Engineering implies organization and discipline.

Whereas Systems Engineering is concerned with an Information System in its entirety, regardless of its physical implementation, Software Engineering pertains only to the programming of the computer. As such, Software Engineering is considered a subset of Systems Engineering. In order to effectively implement Software Engineering, it presupposes a viable Systems Engineering function is in place to provide the necessary specifications required for Software Engineering to implement (such as the work performed during Phases 1, 2 and 3). While Systems Engineering is geared towards business issues and an overall system design, Software Engineering is oriented to technical issues and software design. In situations where the Systems Engineering function is not implemented, Software Engineering is forced to compensate by trying to deduce the business specifications, a talent for which they are not necessarily suited.

THE NEED FOR STRUCTURE

Structuring software is performed more for the human being, rather than the computer. The machine itself does not care. It will process the instructions whether the software is organized or not. Therefore, there are primarily three reasons for structuring software:

1. DESIGN CORRECTNESS - to prove that software reflects its design. This is verified during testing. As such, a design structure becomes a "roadmap" for testing software.

2. MAINTAINABILITY - to simplify the implementation of changes and corrections to the software. A well structured program minimizes the need to depend upon the author of the software.

3. MODULARITY - to divide a complex program into smaller units so that different parts can be designed and developed separately. This is useful from a Project Management perspective. Modularity also promotes software re-usability; where software components can be shared between applications. The exchange of program code expedites project development.

It is a common misconception that the purpose of techniques such as "structured programming" is to produce an efficient program. In reality, many software design techniques sacrifice efficiency for maintainability.

Structuring software can take many forms:

- Graphical - structure charts, block diagrams, decision tables, function charts, or some other programming technique.

- Textual - pseudo-code or organizing source code to reflect a design.

- Specifications - developing and maintaining requirements to suit a specific application development aid, such as a program generator, report writer, or fourth generation language.

- Combinations of the above.

The objective therefore is to select a software engineering tool or technique that produces uniform results and is easy to implement. This may require a compendium of techniques, as opposed to just one.

SOFTWARE IS HARD

Although "software" is considered the opposite of "hardware," it exhibits some very "hard" characteristics. If it were truly "soft" it would be universally applicable to any computer. Obviously it is not. This is why there are different versions of software for different computers. It is highly uncommon to find software that will perform the same way on different computers without any modification. This implies that software is a physical construct, not logical. Therefore, software is hardware dependent.

Before selecting a software approach, the Software Engineer must consider the target platform where the software is to execute. There are many variables about the computer to consider before making a software design decision; for example:

1. Operating System considerations:

   Memory - Virtual or restricted to a specific size.

   Communications - providing either a single-user or multi-user environment.

   Program execution - single serial execution or parallel execution (multitasking).

   Utilities - for sorting, file backup, data base, etc.

   Compilers - languages supported. Also, linker restrictions.

   Application support - screen handling, TP

monitors.

2. Input/Output considerations

Transaction processing speed.

Peripherals used to support data transmission, conversion, and storage.

A. File size limitations.

B. Supported access methods.

3. Application development tools

Programming aids for development and testing.

Performance monitors.

In many cases, the production machine will be the same as the development machine. In other words, the specifications for the targeted platform may be assumed. In any event, the operating environment must be weighed against the application's requirements as detailed in the Phase 3 "Sub-System Design Manual." Depending on the company's overall direction in computing, one of Software Engineering's objectives may be to develop software to accommodate machine portability. This can influence the method of implementation tremendously.

SOFTWARE STRUCTURE

There are basically two types of computer software:

• **SYSTEM SOFTWARE** concentrates on the basic operational needs of the hardware and is represented by the operating system. The expression itself, "System" Software, is perhaps a misnomer in that it has nothing to do with Information Systems. Instead, it is only concerned with operating the computer.

• **APPLICATION SOFTWARE** is concerned with instructing the computer to perform special tasks within an information system.

Although ISEM is oriented to the production of application software, the concepts embodied in the methodology are equally valid for building an operating system (or any other product).

In ISEM terminology, the following definitions are used in relation to software:

1. **A COMPUTER PROCEDURE consists of one or more programs linked together and executed in a prescribed sequence.**

Conceptually, all computer procedures could consist of a single program. But the functionality of the program would probably dictate a program too large for the computer to effectively implement. Because of this, the Computer Procedure represents a collection of programs dedicated to a specific purpose.

2. **A PROGRAM is a set of machine processable instructions that performs a specific task within a computer procedure.**

If desired, a program can be divided into smaller, easier to manage sections of software that are commonly referred to as MODULES, SUBROUTINES, or OBJECTS. These types of components typically represent re-usable software code. The difference between a program and its subordinate components is that a program is executable by the operating system and is controllable at the command level, while modules and subroutines are not. As such, modules and subroutines rely on other modules and subroutines to form a whole program.

METHODS OF IMPLEMENTATION

When designing software, the Software Engineer considers the method of implementation. Obviously, there are many ways to produce a program:

• Manually writing an unstructured program.

• Manually writing a structured program (modular, object oriented, or otherwise).

• Assembling a program through re-usable code.

• Using computer utility programs (such as to perform a sort).

• Through a commercially available software package.

• Generating a program through such tools as code generators, report writers, fourth generation languages (interpreters), etc.

• Combinations of the above.

Any of these approaches are satisfactory. Ultimately, the Software Engineer must consider three things:

1. Project requirements - this includes delivery schedules and operating budget. It also includes an analysis of the skills and proficiencies required to build a certain type of program.

2. Application requirements - batch, interactive, real-time, etc. require different programming tactics. Consideration must also be given to complexity and performance. In addition, hardware specifications as mentioned earlier are considered.

3. Maintenance requirements - to implement changes and corrections to software later on.

Obviously, there will be trade-offs between the three considerations. It is important that the Software Engineer devise a software strategy that is practical to implement. All of this will have a bearing on the techniques and tools required to implement the programs, along with the type of programming languages to be used; either a Machine Language, Assembly Language, Procedural Language, or Non-Procedural Language (specification driven interpreter).

METHODS OF PROCESSING

During Phase 4-II the Software Engineer must be reminded of the basic processing concepts as defined in ISEM:

1. There are three basic processing constructs:

Sequence - representing consecutive steps in processing.

Iteration - representing repetition until a condition is met.

Choice - representing a selected path based on a prescribed criteria.

The Software Engineer will use any one of these constructs, or combinations, in designing the processing flow. This can be used to express concurrent processing, where two programs within the procedure are executing simultaneously. This does not imply a detached process; there must be some form or program dependency. Rather, it means that one program can be used to initiate another, which will then run simultaneously with others.

It must be remembered that as a procedure starts, it continues from program-to-program, until it reaches a conclusion within the specified procedure time frame. This does not mean that a procedure will have only one start or one end; in fact, it can have multiple ways of starting and multiple ways of ending.

2. All processing involves the use of transactions, which may represent:

- How data is collected for storage.
- How a report, screen or other forms of output is requested.

3. Systems communicate through data.

- In Phase 2, primary files were defined and related to the various sub-systems. In Phase 3, external physical files were defined to implement input transaction files and output data files. During Phase 4-II, the Software Engineer specifies the computer working files to pass data between programs.
- These additional file requirements are communicated to Data Base Administration for implementation.

For additional information on "Methods of Processing," consult the Methodology Introduction.

TIMING REFINEMENTS

As the computer procedure is decomposed into programs, timing is refined to express when each program is executed within the procedure and how fast it must be performed. The timing for the overall procedure is expressed in terms of:

- **Offset** - when the procedure is to begin within the time frame of the sub-system.
- **Response Time** - the speed by which the procedure must execute.

However, at the Program level "Offset" and "Response Time" is refined and takes on new meaning:

- **Offset** - when the program is to begin within the time frame of the computer procedure.
- **Response time** - the speed by which the program must execute.

The objective here is to synchronize the timing of the programs. Offset and Response Time, therefore, will not necessarily be the same as specified for the overall computer procedure. As an example, assume that a computer procedure has a response time of one hour. Let's also assume that the procedure can be executed with three sequential programs. The timing of the three programs could possibly be defined as:

PROGRAM-1	PROGRAM-2	PROGRAM-3
Offset - 0 min	Offset - 20mns	Offset - 40 mins.
Resp. - 20 mns	Resp. - 20 mns	Resp. - 20 mins.

This is a simple example; computer procedures can obviously be much more complicated than this. Yet, it demonstrates the synchronization of the three programs. Timing at this level also provides insight into the type of responsiveness which the computer must have during processing.

One last note, the accumulated response time of the programs must not exceed the response time specified for the overall computer procedure.

SOFTWARE RE-USABILITY

When software is decomposed into its basic components, it can be treated like any other re-usable resource. This means that objects such as modules and subroutines can be uniquely distinguished and re-used in other applications. However, sharing must be designed and built into the software. For example, if modules and subroutines are used for nothing more than to subdivide a program, then the opportunity to share software will be missed. But if these software building blocks are constructed properly, they can be re-used elsewhere, thereby saving time during other application development projects.

Designing re-usable components is dependent on the software design technique in use. Some techniques lend themselves to re-usability better than others. A standard and consistent approach obviously provides uniformity.

Regardless of the selected design technique, these components can be cataloged and cross-referenced in the IRM Repository where their relationships can be analyzed. When properly defined, the Software Engineer can:

- Search and locate modules and subroutines for inclusion in a program.

- Perform an "impact analysis" to study the effect of modifying an existing module to incorporate additional functions.

TESTING

The testing of software is just as important as its design. The intent is to validate the design of the product. During ISEM, design is performed top-down, from system down to programs; testing, on the other hand is performed bottom-up, from the program up to the system. During Phase 5, "Software Manufacturing," the program will be produced and tested individually (unit test). In Phase 6, all of the programs in the computer procedure will be tested as a whole (string test). In Phase 7, the procedures are tested for sub-system conformity (parallel test). Test criteria, therefore, should be established during the design phases for use during testing.

During Phase 3, the Test Criteria for the overall sub-system is established. Similarly, in Phase 4-II, the Test Criteria for the computer procedure and each program is established. The test criteria should specify the conditions by which the software can be validated for correctness. Although the Test Criteria is posted to the computer procedure narrative, it is referenced during both Phase 5 and 6.

ENVIRONMENTAL ISSUES

In preparing for Phase 5 "Software Manufacturing" it is important to identify both the Development and Production operating environments, which may or may not reside on the same computer. This is important for change control purposes; when moving from a test mode to production. There are obviously many ways of establishing computer directories/libraries. However, consideration must be given to the following areas:

- TEST versus PRODUCTION versions - Separation can be based on either:

  - Separate directories on the same disk drive.

  - Separate disk drives on the same computer.

  - Separate computers.

- RELEASE number - This can either represent a formal release number, such as REL100, REL110, REL120, etc. or based on the EFFECTIVE DATE of the resource; for example; 19940101 (year/month/day).

• "PRIDE" SYSTEM STRUCTURE NUMBER - Following the System\Sub-System\Computer Procedure\Program format.

NOTE: Since a sub-system consists of only one computer procedure, the need for expressing the computer procedure level may not be necessary.

• SHARED resources - where re-usable resources such as files, modules and subroutines, are truly shared between systems, then they should be separated from the system hierarchy (non-re-usable items should be maintained under the appropriate system directory).

Obviously, the actual setup of a directory/library is based on machine capabilities and performance considerations. However, establishing such a development/production environment offers many benefits:

• It isolates the programs and modules along with associated test files during development.

• Physically locating resources is simplified since it mirrors the contents of the IRM Repository. Although any name can be applied to a file, use of the IRM Repository numbering conventions should be encouraged. For example, PC based files could be expressed as:

TS0102.CMD	MD-90023.OBJ
TS010201.EXE	MD-90023.DLL
TS010201.C	FD-92300.TRN
TS010202.COB	FD-90135.DAT
OD-93020.LST	ID-90032.RC
ID-90021.HLP	ID-90032.IPF

• From a Quality Assurance point of view, this is useful for comparing designs as recorded in the IRM with the actual components.

• It greatly assists change control by being able to copy components from one directory to another (from development to production).

• Although such a directory/library strategy should be in place prior to design, it is confirmed during Phase 4-II.

DEVELOPMENT VERSUS PRODUCTION MACHINES

Developing systems and software on a separate machine from the production environment overcomes the problem of contention for machine time and improves speed (both for developers and production jobs). However, it can present a problem when incompatible operating systems and compilers are used. In this situation, the development organization has a couple of options:

1. Perform Phases 1 through 4-II on the development machine and perform Phases 5 and 6 on the target production machine.

2. Perform Phases 1 through 6 on the development machine. This will result in tested software that complies to the rules of the development machine's compiler. The source code is then moved to the production machine where it may have to be modified to suit the rules of the production machine's compiler. In other words, the testing and debugging steps of Phases 5 and 6 may have to be repeated.

CONTROL LANGUAGE

Between the software design and the development/production environment strategy, the Software Engineer should be able to develop pertinent control language statements for executing the programs. This can take many forms depending on the computer's operating system; some examples:

• Job Control Language (JCL)

• Digital Command Language (DCL)

• Executive Control Language (ECL)

• CLISTS

• CL (Control Language)

• PC Command Files (.CMD and .BAT)

These control language statements are to be prepared during Phase 4-II and made available to the Phase 5 programmer.

SKILLS & PROFICIENCIES

Following the design of the software and the identification of the operating environment, the Software Engineer should be able to define the skills and proficiencies needed to produce the software in Phase 5. For example, the Software Engineer should be able to identify the degree of knowledge regarding:

• Hardware

• Operating System considerations

• Programming Languages

• Techniques and tools for software design and development.

This is useful for selecting and/or recruiting personnel for Phase 5 assignments. Also, it will highlight the need for additional training.

MOVING TO PHASE 5

As mentioned, the formal deliverable resulting from Phase 4-II is the "Computer Run Book." Contained within this document are important design decisions for the programmer to implement in Phase 5, including:

• Program dependencies specifying execution sequence.

• Processing logic for each program.

• Final input/output formats including additional processing messages identified during design of the software.

• File layouts.

• Operating environment considerations including pertinent command language statements.

• Required skills and proficiencies needed to perform the work.

PHASE 4-II REVIEW

Phase 4-II represents a major technical review phase and serves two purposes: to critique the software design, and; to evaluate the software for use in production. Because of this, different types of people participate in the review.

• SYSTEMS ENGINEERING evaluates the software design in terms of whether it will effectively implement the sub-system design as developed in Phase 3.

• OPERATIONS evaluates the design in terms of how it will operate in production. In this respect, operations acts as the end-user who will be responsible for implementing the design.

• DATA RESOURCE MANAGEMENT assures that all of the physical data requirements have been satisfied.

DESCRIPTION OF PHASE ACTIVITIES

Activity A - Design Software

Software Engineering reviews the specifications for the computer procedure from the Phase 3, "Sub-System Design Manual." Based on these specifications, Software Engineering prepares a Detail estimate and schedule to complete Activity A which is reviewed with Project Management for approval.

Software Engineering then studies computer operating requirements and defines what programs will be necessary to implement the computer procedure. This is graphically expressed through the Computer Procedure Flowchart.

For each program, the Software Engineer determines an appropriate method of implementation; either to manually produce software according to the rules of a particular technique, to re-use existing software, to generate programs through an application development aid, or combinations of all three.

Following determination of the method of implementation, Software Engineering prepares a Detail estimate and schedule to complete the remaining activities in Phase 4-II (B through E). This is reviewed with Project Management for approval.

Activity B - Define Software Logic

During this activity Software Engineering:

• Reviews the Input/Output Descriptions. Additional processing messages may be identified during this activity.

• Defines the processing logic for each program in the procedure. This is based on the selected method of implementation. Where pertinent, a Software Structure Chart is produced to reflect the processing logic.

• Defines the need for working files to pass data between programs or for use within programs. This is communicated to Data Base Administration who must implement accordingly.

- Defines the Testing Criteria for the software. This is based on the input, output, file, and program specifications. The Testing Criteria is added to the computer procedure narrative.

Also, Software Engineering considers the skills and proficiencies required to produce each program in Phase 5.

Activity C - Prepare File Layouts

Data Base Administration delivers to Software Engineering the completed file layouts for use in the software. This includes the primary files as expressed earlier in ISEM, along with the external and internal working files for the computer.

Activity D - Define Operating Requirements

Software Engineering defines both the Development and Operating environments for the software. This includes where the software will be produced and tested versus where it will perform when put into production.

Other operating issues are evaluated at this time, such as the operating costs of the software, performance considerations, backup/recovery, etc.

Software Engineering also prepares pertinent control language statements to control the execution of the programs.

Software Engineering then prepares the final Computer Run Book which is reviewed in detail with Quality Assurance for adherence to standards. Revisions are implemented as required.

Activity E - Phase 4-II Review

Software Engineering conducts a review of the "Computer Run Book" with Systems Engineering, DP Operations, Data Resource Management, Systems Resource Management, and Project Management. The run book consists of:

- Phase Cover Page
- Table of Contents
- Computer Procedure Flowchart
- Computer Procedure Discussion
- Test Criteria
- Software Structure Diagram
- Program Specification
- Module Specification (optional)
- Data Resource Layouts
- Phase Review Checklist

Based on this report and subsequent review meeting, management may elect to revise parts of the report, discontinue the project, or approve it for continuation.

The next few pages include sample deliverables resulting from this phase.

SAMPLE PHASE 4-II DELIVERABLES

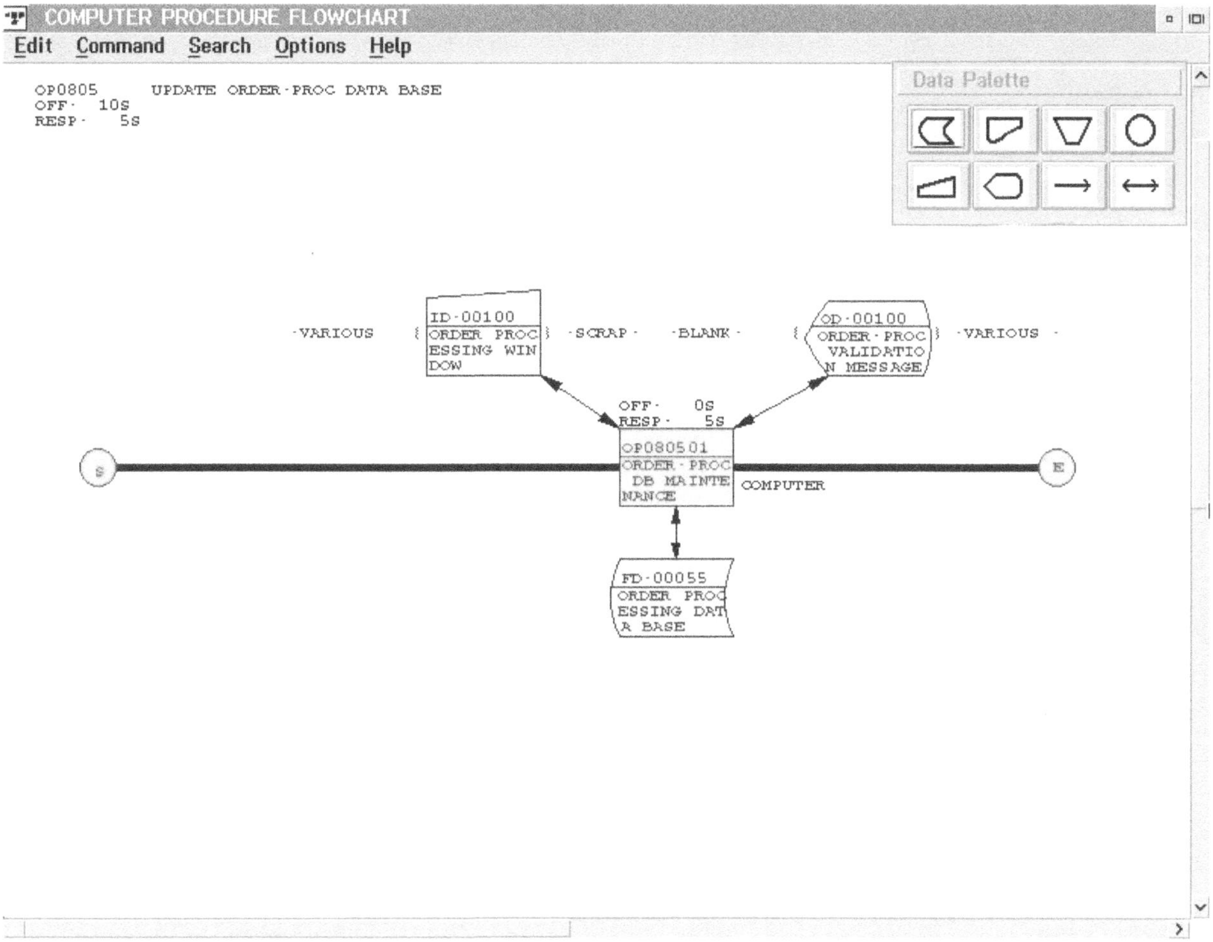

OD-92015 COMPUTER PROCEDURE DISCUSSION "PRIDE"-ISEM

NOV 14, 2004 RELEASE 1.2.0 PAGE 1.2.1

```
*********************************************************************************
* PROCEDURE OP-08-05        UPDATE ORDER-PROC DATA BASE                          *
*********************************************************************************
```

TIMING-

		FREQUENCY	OFFSET	RESP-TIME
OP-08	ORDER-PROC FILE MAINTENANCE	R		15S
OP-08-05		N/A	10S	5S

LOGIC FLOW -

FROM	TO	CONDITION
OP-08-03 ORDER-PROC FILE UPDATE SESSION	OP-08-03 ORDER-PROC FILE UPDATE SESSION	ALWAYS

SETUP-

ID-00100	ORDER PROCESSING WINDOW	/ /R	KEYBOARD
OD-00100	ORDER-PROC VALIDATION MESSAGES	C/ /	SCREEN
FD-00055	ORDER PROCESSING DATA BASE	C/U/R	PRIMARY FILE

ALTERNATE IDENTIFICATION - NONE

PROGRAMS-

OP-08-05-01	ORDER-PROC DB MAINTENANCE	ALT-ID - TMC-O008

DESCRIPTION -

BUSINESS PURPOSE

The purpose of this computer procedure is to validate input transactions and update files in the data base. As such, it operates interactively responding to user requests via on-line entry.

Because of the responsiveness required, the procedure consists of a single program consisting of various software modules that are called as required to fulfill the requested operation.

* * * END OF OUTPUT * * *

SAMPLE PHASE 4-II DELIVERABLES

NOTE: A Software Structure Diagram can take many forms; this is but one.

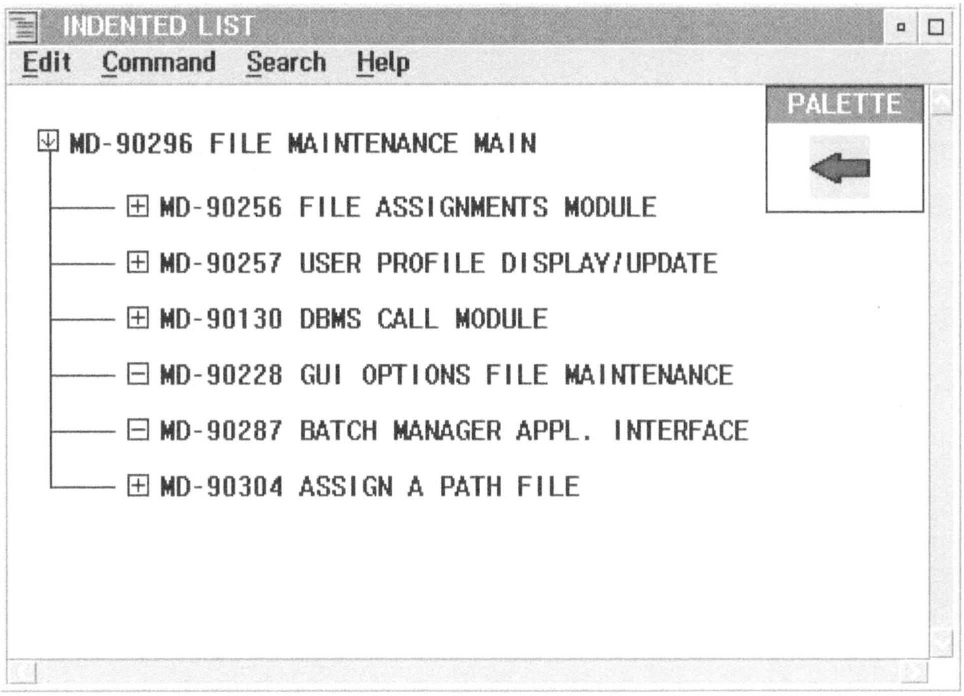

OD-90179 PROGRAM SPECIFICATION "PRIDE"-ISEM

NOV 15, 2004 RELEASE 1.2.0 PAGE 2.2.1

```
****************************************************************************
* PROGRAM OP-08-05-01     ORDER-PROC DB MAINTENANCE                        *
****************************************************************************
```

TIMING-

		FREQUENCY	OFFSET	RESP-TIME
OP08	ORDER-PROC FILE MAINTENANCE	R		15S
OP0805	UPDATE ORDER-PROC DATA BASE	N/A	10S	5S
OP080501	ORDER-PROC DB MAINTENANCE	N/A	0S	5S

LOGIC FLOW-

FROM	TO	CONDITION
-START-	-END-	ALWAYS

SETUP-

ID-00100	ORDER PROCESSING WINDOW	C/ /R	KEYBOARD
OD-00100	ORDER-PROC VALIDATION MESSAGES	C/ /R	SCREEN
FD-00055	ORDER PROCESSING DATA BASE	C/U/R	PRIMARY FILE

ALTERNATE IDENTIFICATION - TMC-O008
LANGUAGE - C
MEMORY - 256K

DESCRIPTION -

BUSINESS PURPOSE

The purpose of this program is to validate input trans- actions and update computer files.

PROGRAM LOGIC

Program is highly modularized and calls modules as required to respond to user input requests. Consult the attached Software Structure Diagram for the hierarchy of software modules.

MODULES-

NUMBER	NAME	ALT-ID	LEVEL
MD-90296	FILE MAINTENANCE MAIN	M90296	01

* * * END OF OUTPUT * * *

OD-90180 MODULE SPECIFICATION "PRIDE"-ISEM

NOV 15, 2004 RELEASE 1.2.0 PAGE 1

****
* MODULE MD-90296 FILE MAINTENANCE MAIN *
****

TYPE- SINGLE USE- PERFORMED RESPONSE TIME- 0005M
ALT IDENT- M90296 LANGUAGE- C MEMORY- 256K

DESCRIPTION-

BUSINESS PURPOSE

This is the root module for updating the Order Processing Files. Start up of the File Manager includes displaying a main panel, and verifying presence of workstation profile and customer profile files.

SUPERIOR PROGRAMS/MODULES-

IF-02-02-00 FILE MAINTENANCE DRIVER PROGRM
ENTRY- M90296

SUBORDINATE MODULES-

MD-90256 FILE ASSIGNMENTS MODULE
ENTRY- M90256

MD-90257 USER PROFILE DISPLAY/UPDATE
ENTRY- M90257

MD-90130 DBMS CALL MODULE
ENTRY- M90130B

MD-90223 AUTO TRACE
ENTRY- M90223A

MD-90228 GUI OPTIONS FILE MAINTENANCE
ENTRY- M90228

MD-90287 BATCH MANAGER APPL. INTERFACE
ENTRY- M90287

MD-90304 ASSIGN A PATH FILE
ENTRY- M90304

CALL PARAMETERS-

RD-03092 ROOT CALL OBJECT

INPUT/FILE/OUTPUT/RECORD-

		COPY	O C	C/U/R	TYPE
FD-90073	CUSTOMER PROFILE FILE	01	O C	/U/R	PRIMARY
FD-90167	WORK STATION PROFILE FILE	01	O C	/ /R	PRIMARY
FD-90124	USER PROFILE FILE	01	O C	/U/R	PRIMARY

* * * END OF OUTPUT * * *

OD-90194 DATA RESOURCE LAYOUT "PRIDE"-ISEM

NOV 15, 2004 RELEASE 1.2.0 PAGE 3.3.1

```
***************************************************************************************************
* FILE LAYOUT FD-00055              ORDER PROCESSING DATA BASE          *
***************************************************************************************************

FILE TYPE - PRIMARY        MEDIUM - DA        ORGANIZATION - R

FORMAT - DBMS              RETENTION - PERMANENT
```

DESCRIPTION

This represents a physical file as used by a relational Data Base Management System (DBMS). The file consists of tables and views containing data about Customers, Orders, Billings, Back-Orders, Shipments, Payments, Territories, and Government Agencies. The master realtional data base is backed up twice a day (noon) and 10 PM.

SORT/ACCESS -

DD-00001	CUSTOMER NUMBER
DD-00003	PRODUCT NUMBER
DD-00008	ACCOUNT NUMBER
DD-00009	GOVERNMENT AGENCY NUMBER
DD-00010	SHIPMENT NUMBER
DD-00011	BILLING NUMBER
DD-00012	ORDER NUMBER
DD-00014	BACK ORDER NUMBER
DD-00042	CONTACT NUMBER
DD-00043	ADDRESS TYPE
DD-00059	TERRITORY ASSIGNMENT NUMBER
DD-00122	TERRITORY CODE
DD-00167	GOVERNMENT TYPE CODE
DD-00192	PAYMENT NUMBER

LABELS -

SQL	ORDER_PROC_DB
COB	ORDER-PROC-DB
C	OrderProcDb
NOM	ORDER-PROC-DB
EZT	ORDER-PROC-DB

RELATIONSHIPS -

(NONE FOUND)

VALIDATION -

(NONE FOUND)

RECORDS -

RD-00500	O-P CUSTOMER IDENT TABLE
RD-00501	O-P CUSTOMER ADDRESS TABLE
RD-00502	O-P CUSTOMER CONTACT TABLE
RD-00503	O-P PRODUCT IDENT TABLE
RD-00504	O-P GOV'T IDENTIFICATION TABLE
RD-00505	O-P GOVERNMENT TAX RATE TABLE
RD-00506	O-P ORDER IDENTIFICATION TABLE

(continued on next page)

SAMPLE PHASE 4-II DELIVERABLES

(continued from previous page)

RD-00507	O-P CUSTOMER/PRODUCT RELATION
RD-00508	O-P BACK-ORDER ID TABLE
RD-00509	O-P ORDER/PRODUCT RELATIONSHIP
RD-00510	O-P SHIPMENT IDENT TABLE
RD-00511	O-P CUSTOMER/PRODUCT RELATION
RD-00512	O-P ACCOUNT IDENT TABLE
RD-00513	O-P BILLING IDENT TABLE
RD-00514	O-P CUSTOMER/PRODUCT RELATION
RD-00515	O-P ACCOUNT/GOVERNMENT RELATN.
RD-00516	O-P CUSTOMER PAYMENT TABLE
RD-00517	O-P CUSTOMER PAYMENT TABLE
RD-00518	O-P TERRITORY IDENT TABLE
RD-00519	O-P SALES ASSIGNMENT TABLE
RD-00520	O-P CUSTOMER/TERRITORY RELATIN
RD-00500	O-P CUSTOMER IDENT TABLE

FORM -

VARIABLE MINIMUM LENGTH - 0040
TYPE - STORAGE RECORD MAXIMUM LENGTH - 0079

DESCRIPTION

This record contains data used to identify a customer.

SORT/ACCESS - DD-00001 CUSTOMER NUMBER

LABELS -	SQL	OP_CUSTOMER_ID
	COB	OP-CUSTOMER-ID
	C	OpCustomerId
	EZT	OP-CUSTOMER-ID
	NOM	OP-CUSTOMER-ID

NUMBER NAME		POSN	LNTH	LEVL	REPT	FL	VD	CL	PR	SC BAS
DD-00001 CUSTOMER NUMBER		2	8		1	Z	Z	N		10
LABELS-	COB	CUST-NUMBER								
	C	CustNumber								
	SQL	CUST_NUMBER								
	NOM	CUST-NUMBER								
	EZT	CUST-NUMBER								
DD-00018 NAME		10	30		1	B	B	X		10
LABELS-	COB	NAME								
	C	Name								
	SQL	NAME								
	NOM	NAME								
	EZT	NAME								
DD-00046 D-U-N-S NUMBER		40	10		1	B	B	N		10
LABELS-	COB	D-U-N-S								
	C	Duns								
	SQL	D_U_N_S								
	NOM	D-U-N-S								
	EZT	D-U-N-S								

*** END OF OUTPUT ***

END SECTION

Section 14

"PRIDE"-ISEM PHASE 5 SOFTWARE MANUFACTURING

Programming is a translation function, going from human understandable specifications to machine processable instructions.
- Bryce's Law

The purpose of Phase 5 is to translate the specifications prepared during Phase 4-II, "Software Engineering," into efficient computer processable instructions. Several events occur during this phase:

- The program is produced according to specifications.
- Test data is developed and a unit test is performed.
- The test results are reviewed and retained for future reference.

METHODOLOGY NAVIGATION

A Phase 5 may be initiated either by following a normal Phase 4-II or, if a modification/improvement, following Phase 1, Activity A. There is a one-to-one relationship between Phase 5 and a program.

Multiple Phase 5's within a computer procedure can be performed in parallel. Each will be used to write a program and perform a unit test. They will then be assembled as a string in Phase 6 and tested accordingly.

There may be unusual circumstances where it may be more practical to create sub-Phase 5's. This usually occurs when a single program is too large and complex to be administered by a single phase. In this situation, it may be necessary to subdivide Phase 5 into its subordinate modules.

Phase 5 represents the bottom of the Standard System Structure hierarchy. The intent is to produce the necessary software and to perform a unit test of the program. Following this phase, testing and implementation begins to revert back up the system hierarchy. In Phase 6, all of the programs in the computer procedure are assembled and tested as a string.

GENERAL DISCUSSION

SPECIFICATIONS

Prior to Phase 5, Software Engineering defined the logic for each program in the computer procedure. The level of definition was based on the method of implementation. If a program generator is to be used, then the specifications were developed to the level where the generator can produce source code in

Phase 5. If the program is to be developed manually, then the specifications were written to a level where the human being can translate them into machine processable instructions. What this highlights is that, at this point, programming is an administrative function.

TYPES OF PROGRAM CODE

There are essentially three types of program code: source, object, and executable. Source code is written according to the specifications established in Phase 4-II. It is then read by the computer's compiler where it is converted into object code. Object code is then linked with other object code and run time libraries to produce an executable program.

Source code can be written either by the human being or generated though an application development aid. However, this does not mean that there will always be source code. Some application development aids have the ability to produce object code directly without having to first write and compile source code. In other words, these types of tools maintain their own internal compilers. However, they are still considered the exception as opposed to the rule in application development. Most development activity still involves creating source code, compiling, producing object code, then executables.

TESTING

Following the production of the executable program, the programmer assembles pertinent test data to test and debug the program. To do so, the programmer consults the Test Criteria, Program Specifications, and pertinent Software Structure Diagrams as stipulated in the Computer Run Book. This becomes the "roadmap" for testing the program.

The IRM Repository is a good source for creating test data. If the data resources were properly defined (inputs, outputs, files, records, data elements), the programmer should be able to assemble the data variables and formats. In fact, a good test data generator should be able to tap into the specifications contained in the IRM Repository to create adequate test data.

There are a variety of testing and debugging tools available on the market. Use of such tools should be encouraged.

When the program has been tested to the satisfaction of the programmer, the results are reviewed with the computer procedure designer and Quality Assurance. This review assures that all standards have been followed, and the program has been tested to conform to specifications. The test data is saved for later reference and the test results are filed with the master copy of the Computer Run Book in the Technical Library.

DELIVERABLES

Unlike the printed deliverables resulting from the previous phases of the methodology, Phase 5 produces an executable program, along with pertinent source code, object code, test data and test results.

DESCRIPTION OF PHASE ACTIVITIES

Activity A - Review Software Specifications

Software Engineering reviews the program specifications as contained in the Phase 4-II "Computer Run Book." Based on these specifications, Software Engineering prepares a Detail estimate and schedule for the activities of Phase 5 (A through E) which is reviewed with Project Management for approval.

Activity B - Produce Executable Program

Software Engineering prepares an executable program based on the specifications. This is normally performed by creating source code, compiling and creating an object program, and linking.

Activity C - Develop Test Data

Software Engineering prepares test data for the program. This is based on the program logic and testing criteria contained in the Computer Run Book.

Activity D - Test/Debug Software - Unit Test

Software Engineering tests and debugs the program accordingly.

Activity E - Phase 5 Review

Software Engineering reviews the test results with the computer procedure designer and Quality Assurance for compliance with installation standards. Revisions are implemented as required.

Note: Not all programs will require sequential execution of these activities. Some programs, particularly interactive, may require Activities B, C, and D to be scheduled and performed in parallel.

NOTE: Because Phase 5 primarily produces code, sample deliverables are not included in this section.

END SECTION

Section 15

"PRIDE"-ISEM PHASE 6 SOFTWARE TESTING

Systems are designed by 'explosion' and implemented by 'implosion'.
- Bryce's Law

The purpose of Phase 6 is to perform a test of all of the programs within the computer procedure. Several events occur during this phase:

- Test data from the various phase 5's are assembled.
- The programs are linked together and a test is performed to assure that the software performs according to the specifications as contained in the Phase 4-II "Computer Run Book."
- Test results are produced and reviewed with the sub-system designer and Quality Assurance. The results are then filed in the master copy of the Computer Run Book in the Technical Library.

METHODOLOGY NAVIGATION

A Phase 6 may be performed only after all of the Phase 5's in the computer procedure have been completed. There is normally a one-to-one relationship between Phase 6 and a computer procedure.

The formal deliverables resulting from Phase 6 will be the test data and test results, which is reviewed by the sub-system designer and Quality Assurance.

Phase 6 will then proceed to Phase 7 where it will rejoin the administrative procedures in a test of the entire sub-system.

GENERAL DISCUSSION

During the preceding phases of ISEM, primary emphasis was placed on the design or "explosion" of the system into its detail components. This phase and the succeeding phases are concerned with the "implosion" of the detail components into operating units for implementation, e.g., programs into procedures, and procedures into sub-systems.

The purpose of this phase is to test and correct errors in the computer procedure. This is sometimes called a "string test," "job stream test," or "stress test." Whereas the "unit test" performed in Phase 5 was concerned with data format, the emphasis in Phase 6 is on volume testing.

Testing within ISEM is performed in three parts: Unit testing of individual programs is conducted in Phase 5 by the programmer; the computer procedure test is conducted in Phase 6, normally by the computer procedure designer; and the sub-system test, of both administrative and computer procedures, is conducted in Phase 7 by the sub-system designer. Testing performed in this manner assures a continuity of specifications throughout the system. Each designer tests the implementation of their design and specifications. The Computer Procedure Designer, also tests and corrects the individual programs, which assures that the various programs in the procedure are maintainable and completely documented.

It is the objective of the Computer Procedure Designer to assure that the procedure meets the sub-system specifications. In addition to testing for correct processing logic, they also check the operational considerations of timing, re-start, etc.

Following testing, the results are reviewed with Quality Assurance and the Sub-System Designer. This assures compliance with all standards and that the Sub-System requirements are met. When the testing is satisfactorily accomplished, the test data and results are saved for use when any modifications/improvements are executed in the future.

As in Phase 5, program testing/debugging aids should be encouraged.

DESCRIPTION OF PHASE ACTIVITIES

Activity A - Prepare Test Data

Software Engineering prepares a Detail estimate and schedule for the activities of Phase 6 (A through C) which is reviewed with Project Management for approval.

Software Engineering (the Computer Procedure Designer) assembles the test data from the various Phase 5's and in accordance with the Test Criteria defined in the Computer Run Book.

Activity B - Test/Debug Software - String Test

Software Engineering tests and debugs the computer procedure in its entirety, including all programs.

Activity C - Phase 6 Review

Software Engineering reviews the test results with the sub-system designer (Systems Engineering) and Quality Assurance for compliance with installation standards, and to assure that the Sub-System Design requirements are satisfied.

NOTE: Because Phase 6 primarily produces test data and test results, sample deliverables are not included in this section.

END SECTION

Section 16

"PRIDE"-ISEM
PHASE 7
SUB-SYSTEM TEST

How a system is implemented is of little importance if it solves the problem effectively.
- Bryce's Law

Sub-System testing represents a test of both the administrative and computer procedures in a sub-system. Several events occur during this phase:

- Test data from the previous phases are assembled.
- The procedures are tested in accordance with the Test Criteria of the Phase 3 "Sub-System Design Manual."
- Test results are produced and reviewed with Quality Assurance. They are then filed with the master copy of the Sub-System Design Manual in the Technical Library.

METHODOLOGY NAVIGATION

A Phase 7 may be performed only after Phase 4-I and Phase 6 have been completed. There is a one-to-one relationship between Phase 7 and a sub-system.

The formal deliverables resulting from Phase 7 will be the test data and test results, which are reviewed by Quality Assurance.

Phase 7 will then proceed to Phase 8 where it will rejoin other sub-systems and put into production.

GENERAL DISCUSSION

Testing in Phase 7 is performed in a similar manner as in Phase 6, except at a higher level. Whereas Phase 6 "Software Testing," is performed by Software Engineering for all of the programs within the computer procedure, Phase 7 is performed by Systems Engineering for all of the procedures in the sub-system, both administrative and computer.

During Computer Procedure Design, Software Engineering prepared the computer procedure that will be used by Operations. In order to test this procedure, it is recommended that DP Operations actually perform the tests under Software Engineering's direction. In this way, the operating procedures are tested along with the programs used to implement the procedure. Often the operating people will point out areas for procedure improvement.

During Phase 3, Systems Engineering specified a testing criteria for the overall sub-system. This included the validation of input/output processing, along with anticipated transaction volume for the sub-system. Phase 7, therefore, is used to verify the test criteria and benchmarks.

Following Phase 7, the sub-system is essentially ready for operation. However, considerable care should be exercised to avoid a premature installation of the sub-system for day-to-day operation. This situation could cause panic conditions if other parts of the system are not ready. This is why Phase 8 is used to release sub-systems.

DESCRIPTION OF PHASE ACTIVITIES

Activity A - Prepare Test Data

Systems Engineering prepares a Detail estimate and schedule for the activities of Phase 7 (A through C) which is reviewed with Project Management for approval.

Systems Engineering (the Sub-System Designer) assembles the test data from the previous phases and in accordance with the Test Criteria defined in the Sub-System Design Manual.

Activity B - Test/Debug Sub-System

Systems Engineering tests and debugs the sub-system in its entirety, including all procedures. Software Engineering provides assistance as required.

Activity C - Phase 7 Review

Systems Engineering reviews the test results with the system designer and Quality Assurance for compliance with installation standards, and to assure that the system design requirements are satisfied.

NOTE: Because Phase 7 primarily produces test data and test results, sample deliverables are not included in this section.

END SECTION

Section 17

"PRIDE"-ISEM PHASE 8 SYSTEM OPERATION

Systems are built by evolution, not by revolution. The day a system is installed is the day it begins to undergo change.
- Bryce's Law

Phase 8 "System Operation" represents the point in the methodology where a system or groups of sub-systems are put into production. Several events occur during this phase:

- The system and data base are installed and a final test of the sub-systems is performed.
- Users are trained in the effective use of the system. This includes Data Processing Operations.
- The system is put into production.

METHODOLOGY NAVIGATION

A Phase 8 may be performed for either an entire system or for a group of sub-systems. It ultimately is based on the magnitude and complexity of the system, along with the delivery schedule of the project. Normally, one Phase 8 is performed for a single project. However, It is not unusual to have multiple Phase 8's; one for each grouping of sub-systems.

There is a one-to-one relationship between Phase 8 and a system.

The formal deliverables resulting from Phase 8 will be training aids which are reviewed by Quality Assurance.

From Phase 8 the project will proceed to Phase 9 where the system/project audit will be performed.

GENERAL DISCUSSION

The installation of a new information system can either be a frustrating, time consuming, traumatic endeavor, or it can be an orderly, well planned, rewarding experience.

NOW IS NOT THE TIME TO GRASP DEFEAT FROM THE JAWS OF VICTORY!

The more preparation and organization, the greater the chances are for a smooth installation and success.

INSTALLATION

A system can be installed either in its entirety or in groups of sub-systems. This permits the sub-systems to be implemented in a sequence that is convenient to the Users, thus reducing the amount of interference that typically occurs in the user area when complete systems are implemented at one time. In addition, the user will benefit from using some parts of the system earlier than waiting until the entire system is implemented.

Selection of sub-system groupings is important. Normally, sub-systems accommodating operational requirements are implemented before those satisfying control and policy requirements. This is because the operational sub-systems typically represent those processes where data is collected and stored (File Maintenance type of sub-systems) for use by other processes. Also, special file conversion sub-systems are implemented to move data from old files formats to new formats. This implies that Data Base Administration participates in the implementation of the data base.

TESTING

A final test of the system is performed where the various sub-systems are tested in parallel. This is similar to the testing in Phases 6 and 7, but at the highest level. It is not uncommon to invite user personnel and Operations to participate in the test. The Administrative Procedure Manuals and Computer Run Books, as created in the various Phases 4-I's and 4-II's, are used as part of the formal walk-through. The test data as created in previous phases is assembled and used as the basis for the test. Where available, actual data should be tested to simulate actual system situations. If any errors are discovered, they are corrected immediately and testing is resumed. Changes to the sub-systems may also be proposed; however, these should be considered carefully. In some instances, it may be more appropriate to suspend any modifications or improvements temporarily until use of the new system has settled into routine operation.

EDUCATION

Even the best designed system can fail unless it is properly presented to the users. Educating and training the users should be handled carefully and diplomatically. Good presentation skills are essential.

Users must be convinced that the system is a viable business solution and they must know how to effectively use it. The intent of training, therefore, is to communicate the scope of the system, along with the business problems/opportunities it addresses. All parties involved with the system must understand WHO is responsible for performing WHAT, WHEN, WHERE and WHY. This must be communicated in simple business terms that the user will understand. Technical jargon should be avoided since it tends to alienate users. In other words, the trainer should be promoting business benefits as opposed to the technical details of the system.

Prior to conducting the actual training of the users and DP operations, Systems Engineering assembles presentation aids, such as:

• Transparencies/Slides - using graphics developed during the methodology, for example:

- System Concept Diagram
- Sub-System Concept Diagrams
- System Flowchart
- Sub-System Flowcharts
- Illustrative Input/Output Examples
- pertinent matrices

• Administrative Procedure Manuals (from the various Phase 4-I's) and Computer Run Books (from the Phase 4-II's).

• Reference Cards, Templates, Promotional items, etc.

• A demo or training version of the product.

OPERATIONS

Following testing and education of the users, the system is put into production on an agreed upon date. During system start-up it may be necessary to make systems personnel available to users and operations in order to answer questions and solve problems.

DESCRIPTION OF PHASE ACTIVITIES

Activity A - Install/Test System

Systems Engineering prepares a Detail estimate and schedule for the activities of Phase 8 (A through C) which is reviewed with Project Management for approval.

Systems Engineering assembles the test data from the prior Phase 7 tests.

The system and data base is installed and a parallel test of all sub-systems is performed.

Activity B - Educate/Train Users

Systems Engineering prepares pertinent training aids and conducts educational training for users and DP Operations.

Activity C - System Operations

The system is then put into production. Systems personnel are made available during start-up to expedite problems.

NOTE: Because Phase 8 primarily produces test data and test results, along with educational aids, sample deliverables are not included in this section.

END SECTION

Section 18

"PRIDE"-ISEM PHASE 9 ISEM EVALUATION

Systems do not have a 'life cycle.' They may go on forever if kept viable with change. The only thing that has a 'life cycle' is a project which has a beginning for planning, a middle for execution, and an end for review.
- Bryce's Law

The purpose of this phase is to evaluate the system and project in terms of how well they were delivered as compared to what was proposed in the early phases of the methodology. Several events occur during this phase:

• The implemented system is analyzed in terms of how well it satisfies its requirements. Also, performance issues are evaluated. From this, strengths and weaknesses are evaluated.

• Project performance is evaluated in terms of how well the project met estimates, schedules and proposed costs. Further, an analysis of the project is prepared.

• A system evaluation report is prepared summariz ing the phase findings and reviewed by management.

METHODOLOGY NAVIGATION

This represents the concluding phase of the ISEM project. It is performed either after Phase 8, or if a project has been discontinued earlier on (such as after Phase 1, 2 or 3).

The formal deliverables resulting from Phase 9 will be the "System Evaluation Report" which is reviewed by Quality Assurance prior to being issued to management.

GENERAL DISCUSSION

During Phases 1, 2 and 3, Systems Engineering prepared the "System Study and Evaluation Report," "System Design Manual" and "Sub-System Design Manuals." On the basis of these reports, management either approved or disapproved the project for design and implementation of the system.

If the project was disapproved, the reasons are outlined during this phase and placed with the original documentation for future reference. If the project was approved, this approval was given as a result of certain expectations outlined within the reports. The purpose of this evaluation is to determine if these expectations were realized.

THE BENEFITS OF PHASE 9

"Unless we learn from history, we are doomed to repeat it." This may be an old adage, but it is certainly true. Many organizations tend to regard project reviews as a waste of precious time and resources. Consequently, systems are installed without any type of follow-up. Because of this, many of the benefits were never realized. For example, clerical savings may have been part of the justification for a new system, but operating management did not reduce or reallocate the staff to realize the benefits. Without follow-up, "Parkinson's Law" may occur where "work expands to fill the time available."

As another example, a new operating management, created as a result of organizational changes, may not be utilizing the information provided by the system. They may be unaware of, or lack knowledge about the system. In order for management to have confidence in their systems organization and the systems installed, they should have tangible evidence that their significant investments are producing the promised results.

The systems organization can realize many benefits from an "ISEM Evaluation." It is not just a platitudinous statement to say that *"we learn from our mistakes."* It is a clear and established fact. A detailed audit provides Systems Engineering and Software Engineering with an opportunity to review their estimating and design skills. This knowledge, along with historical data, can be gainfully applied to new systems and projects. As such, estimating guidelines can be updated.

A carefully executed audit can also add to the credibility of the systems organization by showing how well they performed and that they can account for their actions.

Finally, as a result of an audit, the systems organization may find that the new system is simply not functioning in spite of systems maintenance and revisions. In this event, the audit may indicate an entirely new approach should be taken as opposed to continually fighting the problem.

"WHEN" SHOULD YOU CONDUCT PHASE 9?

The System Audit should be scheduled for execution after the system has operated for an adequate period of time. Typically, this will be between thirty (30) and ninety (90) days after Phase 8. In some instances, it may be necessary to conduct more than one audit, depending on the timing of implementation and execution of the various sub-systems within the system. For example, User Management may want a detailed accounting of project costs immediately after each Phase 8. The actual system evaluation can follow thereafter.

"HOW" SHOULD YOU PERFORM PHASE 9?

Ideally, the audit should not be performed by the same individual(s) who developed the system. A neutral third party should be involved who can audit the system and project objectively.

The steps required to execute the phase are similar to those in Phase 1, Activities B and G.

- Phase 1, Activity B "Current System Analysis" - the intent of this activity is to study the existing system, sample work, and evaluate strengths and weaknesses. This same type of work is performed in Phase 9. Here, the Systems Engineer uses the systems documentation from the various phases to walk the engineer through the system.

- Phase 1, Activity G "Prepare System Evaluation" - the intent of this activity is to estimate and schedule the remaining work effort, and perform a Cost/Benefit Analysis. During Phase 9, the Project Manager examines actual time reported, costs incurred and delivery dates. A final Cost/Benefit Analysis is calculated based on actual data (not estimated).

AUDIT CONSIDERATIONS

During the system audit the Systems Engineer may identify errors, omissions and severe weaknesses in the new system. In this event, the Systems Engineer may initiate a Work Request/Objective to document the modification/improvement. This will then go through the normal process of evaluation and priority calculation as provided for under the Enterprise Engineering Methodology (EEM).

A Project Management (PM) system can assemble all pertinent project data for analysis. This data can also be exported to other financial packages and spreadsheets for further analysis as required.

In addition to controlling information resources, the IRM Repository provides auditors with the "roadmap" for locating the control points in a system. Further, the "impact analysis" feature is useful for considering enhancements.

PHASE 9 DELIVERABLES

The formal deliverable resulting from Phase 9 is the "System Evaluation Report" which details the findings of the System Audit. This is then reviewed with management.

DESCRIPTION OF PHASE ACTIVITIES

Activity A - Perform System Audit

Systems Engineering reviews the original system design documentation to become familiar with how the system was intended to operate. Next, a survey of the system is conducted to identify how the system is actually operating. Any differences are identified and noted as part of the survey. If any problem areas are identified, Systems Engineering works with the Users to determine the problem and possible solutions. If necessary, a Work Request/ Objective is initiated to resolve the issue.

Activity B - Prepare System Evaluation

Project Management prepares the formal Phase 9 **"System Evaluation Report"** consisting of:

- Phase Cover Page

- System Evaluation - a textual analysis of the completed system (as prepared by Systems Engineering).

- Project Evaluation - a textual analysis of the completed project. This should include sections describing "Lessons Learned" and the final "Cost/Benefit Analysis."

- O-O-M Estimate/Actual - detailing all project estimates, schedules and expenditures.

- Phase Review Checklist

Quality Assurance checks this report to assure conformance with standards. User Management is then provided a copy of the report for review, along with Information/Systems/Data/Enterprise Resource Management.

The next few pages include sample deliverables resulting from this phase.

SAMPLE PHASE 9 DELIVERABLES

OD-90400 SYSTEM EVALUATION "PRIDE"-ISEM

APR 14, 2005 RELEASE 1.2.0 PAGE 2.1

AS DEFINED ON PD-00132 ORDER PROCESSING SYSTEM

The new Trundle Order Processing System (TOPS), which was recently completed, has been reviewed in detail. A complete walkthru of the system and interviews with key users (Sales Management, and Order Processing) was performed. First, Sales Management expressed great enthusiasm and satisfaction with the new system. Miscalculation of generated data values, which plagued the old system, has been eliminated. The Order Processing clerks have indicated that the new system has alleviated a considerable amount of their time and processing. Communications between Order Processing and Production has greatly improved due to the new system and, consequently, orders are processed and filled in a much more timely manner. Further, customer relations has improved due to timely responsiveness to the status of their orders.

The only suggestions for improving the new system came from the Sales Manager who would like to study customer buying trends for forecasting purposes. This suggestion has been turned into a Work Request (MI) for consideration. Further, backup/recovery procedures of the system need to be improved and a mod/imp has been initiated for this purpose.

- Art Powell
Systems Engineering
April 13, 2005

* * * END OF OUTPUT * * *

SAMPLE PHASE 9 DELIVERABLES

OD-90392	PROJECT EVALUATION	"PRIDE"-ISEM
APR 14, 2005	RELEASE 1.2.0	PAGE 2.1

AS DEFINED ON PD-00132 ORDER PROCESSING SYSTEM

The TOPS Order Processing System project was recently concluded. The new system was installed on time with a minimal overrun of costs.

Although the project was successfully completed, the learning curve used to implement the methodology resulted in a modest 3% overrun ($4,550) from the Phase 1 Order-of-Magnitude Estimate. The project took less than eight (8) elapsed months to perform at a total cost of $145,750.00. Consult the attached Project Management reports for details on estimated versus actual figures.

LESSONS LEARNED

1. Programming standards need to be enhanced, particularly in the areas of coding and testing.

2. User cooperation and participation significantly expedited project development for which we give our thanks.

3. Estimating guidelines need to be updated to reflect the results produced from this project.

- Ernest Bilko
  Project Manager
  April 14, 2005

* * * END OF OUTPUT * * *

END SECTION

Section 19

"PRIDE"-DBEM DATA BASE ENGINEERING METHODOLOGY

A Data Base is all of the data needed to support the information requirements of an enterprise, regardless of where used or how stored. By this definition, all companies have a data base; some are managed, most are not.
- Bryce's Law

The "PRIDE"-Data Base Engineering Methodology (DBEM) is an important part of an overall philosophy of Information Resource Management (IRM) as defined by "PRIDE". This involves the development and control over all of the resources required to produce information. Whereas the "PRIDE"-Enterprise Engineering Methodology (EEM) is principally concerned with developing an Enterprise Information Strategy, methodologies such as DBEM and the "PRIDE"-Information Systems Engineering Methodology (ISEM) are concerned with actually creating the data and system resources needed to produce information.

The intent of any of the "PRIDE" methodologies, including DBEM, is to define the business environment as to "Who" is to perform "What," "When," "Where" and "Why" (the "5-W's"). As a result, it is used to convert a heterogeneous operating environment into a homogeneous environment. This improves communications and promotes cooperation and teamwork throughout an enterprise. Better organization and discipline also enhances the ability to build quality products and make effective use of resources. In addition, to the 5-W's, the methodology provides "How" to perform the work by providing a variety of techniques and tools deployed throughout the methodology. A methodology, therefore, resembles an assembly line where work is performed in manageable stages.

DBEM is a generic and universally applicable approach for building any type of data base, regardless of industry, type of application, software language/ technique, or data base tool. A Data Base Management Systems (DBMS) is not a prerequisite for DBEM. The methodology is based on tried and proven approaches that are so fundamental to sound data base design that tailoring to individual development requirements is not only unnecessary but highly undesirable.

CONCEPTS & PHILOSOPHIES

INTRODUCTION

Data is one of the resources needed to produce information. This implies that Data exhibits distinctively different characteristics than Information. Information represents the intelligence or insight needed to support business actions and/or decisions. A data element by itself is meaningless. It is used to define the facts and events about a business. Data identifies and describes the objects of importance to the enterprise, such as products, orders, customers, vendors, parts, billings, payments, shipments, etc. It is also used for quantitative purposes in measurements and calculations. A single information requirement, thereby, represents an assemblage of these business facts presented in a specific context and time frame. In this respect, data represents the raw material needed to produce information. Obviously, one data element can support many information requirements. Because of this, it is necessary to manage data like any other resource.

Data is the binding force behind information systems. The only way systems communicate, either internally or externally to other systems, is through data. For this reason alone, data must be controlled to promote sharing between information systems.

The management of data as a resource begins with a corporate attitude and disposition, not with an elaborate set of tools. Data Resource Management requires the same perspective as managing the parts department of a manufacturing company. The objective is twofold: to classify and standardize resources so they may be shared by multiple applications, and; to control the collection, storage and distribution of resources to minimize overhead. Both are concerned with the efficient and cost effective use of resources.

Under DBEM, a data base is defined as all the data required to produce information, regardless of where the data is used or how it is stored. From this perspective, all companies have a data base. In fact, the day a company begins to conduct business is the day its systems and data base are born. Both evolve with the business over time as the company's information needs change.

Data Resource Management is another area where corporate management has abdicated its responsibilities to technicians who have turned a simple concept into an esoteric technical practice. As computer technology evolved, several physical file management techniques were introduced to manage data on the computer, most notably the Data Base Management System (DBMS). The intent of the DBMS is to physically store and retrieve data for use in computer programs. In fact the term "DBMS" is a misnomer since it only deals with data on a direct access device (disk). It does not deal with data on other devices, such as tape files, card files, manual files, etc. Therefore, it does not manage the entire data base, only a portion of it. Nor does it do any logical file management which is more important than the physical file management.

Although the DBMS was originally designed to permit the sharing of data among applications, this has seldom been implemented. Due to a lack of management discipline, the DBMS is one of the most abused and misapplied products in the industry. It is typically used as nothing more than an elegant file access method, not as a tool for integrating systems.

What this points out is that companies have been taking a tool oriented or physical approach to managing data. Despite the considerable investment in DBMS technology over the last 25 years, very few companies have realized a managed data base environment. Why? Primarily due to management's failure to recognize and treat data as a reusable resource. Data Resource Management is a "materials management" issue, nothing more, nothing less.

Imagine a manufacturing company without a materials management function. Under this scenario, engineers would design products without consideration for the other products marketed by the company. Each product would be designed with a unique set of parts. Inevitably, many duplicate parts would be designed. Without some form of coordination, there would be significant overhead and waste from collecting and storing redundant parts. Also, because no formal control mechanism existed to track parts, implementing changes to parts consistently throughout a product line would be a haphazard endeavor.

This is exactly the situation that occurs during traditional systems development. Each analyst and programmer is permitted to design data bases unique to their application. The result: rampant data redundancy throughout the organization. There is a natural tendency for an analyst to do only what is best for their individual assignment, not necessarily what is best for all corporate applications. This is why a neutral third party is required, to coordinate and standardize data resources on a corporate basis. This is the mission of the Data Resource Management function.

DATA BASE CONSTRUCTS

Data resources can be organized into a generic and universal structure. The basic building block is the data element itself, the representation of an individual fact or an event. A collection of one or more data elements is a record and one or more records make up a file. As previously mentioned, a data base represents all of the data used to produce information, regardless of where used or how stored.

BASIC CONSTRUCTS

All data resources are structured in this generic manner. Terms such as "schema," "sub-schema," "segments," "tracks," "cylinders," "sectors," "tables," "arrays," "tuples," "data stores," etc., (all of which deal with particular computer techniques and tools), can all be translated into the basic constructs mentioned above.

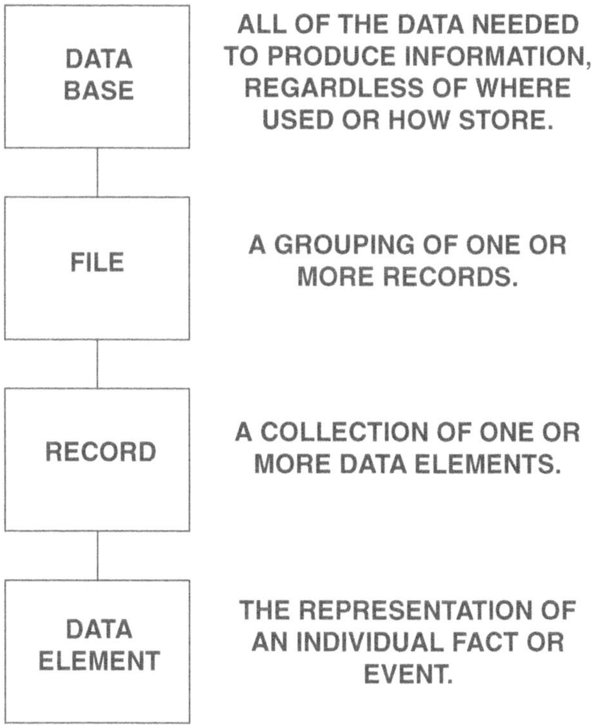

ALL DATA STRUCTURES CONFORM TO THIS MODEL.

The organization of data serves two purposes: one is to logically describe the "objects" used to manage and operate the business, and; to express how data will be physically stored. The differences between logical and physical are substantial; there will not necessarily be a direct relationship between the two.

Physical files may differ considerably from logical files. Here, the file represents a particular way of physically storing data. Data may be physically stored in a variety of files, such as an indexed file, a "flat" file, a DBMS file, etc. Even manual files follow this model with the exception they also store inputs and/or outputs (both of which consist of records and data elements).

Unlike the logical file that is organized according to a unique data identifier ("primary basic grouping"), the physical file does not require any specific organization and can use any sort/access key desired. Ultimately, it depends on the file management technique or tool being used.

The logical view of data is the basis for all physical data base design, regardless of the file management technique or tool selected. The physical files must ultimately carry out the intentions of the logical files in terms of what data must be stored, the dependencies between data, and volume. As a matter of fact, all DBMS packages can implement these logical views, regardless of whether they have a hierarchical, network, relational, or object-oriented structure.

Again, there are substantial differences between logical and physical files. Perhaps the most noticeable difference is the logical file will remain relatively static while the physical file will change dynamically, based on advances in technology. One of the most important reasons for defining data resources logically is to seek data independence from the physical environment, thus allowing any physical implementation without disrupting systems.

OBJECT CONCEPT

In its simplest terms "objects" represent "things" used in the operation of an enterprise, such as Products,

LOGICAL/PHYSICAL CHARACTERISTICS

LOGICAL		*PHYSICAL*
REPRESENTS AN 'OBJECT' OF THE BUSINESS	FILE	REPRESENTS HOW DATA IS STORED
REPRESENTS A 'VIEW' OF AN OBJECT	RECORD	REPRESENTS AN AREA WITHIN A FILE
REPRESENTS AN INDIVIDUAL ELEMENT ABOUT AN OBJECT	DATA ELEMENT	REPRESENTS AN INDIVIDUAL ELEMENT WITHIN A RECORD

There is not necessarily a one-to-one relationship between logical and physical

LOGICAL TO PHYSICAL RELATIONSHIPS

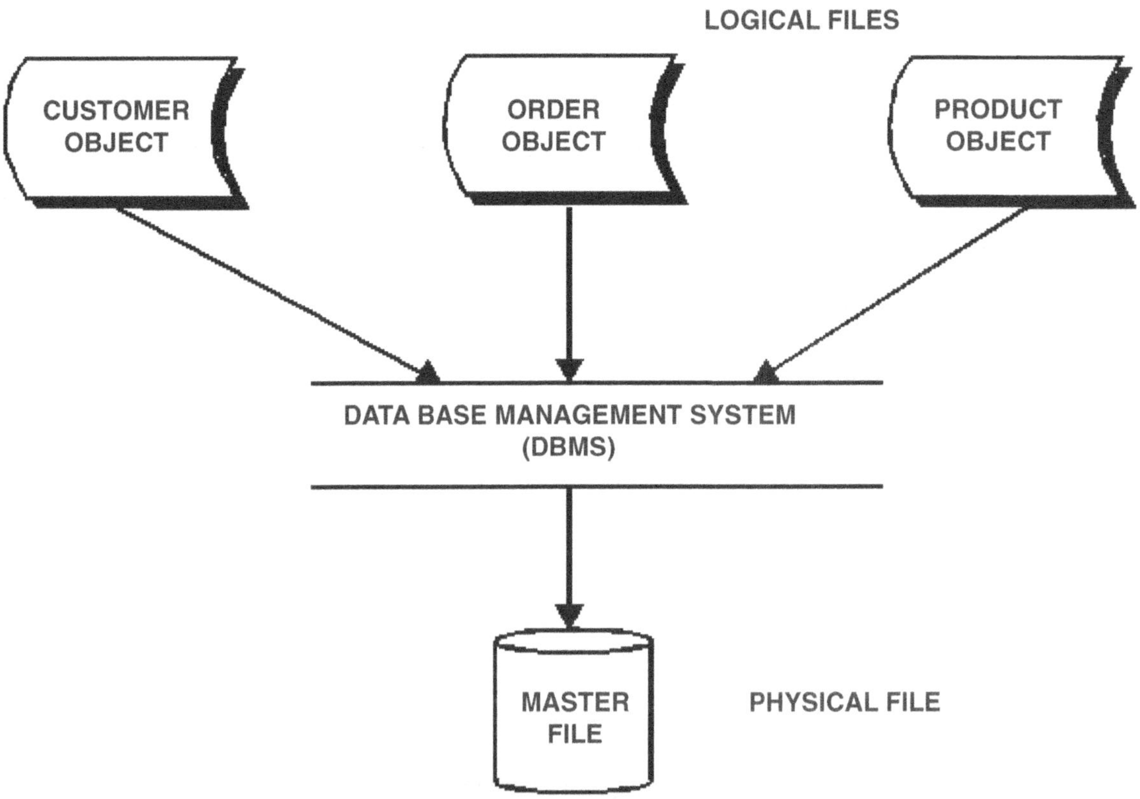

Parts, Customers, Employees, Vendors, Orders, Shipments, Billings, Payments, etc. Objects are initially identified when the business is defined under the Enterprise Engineering Methodology (EEM) and passed to DBEM for further definition.

Objects are uniquely identified by a single data element which is referred to as the "primary basic grouping." For example, "Customer Number" is used to uniquely identify a "Customer" object; "Shipment Number" identifies a "Shipment"; "Order Number" identifies an "Order"; etc. This means that **AN OBJECT IS AN OBJECT WHEN A UNIQUE DATA ELEMENT HAS BEEN ESTABLISHED TO IDENTIFY IT.** Further, the "primary basic grouping" becomes the common bond that relates data to one another. For example, all customer related data will be "grouped" around the "Customer Number," shipping related data around "Shipment Number," etc.

Not all numbers and codes will necessarily represent objects. For example, to a post office or shipping company, a Postal Territory is extremely important and requires management. As such, a "Zip/Postal Code" data element is created to uniquely identify each territory. However, to other companies, a Postal Territory is of little concern or importance; it is not an object they must deal with in the operation of their organization. In this situation, "Zip/Postal Code" is used for nothing more than descriptive purposes about "Customers," "Vendors," "Employees," etc. This means **AN OBJECT MUST BE RELATED TO AT LEAST ONE BUSINESS FUNCTION THAT IS RESPONSIBLE FOR ITS CONTROL. IF SUCH A BUSINESS FUNCTION DOES NOT EXIST, IT IS NOT AN OBJECT.**

As another test of the validity of an "object," consider how the data element is assigned (its "source"). If the data element is assigned from an external source, then in all likelihood the object is not valid. To illustrate, "Zip/Postal Code" is assigned by the Post Office (not by the average company). "Postal Territory," thereby, is a pertinent object for the Post Office, but not for the average company.

TYPES OF OBJECTS: FACTS AND EVENTS

There are basically two types of "objects": facts and events. Factual objects represent tangible things such as Products, Parts, Employees, and Vendors. In contrast, event related objects are intangible things that are date/time related, for example: Order, Shipment, Billing, Purchase, etc. An event represents some form of interaction between two or more factual objects. To illustrate, a "Customer" (fact), places an "Order" (event), for a "Product" (fact).

One nuance that distinguishes facts from events is the type of data associated with each. "Names" and "Locations" can be associated with facts; "Dates" and "Times" with events. For example, "Product Name" makes sense, "Product Date" does not. "Order Date" makes sense, "Order Name" does not.

Since logical files are used to model "objects," they contain only primary data elements to identify, describe and quantify the object. Generated data is not included in logical files since it can be generated based on the primary values. (This is another example of the differences between logical and physical files; in some situations, it is practical to store generated data elements).

UNDERSTANDING "VIEWS"

An "object" is divided into "views" to represent different perspectives about an object. Data elements could conceivably be grouped into one large record, but this would not provide any significant insight into objects and data. Grouping data into distinctly separate views, simplifies our perspectives on data. The objective is to uniquely identify an occurrence of data so that it is not confused with another (for example, "Quantity Shipped" versus "Quantity Ordered.") There are three types of "views" (logical records) associated with "objects":

1. **IDENTIFICATION VIEW** - normally consists of a control number or code to identify the object followed by a name or date; for example:

```
EMPLOYEE OBJECT - Employee Number & Name
PURCHASE OBJECT - Purchase Order Number & Date
```

There may be additional data elements used in this view. However, the point is, all objects have one identification view, regardless if they are facts or events. This implies that if the Identification View is deleted, then the other views in the objects are also deleted, as well as pertinent relationship views in other objects.

2. **CHARACTERISTICS VIEW** - consists of data that is not used to identify a separate object, nor create a relationship to other objects. Instead, it is used to describe the internal characteristics of an object; for example, a "Customer" object may have two types

of characteristic views:

ADDRESS - Including "Customer Number," "Type Address" (a code to identify an address type for Shipping/Billing/Mailing purposes), "Address," "City," "State," "Zip/Postal Code," and "Country."

CONTACT - Including "Customer Number," "Contact Number" (to identify each customer contact), "Name," "Telephone Extension," "Internal Mail Code/Drop."

This implies that an address and a contact are not strong enough to represent separate objects by themselves. Rather, they are describing various aspects of the customer object.

An Object can have one or more of these views (or none). Factual objects will typically have Characteristics Views; Event related objects normally do not.

3. **RELATIONSHIP VIEW** - consists of data that is used to establish relationships between objects. For example, in a Shipping Record, there may be data elements to relate "Quantity Shipped" to a Shipment ("Shipment Number"), to a Product ("Product Number"), and to a Customer ("Customer Number"). This implies relationships between..

```
CUSTOMER ------ SHIPMENT ------ PRODUCT
```

Event related objects will have at least one Relationship View; Factual objects normally have none. There cannot be more than two facts associated with a single event/view.

The relationship view connects with identification views (not characteristic views) through record-to-record relationships. To illustrate:

These relationships require clarification to describe the nature of each relationship, as denoted by the arrow notation. In the example above, a Customer may have many Shipments, but a Shipment pertains to a specific Customer (this is a one-to-many relationship (1:M)). Also, a Shipment may consist of many Products, and a Product can be used in many Shipments (many-to-many relationship (M:M)).

These relationships and the basic grouping of the view establish the constraints of the data base and are essential in physical data base design.

BASIC GROUPING CONCEPT

What distinguishes the different views of an object is the "basic grouping" of the logical record. The term "basic grouping" refers to the indicative data elements used to uniquely identify a view. It also represents a dependency between data elements in a particular context (it is how they are "grouped" into separate views). For example, in a Customer Object, it is used to segregate address data, from credit data, from customer contact data, etc. Perhaps it is easier to think of the "basic grouping" as the key to a logical record (not a physical record). Because the intent is to uniquely identify data, the "basic grouping" consists only of "Indicative" data (not "Descriptive" or "Quantitative").

There may be up to two parts in a single basic grouping:

PRIMARY BASIC GROUPING - Since views are used to describe objects, they must all be defined with the one data element used to uniquely identify the overall object; this will be a primary/indicative/object-oriented data element. The Primary Basic Grouping will be the basis to sort views into objects (e.g., all Product related Records will be put in the Product File, all Customer related Records in the Customer File, etc.).

SECONDARY KEYS are used to either distinguish Characteristic Views or Relationship Views.

```
   CUSTOMER                      SHIPMENT                    PRODUCT
IDENTIFICATION <---------->> RELATIONSHIP <<---------->> IDENTIFICATION
    RECORD                        RECORD                      RECORD

              (1:M)                          (M:M)
```

For Relationship Views where it is necessary to relate one object to another (such as an Order to a Customer and to a Product) additional primary/indicative/object-oriented data elements may be added to the basic grouping. These data elements are sometimes referred to as the "Foreign Keys" to another object.

The concept of *"Referential Integrity"* (as commonly referred to in the industry) is concerned with the logical consistency between views or records. It requires that every occurrence of a foreign key has a corresponding occurrence as a primary basic grouping in an identification view. As in the "Customer/Shipment/Product" example previously mentioned, "Customer Number" and "Product Number" can only be used as a Secondary Key as long as they are also used as a Primary Basic Grouping in other identification views.

As an aside, since a "Relationship View" typically applies to event related objects, the secondary key should normally consist of object-oriented identifiers related to factual objects. This will be used to bridge factual objects through the event object (see "Other Object Considerations").

Characteristic Views also require Secondary Keys. However, the intent here is not to establish a relationship to another object, but rather to establish a separate view within an object (an internal relationship). Under this approach, "view identifiers" are used to segregate data into separate records. For example, "Type Address" is used to distinguish address related data from other data.

The sequencing of the basic grouping is extremely important for three reasons:

1. It is the principal criteria for combining logical records into logical files (based on the primary basic grouping data element).

2. It is the principal criteria for establishing relationships between logical records (based on secondary keys).

3. As we will see when we discuss "Descriptive" and "Quantitative" data, the "basic grouping" gives meaning to the data.

Because of its importance, the "basic grouping" must be established in a prescribed format. Unlike a key in a physical record, which can be set to whatever is convenient, the basic grouping must be assigned as:

1. The Primary Basic Grouping first.

2. Secondary Keys last.

OTHER OBJECT CONSIDERATIONS

Factual Objects will typically relate to other Factual Objects through Events. This is quite common and natural. As facts or events are being defined, the analyst should challenge what peripheral facts and events are involved.

The "PRIDE" concept of "Objects" is an advanced refinement over the "Entity/Relationship" model as commonly referred to in the industry. The differences are subtle but significant:

1. "Entities" typically represent only "facts" and have trouble depicting "event" relationships. Under DBEM, an "event" is just another "object," thus simplifying the establishment of relationships.

FACTS & EVENTS

OBJECT TYPE	*DIFFERENCE*	*EXAMPLES*	*ASSOCIATED WITH*
FACT	TANGIBLE	PRODUCT EMPLOYEE PART	NAME LOCATION
EVENT	INTANGIBLE	ORDER PURCHASE BILLING	DATE TIME

EVENTS BRIDGE FACTS

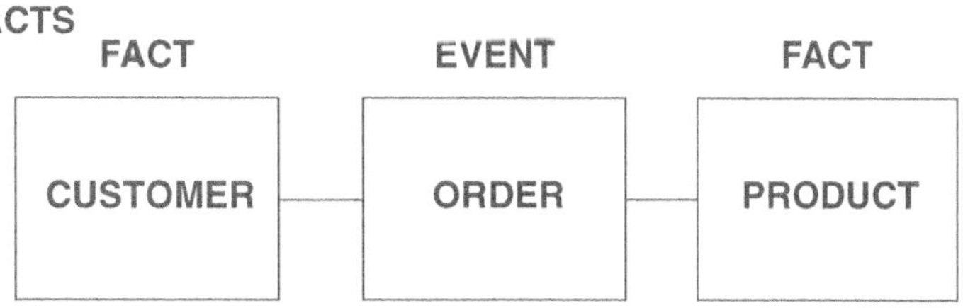

2. "Entities" are typically scattered with no cohesive bond to unite common entities. For example, there is no one point where a global view of a customer can be found. In contrast, the "Object" concept uses the primary basic grouping of the various views to group compatible logical records into a single logical file, thus providing a total picture of the object.

Coincidentally, the "PRIDE" concept of "Objects" is compatible with what the industry refers to as "object-oriented" data bases and programming. These are techniques that were developed independent of DBEM, yet are complementary.

APPLICATION VERSUS ENTERPRISE

It would be easy to say there are just two types of data base models, logical and physical. However, there is another perspective that adds a different dimension to this, and that is how data is viewed from an "enterprise" versus an "application" perspective.

An "application" view refers to the data used in a specific system. In terms of the logical model, it represents the "local" data used to describe objects for a particular Information System. It represents only those data elements required to satisfy the information needs for a particular application. Obviously, this will not necessarily be the "global" view of the object, which is the intent of the "enterprise" view. In other words, the "application" view will usually be a subset of the "enterprise" view of data.

"APPLICATION VERSUS ENTERPRISE"

APPLICATION VIEW OF A CUSTOMER	ENTERPRISE VIEW OF A CUSTOMER
Customer Number	Customer Number
Name	Name
Credit Rating	Credit Rating
	Contact Number
	Title
	Telephone
	Address Code
	Address
	City
	State/Province
	Zip/Postal Code

The "enterprise" view represents a complete picture of the object, with all of the data required to satisfy all applications, not just one. Under this arrangement, there may be multiple "application" views of objects, but only one "enterprise" view of an object. In fact, it is quite common to have many different "application" views of an object. One system may require certain data elements about a customer object while another requires a totally different set of data elements to describe a customer. These legitimately separate views of the customer, as defined by Systems Engineering during design, are coordinated through the enterprise view of the customer as controlled by the Data Engineering function.

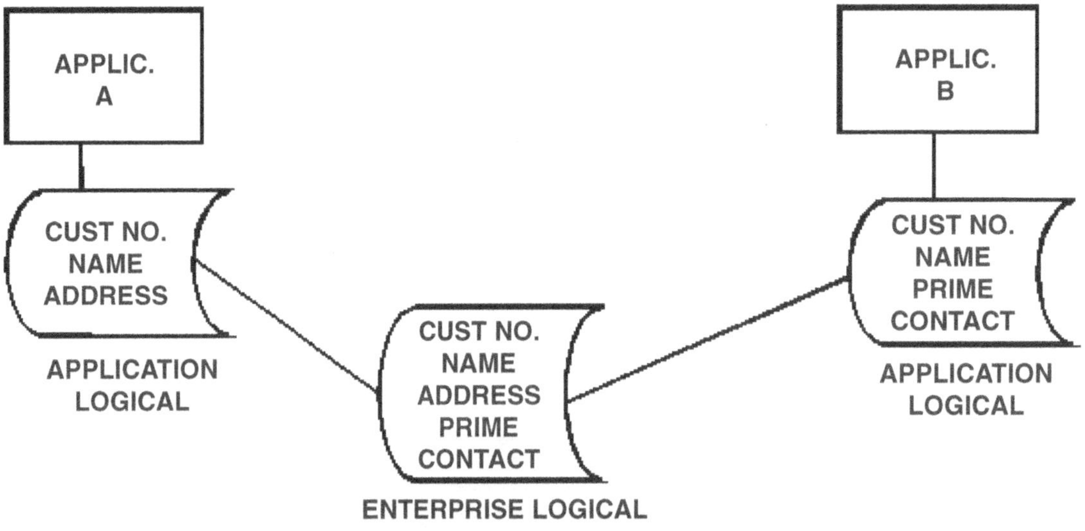

When a system is designed into sub-systems with logical files, the "enterprise" data base is adjusted to accommodate the "application" data base. If the objects encountered in the system are new to the enterprise, then new enterprise views must be defined. Initially, the application and enterprise views of an object are identical. As new applications are introduced with different views of the same object, then the enterprise view is modified accordingly by Data Engineering.

This application/enterprise relationship highlights the fact that data base design is an evolutionary process. Other data base design techniques typically take a "revolutionary" approach by trying to identify all of the data requirements for the entire company at one time. Obviously, the problem with this approach is that it becomes an enormous and unmanageable data base design project with questionable results. Whereas the evolutionary approach naturally synchronizes the data base with all of the applications, the revolutionary approach develops a data base that will not necessarily match the applications.

———→ EVOLUTIONARY DESIGN ———→

BUSINESS INFO. REQUIREMENTS	SYSTEM APPLICATION	CORPORATE DATA BASE

←——— REVOLUTIONARY DESIGN ←———

Under the evolutionary approach, the corporate data base will expand and contract naturally as the business and applications change. Consequently, excessive or unnecessary data definitions will be avoided.

THE FOUR DATA BASE MODELS

The variables of logical versus physical and application versus enterprise results in four data base models:

• The **Application Logical Data Base Model (ALDBM)** represents all of the primary data elements needed to satisfy the information requirements of a single application. In other words, all of the data needed to describe the objects pertinent to a given information system. The ALDBM defines the logical files used in a single system. It also represents a subset of the Enterprise Logical Data Base Model.

• The **Enterprise Logical Data Base Model (ELDBM)** represents the primary data elements used to describe all objects in an enterprise, not just that data used in a single system. It represents all logical files in the corporate data base.

• The **Enterprise Physical Data Base Model (EPDBM)** represents how the data in the ELDBM is physically stored in files. The corporate data base can be either centralized or distributed. A variety of file management techniques can be used to store the data, e.g., computer files, manual files, etc. The EPDBM, therefore, defines all of the physical files in the corporate data base.

• The **Application Physical Data Base Model (APDBM)** represents subsets of the EPDBM used to fulfill a specific application. It satisfies the data requirements of the ALDBM and denotes the physical files used in the system.

As can be readily seen, there are corresponding relationships between the four data base models; the enterprise view must satisfy the application view and the physical must implement the logical. This type of data resource definition provides for the integrity of the data base by assuring that only those resources required to serve changing business information needs are maintained. In summary, it assures the corporate data base will be correctly synchronized with all of the various applications.

The relationships between the four data base models can be rather extensive. Diagrams using boxes and arrows are fine for expressing simple relationships, but this is seldom the case. Instead, a set of matrices with horizontal rows and vertical columns is a much more convenient and simpler approach for expressing these relationships. These data base relationship matrices are similar in intent and format to those mentioned in Enterprise Engineering. Data base matrices are used to express relationships between:

- Application Logical Records
- Enterprise Logical Records
- Enterprise Physical Records
- Application Physical Records
- Application Logical Records to Enterprise Logical Records
- Enterprise Logical Records to Enterprise Physical Records
- Enterprise Physical Records to Application Physical Records
- Application Logical Records to Application Physical Records

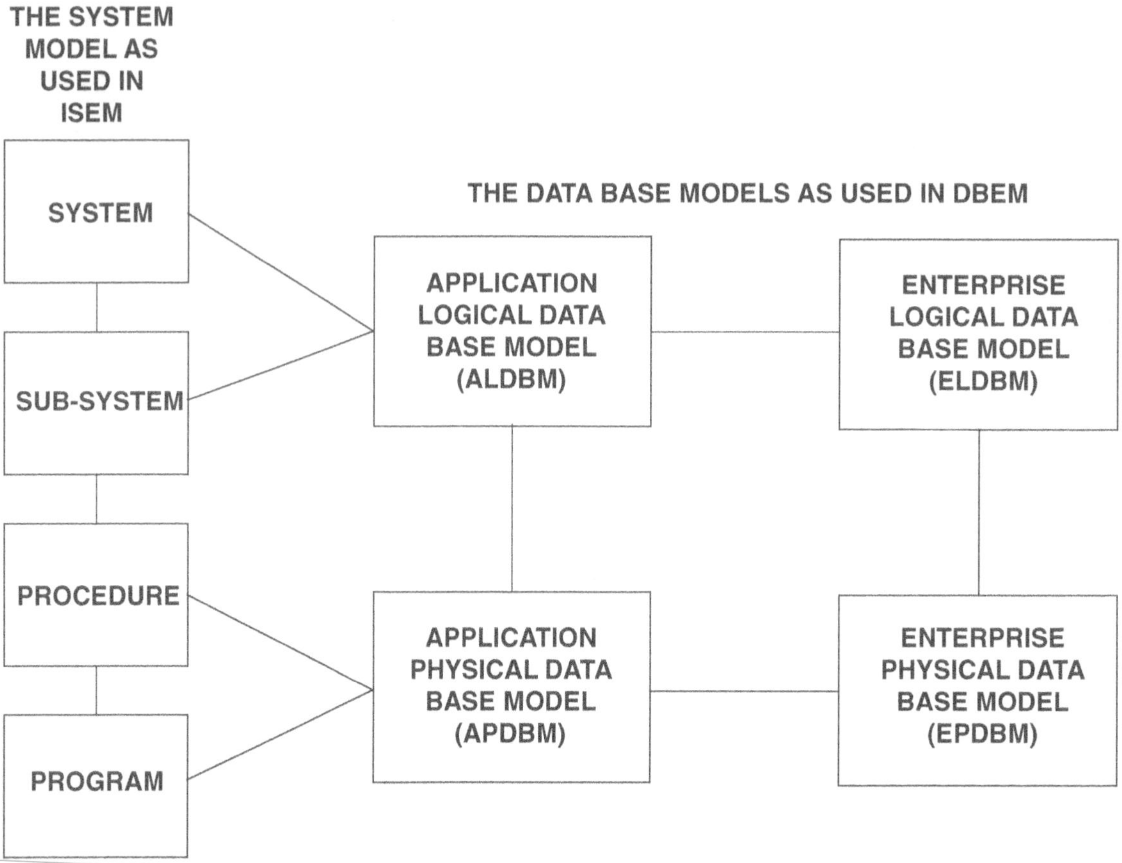

One by-product of data base modeling is that it provides for a complete description of how data is used throughout a company, regardless of where used or how stored. It tracks where data is collected, stored and retrieved. This type of inventory control implements one of the basic missions of resource management.

DATA AS A RESOURCE

Like a component in a product, data is a reusable resource that can be shared between applications. To avoid redundancy and to verify the integrity of the component, data must be defined with a high degree of precision. Otherwise, it will be virtually impossible to check for duplication. In order to maintain its "cleanliness," data must be specified and classified in the same manner as any other part in a product.

How data is defined will dictate how it is used. If management cannot see what constitutes an element of data or how it is derived, its validity, currency and accuracy will be highly suspect. Ultimately, users will not be able to trust it. They will not be able to truly tell if they can base decisions and actions on information created by using data they do not understand or know its source. This type of specification traditionally is defaulted to obscure and inflexible program source code. If it is not visible, it is not reusable. This is why data elements like "Net Pay" are difficult to maintain, because their logical calculations reside within source code and are not maintained separately. The definition of data should be used as the specification for programming. A program, therefore, is nothing more than a mechanism to carry out intentions of how the data should be manipulated and processed.

LOGICAL VERSUS PHYSICAL DATA

There is more to defining a data element than providing a cryptic program label, yet this is all that is commonly considered by the average programmer. This is far too vague and inconsistent to assure any precision in data definition. There are actually two aspects to be considered when defining data: its logical meaning and its physical implementation.

A data element can have only one logical definition but can have one or more physical implementations. If a data element is an expression of a single fact or an event, it is important that it be explicitly defined so it will not be confused with another. If there is a genuine difference in interpretation of the meaning of data between users, then more than one data element is involved.

Although standardization of data's physical characteristics is an objective, there can be multiple physical representations of data. For example, there can be several legitimate ways to represent "Calculated Delivery Date":

December 11, 2004
11 DEC, 2004
DEC-11-04

In this example, the data element has a singular logical definition, *"The calculated date when a delivery is due."* All that differs is how the data element is physically represented. What this points out is the physical characteristics of data may vary from one application to another.

Obviously there are significant differences between how a data element is logically and physically defined. Systems Engineers and Data Engineers primarily deal with the logical. Software Engineers and Data Base Administrators work with the physical definition. Data Resource Management must govern both.

THE ANATOMY OF A DATA ELEMENT

A well defined data description contains vital intelligence for establishing relationships between data elements, and for constructing logical records and files. Superficial or inaccurate data definitions will produce erroneous results. For example, logical data base design is totally dependent on the precise and accurate definition of data. The objective when defining data, therefore, is to prove that the data element is unique and non-redundant, thus promoting sharing and re-using resources. This, in turn, leads to system integration.

As mentioned, there are two aspects to data definition: Logical and Physical. This narrative will describe both with emphasis on Logical Characteristics.

LOGICAL CHARACTERISTICS

A. NAME AND DEFINITION

Each data element has a proper name that it is commonly referred to in business (not necessarily how it is used in programming), such as "Employee Number," "Address," or "Net Pay." To eliminate confusion, the name should not be redundant with another item.

The data element also has a textual description expressed in a Webster or Oxford style dictionary format. The description provides the meaning of the data element and should be expressed in the terminology of the business.

B. TYPES OF DATA

There are three types of data elements: Indicative, Descriptive, and Quantitative.

INDICATIVE data is used to uniquely identify an object in part or in full. "Uniqueness" is an inherent property of indicative data so that it can be used to clearly differentiate occurrences of an object. This is why control numbers and codes are typically used as indicative data, as opposed to names. Names can be too vague. For example, there may be more than one employee named "John Smith." Without some form of qualifier, it is virtually impossible to distinguish one "John Smith" from another. Consequently, an "Employee Number" is assigned to uniquely identify each employee.

In most major corporations, "names" are treated as descriptive data due to the volume of occurrences (many employees, many products, many customers, many parts, etc.). However, in smaller enterprises, where there is not a high volume of occurrences, "names" are a much more effective means for controlling occurrences. For example, at a small produce market, the "Produce" object is uniquely identified by name, not by number.

The point is, numbers and codes offer better control in a major business, yet in a smaller organization, names may be more practical and easier to use.

Indicative data is used to identify either a whole Object, or a View within an Object. The difference between the two is significant. Data elements such as "Part Number," "Product Number," "Shipment Number," and "Billing Number" are strong enough to identify a whole object by themselves (facts or events).

Other data elements may not be strong enough to represent a separate object and are subordinate to the object-oriented identifiers. For example, "Contact Number" may be used to represent an individual person within a "Customer." In this situation, a "Contact" object is not strong enough to be independent from a "Customer" (the company does not manage "Contacts," it manages "Customer Contacts" instead). Under this scenario, "Contact Number" is subordinate to "Customer Number" and, as such, is used to represent a view of a customer.

"View Identifiers" are typically encountered on "Characteristics Views" and are used in situations where there are multiple occurrences (repeating groups) of the same data element. As in the "Customer Contact" example, a customer may have many people within it to contact. To distinguish each person, a "Contact Number" is devised to uniquely identify each person. This is similar to a body of text where a "Line Number" is used to differentiate each line of text.

A "View Identifier" is not used on an "Identification View" of an object since the view implies a single occurrence of a data element, not multiple. For example, an Employee will only have one name, an Order will only have one date, etc.

A "View Identifier" is typically not used on a "Relationship View" since object-oriented identifiers should be strong enough to uniquely identify each occurrence of a data element.

DESCRIPTIVE data consists of alphanumeric characters that are not strong enough to identify an object, but convey important business facts about an object, such as names, addresses, text, codes, etc.

QUANTITATIVE data deals with numeric values that are either calculated or are calculable. Measurements and computations are typical examples: "Net-Pay," "Quantity Ordered," "Elapsed Time," "Percent of Gross," etc.

It is sometimes difficult to differentiate Quantitative data from Indicative data. Indicative data will often use numeric values for identification purposes, such as "Invoice Number," "Purchase Order Number," "Customer Number," etc. However, it would be a mistake to use these numbers for quantitative purposes (aside from counting the number of occurrences).

Descriptive data should be expressed in the simplest of terms, allowing their dependency to Indicative data (the basic grouping) to express their meaning. For example:

INDICATIVE DATA +	DESCRIPTIVE DATA =	MEANING
Customer Number	Name	Customer Name
Product Number	Name	Product Name
Vendor Number	Address	Vendor Address
Employee Number	Address	Employee Address

This approach results in a minimal number of primary data definitions. The alternative would be to define many names, many addresses, and many dates. Data elements such as "Employee Name" should be defined only in the absence of a control number/code (such as "Employee Number"); in this situation, "Employee Name" is indicative, not descriptive. Names should automatically trigger the analyst to consider the facts being represented, and the indicative data elements related to them.

This discussion reflects the fact that the nature of the business will dictate "data type." Not all numbers and codes will necessarily be indicative; it is based on the "object" or "view" being identified. This also highlights the fact that "type" refers to the logical nature of data exclusively; not how it will be physically used. After all, descriptive and quantitative data can be used as a physical sort/access key just as well as any indicative data element can.

C. FORMS OF DATA

Data comes in two forms: Primary and Generated.

PRIMARY data refers to data in its virgin state; as introduced to the system from an external source (such as a person or department). "Source" defines who is responsible for entering the data to a system, and who has ultimate authority for the definition of the data element. Depending on how well the data element is defined, it may have either one or many "sources." For example, "Customer Name" may have one source, the Customer Services area. Conversely, a generalized data element, such as "Name," may have many sources depending on circumstances.

GENERATED data refers to data that relies on other data elements in order to produce the necessary result. This type of data can involve elaborate calculations and algorithms (e.g., DD-1 + DD-2 = DD-3). "Net Pay," "Balance Amount" and "Percent Complete" are some examples of calculated data.

Another form of generated data is a "Group" data item that represents a specific string of data elements in a assigned format. For example, "Credit Card Number" typically consists of "Financial Institution ID," "Bank Region Number," "Bank Branch Office ID," and "Account Number." There are many other examples of "group" data: such as telephone numbers, product identification codes, public utility account numbers, etc.

It is a common misconception that group data elements should be used for basic groupings in logical records; THEY SHOULD NOT! Group data is used as a convenient means to describe dependencies between primary data elements. As such, a group data element provides tremendous insight into objects and views. For example, consider the objects and views associated with "Telephone Number."

Observe the dependencies between the three views in the exhibit below. Each has an impact on the others. Should the first view be deleted, the second and third views will also be deleted. From this perspective, the basic grouping defines dependencies and eliminates the problem of multiple occurrences.

Data elements such as "Telephone Number" and "Credit Card Number" should only be defined as group items if they truly represent a concatenation of indicative data elements representing objects. For example, "Telephone Number" is a valid group item to identify a "Communications Area" and its views for a telephone company. But if "Communication Area" is not a pertinent object to your business, there is little point in defining it as a group item. Instead, it is a simple primary value.

Group data may not be pertinent for logical records and files, but in certain situations it can be a convenient sort/access key for physical files.

D. DATA DEPENDENCIES

Aside from data-to-data relationships required to produce generated data, another form of data dependency is required when defining primary data. The purpose for defining this dependency is to supply additional meaning to subordinate descriptive and quantitative data definitions. For example, it is reasonable to assume that there is a relationship between a "Customer Number" data element and a "Name" data element - "Name" DEPENDS on "Customer Number" to give it meaning ("Customer Name").

HOW A PRIMARY DATA ELEMENT IS DEFINED IS BASED ON THE DATA ELEMENTS IT DEPENDS ON.

Logical data dependencies must be explicitly defined. This intelligence will be required for creating logical records ("Views").

A data element may require a dependency on more than one data element. For example, a "Quantity Ordered," may depend on "Order Number," "Product Number," and "Customer Number" to give it meaning and uniquely identify a single occurrence of "Quantity Ordered."

Descriptive and quantitative data have dependencies with indicative data elements:

GROUP DATA ELEMENT

TELEPHONE NUMBER = AREA CODE + EXCHANGE + ACCOUNT
727/786-4567

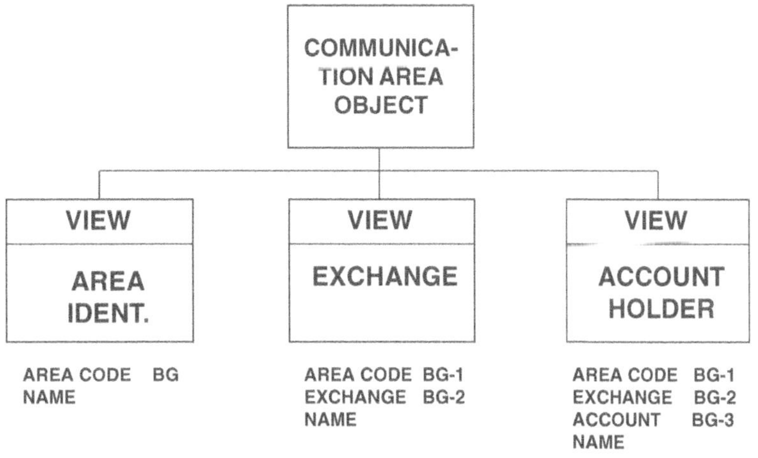

• They will require one or more "object" type data elements.

• They may require one "view identifier" type data element to uniquely identify an occurrence of data.

Dependencies between indicative data elements should also be established to reflect how "view identifiers" depend on superior "object identifiers" (for example, "Contact Number" to "Customer Number").

Dependencies between primary indicative data elements that are prohibited include:

1. An "object" to "object" data element relationship is prohibited since this will be defined through logical records.

2. A "view identifier" to "view identifier" data element relationship is prohibited since this would create an abnormal logical data base design.

E. PROBLEMS IN DATA DEFINITION

Defining data elements is not always easy. Quite often a data element's assignment may be different than its business purpose. For example, a financial institution may require "Mother's Maiden Name" from an account holder. In reality, the financial institution is not really interested in the mother as it is in establishing a unique "Security Password" to validate the account holder in case of an emergency. "Security Password," thereby is the data element, "Mother's Maiden Name" is how it is assigned. Another common example is "Social Security Number" as used in the United States. This is a number used by the federal government to identify each citizen for retirement benefits. Many companies will use the number to uniquely identify each employee as opposed to inventing a separate numbering convention. In this situation, the data element is actually "Employee Number," but the number is assigned by the U.S. Social Security Administration.

Another common problem is to create an indicative item that is bound to a physical input/output as opposed to an object. The classic examples here are "Check Number" and "Deposit Slip Number" as used in banking. In reality, a check or deposit slip are physical inputs for recording a "Debit" or "Credit." Obviously, there are other ways of creating "Debits" and "Credits," particularly with electronic banking (automatic funds transfer for example). Under this scenario, checks and deposit slips are not used; therefore, "Check Number" and "Deposit Slip Number" are invalid. The actual data elements are "Debit Number" and "Credit Number."

PHYSICAL CHARACTERISTICS

The physical definition of data is perhaps easier to comprehend by the average programmer and data base administrator. It includes such things as:

• **Length** - defines the maximum number of characters that can be assigned to a data element.

• **Class** - defines the type of characters used to express a data element, e.g., alphabetic, numeric, alphanumeric, signed numeric, etc.

• **Justification** - defines the alignment of data within a field when the number of characters is less than the length of the receiving field, e.g., left, right, around the decimal point.

• **Fill Character** - defines the character to complete a field when the data item is shorter than the maximum length, e.g., blank, zero, X, etc.

• **Void Character** - defines the character to be used when a data item's value is unknown or non-existent, e.g., blank, zero, X, etc.

• **Unit of Measure** - defines the representation of numeric data, e.g., area, volume, weight, length, time, energy rate, money, etc.

• **Precision** - defines for numeric data the number of significant digits in a number.

• **Scale** - defines for numeric data the placement of the decimal point.

• **Base** - defines for numeric data the radix used for representing the number in programming, e.g., decimal, binary, octal, hexadecimal, etc.

• **Mode** - defines the format (and type) of a data element for programming, e.g., fixed point integer, floating point, double precision floating point, complex, binary, packed decimal, polar coordinates, etc.

• **Picture** - defines how the data element is expressed for programming. It is typically based on length, class, precision and scale.

• **Program Label** - defines the proper name of the

data element as it will be referred to in a programming language, such as COBOL, FORTRAN, PL/1, Assembler, C, Pascal, ADA, etc. One data element may have many program labels.

• **Validation Rules** - defines specific values which the data element may assume. For example, Yes/No, specific codes or numbers to be used, editing/syntactical rules, etc.

Although Systems Engineering is primarily concerned with logical specifications, they will also provide assistance in gathering physical specifications, particularly when they format inputs and outputs.

DATA TAXONOMY & DOMAINS

The management of any resource requires the development of a classification system. Financial resources are typically arranged according to a chart of accounts; material and human resources are categorized by type. In science, everything from chemical elements to the animal kingdom are organized according to a class structure. There obviously is a purpose to uniquely identify common elements; to provide for the ability to distinguish one from another, and eliminate redundancy. In all instances, classification is based on the inherent characteristics of the element.

A Data Taxonomy is a hierarchical structure that separates data into specific classes of data based on common characteristics. The taxonomy represents a convenient way to classify data to prove that it is unique and without redundancy. This includes both primary and generated data elements.

The lowest level in the classification hierarchy represents what is commonly referred to as the "domain" of a collection of data elements, one or more, with common characteristics. For example, "text" related data elements would be in one domain, "weights" in another, "percentages" in another, "monetary values" in another, etc.

The domain also defines the standard physical characteristics and values the data may assume. For example, we could establish that all "location" values are alphanumeric, left justified, with blank fill and void characters. In other words, data elements such as "Address," "City," and "State" should assume these physical characteristics for consistency.

If a data element does not have the standard logical and physical characteristics, it must belong to another "domain." In the situation where a data element may have only one logical definition, but multiple physical definitions, its primary physical definition must first conform to the Domain standards before it

CLASSIFYING DATA
The objective is to eliminate redundancies and promote sharing/integration

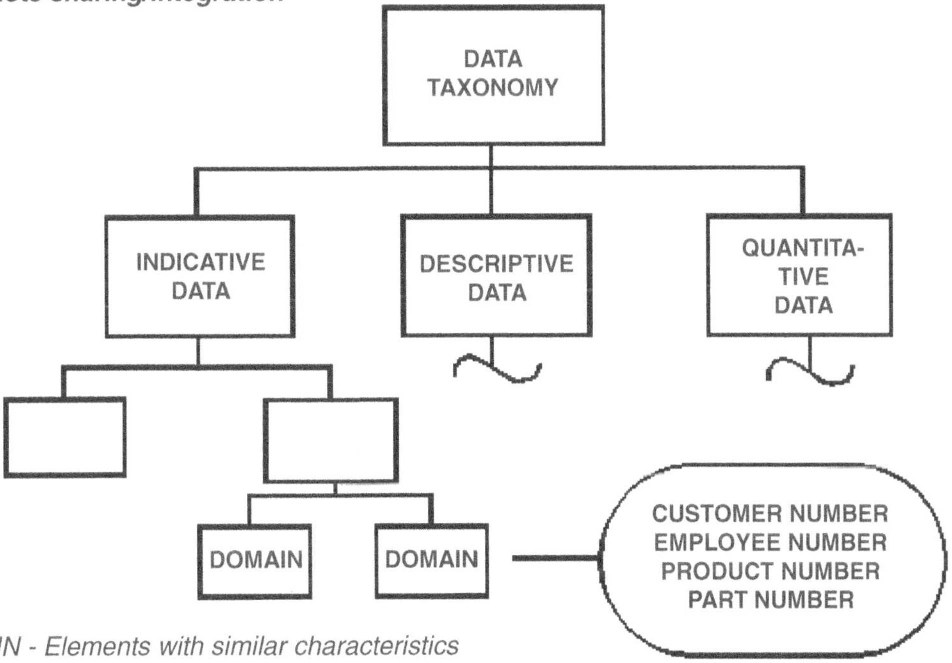

DOMAIN - Elements with similar characteristics

can be deviated from in an application record. In other words, the primary physical representation of "Unit Cost" is expressed as an eight character numeric to conform to the "currency" domain. However, in one application, a user desires the data element be expressed as a ten character numeric. It is the same logical data element with just another form of physical expression.

With a classification system in place, data elements can then be uniquely and consistently defined. When this is done, we then have a basis for checking data redundancy. Also, when a data element has been properly specified in this manner, it becomes rather simple to locate it again for use in other applications.

Classifying data helps to fulfill one of the the major objectives of Data Resource Management: to eliminate redundancy and promote the re-use of resources in applications.

SUMMARY OF MAJOR DBEM CONCEPTS

• DATA IS A RESOURCE THAT MUST BE MANAGED AND CONTROLLED LIKE ANY OTHER RESOURCE. SYSTEMS COMMUNICATE THROUGH DATA.

• THE MISSION OF DATA RESOURCE MANAGEMENT IS TO STANDARDIZE AND CONTROL DATA RESOURCES IN THE MOST COST-EFFECTIVE MEANS POSSIBLE.

• BASIC CONSTRUCTS - Data Base, Files, Records, Data Elements

	LOGICAL	PHYSICAL
FILE	Represents an "object" of the business.	Represents how data is stored.
RECORD	Represents views of the "object."	Represents an area within the File.
DATA ELEMENT	Represents an individual element about an object.	Represents an individual element within a record.

There is not necessarily a one-to-one relationship between logical files and physical files. However, the logical is used to design the physical.

• THERE ARE TWO TYPES OF "OBJECTS": Facts and Events.

-FACTS are Name/Location oriented (tangible things).

- EVENTS are Date/Time oriented (intangible actions) and represent some form of interaction between two or more factual objects.

• AN OBJECT HAS ONE DATA ELEMENT USED TO UNIQUELY IDENTIFY THE OVERALL OBJECT, THIS IS REFERRED TO AS THE "Primary Basic Grouping."

• A FACTUAL OBJECT TYPICALLY RELATES TO ANOTHER FACTUAL OBJECT THROUGH AN EVENT:

```
CUSTOMER ------- ORDER ------- PRODUCT
```

• THREE TYPES OF VIEWS WITHIN AN OBJECT:

- IDENTIFICATION VIEW - all objects will have one.

- CHARACTERISTIC VIEW - describes an object and is typically associated with a factual object.

- RELATIONSHIP VIEW - establishes a relationship between two or more objects. Will typically apply to event related objects.

ONLY PRIMARY DATA IS STORED IN A LOGICAL RECORD; GENERATED DATA CAN BE DERIVED FROM PRIMARY DATA.

• BASIC GROUPING: The key to a logical record. It is used...

- As the principal criteria for combining logical records into logical files (based on the primary basic grouping data element).

- As the principal criteria for establishing relationships between logical records (based on secondary keys).

- To give meaning to descriptive and quantitative data.

• THE ASSIGNMENT OF THE BASIC GROUPING INCLUDES:

- The Primary Basic Grouping; a primary/indicative data element used to identify an object in its entirety.

- Secondary Key, to either:

1. Establish a relationship to other objects (a Foreign Key); object-oriented data elements are thereby used.

2. To identify a specific view within an object (a qualifying key to note a single occurrence of data). A view-identifier data element is used.

• THERE ARE FOUR DATA BASE MODELS;

1. ALDBM - Application Logical Data Base Model - all of the primary data needed to satisfy the information requirements of a system.

2. ELDBM - Enterprise Logical Data Base Model - all of the primary data needed to satisfy all of the applications (the global view).

3. EPDBM - Enterprise Physical Data Base Model - represents how all corporate data is physically stored.

4. APDBM - Application Physical Data Base Model - represents how data for a single system is physically stored. It also represents a subset of the EPDBM.

• DATA HAS ONE LOGICAL DEFINITION, BUT CAN HAVE MORE THAN ONE PHYSICAL REPRESENTATION.

• TYPES OF DATA:

- INDICATIVE - to uniquely identify and control objects, in part or in full. This will include data elements to either identify a whole object or a single view.

- DESCRIPTIVE - to describe objects.

- QUANTITATIVE - numeric values used in calculations.

• FORMS OF DATA:

- PRIMARY - data assigned from a user area; outside a system.

- GENERATED - data derived from other data values; either from calculations or group (concatenated data).

• ONLY PRIMARY/INDICATIVE DATA CAN BE USED IN THE BASIC GROUPING OF A LOGICAL RECORD. Group data elements cannot.

• DATA TAXONOMY - a hierarchical structure used to classify data elements. The intent is to eliminate redundancy.

• DOMAIN - the lowest level in the Data Taxonomy. A collection of data elements exhibiting common characteristics.

DATA RESOURCE MANAGEMENT: THE FUNCTION

The scope of Data Resource Management is much more encompassing than most people envision. It represents a large investment by a company. This is partially the reason why few companies have succeeded in this area, they simply fail to comprehend the magnitude of the function and its importance.

One of the mistakes made when implementing Data Resource Management historically has been that the task has been delegated to technicians. The "tool approach" to managing data has emphasized the physical data base considerations but little has been accomplished in regard to the logical side. This is the principal reason why the proliferation of DBMS technology has been so widespread. Another symptom of the problem is the use of the term 'Data Base Administrator' which is clearly a technical task.

The management of data requires centralized control to coordinate the use of resources. This is not to suggest a centralized data base. A company could have a distributed data base spread throughout various locations. It simply means the function is best served through a focal point. This is no different than the function of materials management which is concerned with coordinating the use of all parts, regardless of where used or how stored.

A centralized Data Resource Management function can better promote and control the exchange of data between applications than a decentralized function which would only have a partial view of the corporate data base. Centralization would be able to maintain the data base models more effectively than if they were left to separate operating units. It would also be able to enforce data base design standards on a more consistent basis.

One of the more controversial subjects in Data Resource Management pertains to the "ownership" of data. There are those who suggest that data "belongs" to the various users or departments of a company. This is like saying money is the property of the sales or accounting departments, not the company. Data belongs to the enterprise as a whole and not to any single person or department. Of course, how data is accessed should be controlled and safeguarded, just as we would do with any other resource. This is a responsibility for Data Resource Management to perform.

The concepts of "end-user computing," "Data Mining" and the "Information Center" are totally dependent on the integrity of the data base. Without effective Data Resource Management, they would not be possible. Users could access erroneous or unauthorized data which could present serious financial and security problems to a company.

RESPONSIBILITIES OF DATA RESOURCE MANAGEMENT

There are essentially seven responsibilities associated with the Data Resource Management function:

1. To eliminate data definition redundancy.

This does not mean the elimination of duplicate data, only the elimination of duplicate definitions. In many instances, it may be more practical to physically store redundant data in various locations. This decision is based on what is effective for fulfilling the needs of an application.

Data should be defined properly one time and then re-used as often as there is an application need for it. This is accomplished by classifying data resources and controlling the four data base models.

2. To satisfy the data needs of all applications and promote data sharing.

Here again, data base is defined as all of the data in an organization required to produce information, regardless of where used or how stored. In any given organization, there is a finite number of primary data elements, not infinite. The logical corporate data base will keep expanding until this objective is reached. When this finally occurs, the company is in an enviable position. Users and the systems staff will be able to implement new information requirements simply by adjusting timing and creating new combinations of data.

3. To design the data base to be easy to maintain and modify.

The four integrated data base models provides invaluable assistance in this regard. They have the ability to expand and contract as the business and applications change. It also provides the means to scope and isolate problem areas in the corporate data base.

Mapping the four models can be best served by using matrices to express the extensive relationships.

4. To design the physical data base in the most efficient and cost effective means possible.

Data Resource Management should be concerned with all file management tools and techniques, not just one. Indexed sequential files, sequential files, etc. are just as important as any DBMS file. The function should be just as concerned with the organization and filing techniques of manual files as they are with computer files. Unfortunately, this is not the situation in most companies today. The tools and techniques selected should be based on such things as required processing speed, anticipated transaction volume, security, and other performance considerations.

5. To design the data base to be independent from applications and physical environments.

Data Resource Management must be careful not to impose a dependency on particular hardware or software. In some situations, this may not be avoidable. In this event, conversion options should be explored and planned. Obsolescence in the areas of storage devices and techniques can mitigate against data base planning and create future problems. Also, Data Resource Management should avoid any short term solutions that may create long term problems. This applies to designing a Data Base to meet some particular application need without consideration for the overall needs of the other applications. The Data Resource Manager should have as objectives the complete independence of the data base from hardware, system software and applications.

6. Cooperate with other IRM functions.

Data Resource Management is a function that does not operate in isolation. It works closely with the other functions associated with the disciplines of information systems engineering and enterprise engineering. For example, the application logical data base design developed by Systems Engineering is checked for accuracy.

The specification of the logical designs will result in a physical implementation requiring hardware and software which may not exist or will require modification. Data Resource Management must consider this carefully and advise Systems Engineering accordingly. When determining project costs, Project Management must be advised by Data Resource Management of the pertinent data base costs. Also, the application physical data base design must be delivered to Software Engineering prior to programming.

For Enterprise Engineering, Data Resource Management must create skeletal definitions of the objects required to operate and manage the enterprise.

7. To control all data resources.

In order to provide an accurate accounting of all resources, a "bill of materials" type of system is required to catalog, classify, and cross-reference components to where they are used. This is the intent of an Information Resource Manager (IRM), a software tool used to inventory and track the use of organizational resources, systems resources, and data resources. This type of tracking could be performed manually, but this would create a large and cumbersome trail of paper.

Acting as a Bill of Material Processor (BOMP), an IRM Repository provides tremendous analytical capabilities, particularly in the area of "impact analysis," which permits a user to evaluate the effect of a change to a resource as it applies to the other information resources in an enterprise. It can also be used to maintain documentation on the various resources. As changes are made, the documentation is automatically updated. Because of the extensive resource intelligence contained in the IRM, it can be used to drive multiple physical data base management systems.

Because the IRM Repository represents the central location for a company's information knowledge, it becomes a tool used in enterprise engineering and information systems engineering, as well as data base engineering.

Selling the concept of Data Resource Management to corporate executives is a relatively simple task, as long as it is communicated in management terms, not technical jargon. Data Resource Management will not be successful as long as executives view it as a technical function. However, if management understands and accepts the fact that Data Resource Management is simply another form of materials management, a vital part of the "Information Factory" concept, they will understand and support an effective Data Resource Management function.

METHODOLOGY CONSTRUCTION/NAVIGATION

The Data Base Engineering Methodology (DBEM) consists of an assembly of six phases detailing what is to be accomplished and by whom. Each phase consists of a defined set of activities (a total of 24); each activity consists of a series of operations or tasks to be performed. All phases, activities and tasks produce tangible deliverables that can be reviewed and checked. These deliverables substantiate adherence to the methodology and permits the measurement of progress. Both formal and informal review points are contained throughout the methodology which provides for the effective dialog between management and data base engineers.

DBEM emphasizes design correctness and the production of a quality product. The first phase is essentially used to plan the DBEM project. The remaining phases map the four data base models as mentioned earlier. The final phase (6) is used to evaluate the DBEM project.

PHASE 1 - "Data Base Study & Evaluation"

A DBEM project is normally initiated in conjunction with an EEM or ISEM related project. During the Enterprise Engineering Methodology, objects are initially identified that relate to business functions. DBEM can use these objects as the basis for initiating a tentative set of Enterprise Logical objects. EEM may also identify maintenance projects where certain parts of the data base may need to be documented. However, most DBEM projects are performed in support of the Information Systems Engineering Methodology. Although DBEM projects are normally performed separately from ISEM projects, it may be desirable from a manageability point-of-view to conduct a combined ISEM/DBEM project, where phases from both methodologies are used.

In all instances, Phase 1 is used to plan the project and identify the parts of the four data base models affected by the project, either new or existing.

Information requirements and preliminary object definitions resulting from EEM and/or ISEM are used as the basis for creating or modifying a preliminary design of the data base. Any existing data base models are referenced at this time.

The rough design becomes the basis for developing a project plan, estimate and schedule for the DBEM project. This type of analysis is used not only in relation to DBEM, but also any ISEM projects it may support.

The formal deliverable resulting from this phase is the "Data Base Study & Evaluation Report" containing the project scope, requirements, approach and project evaluation considerations. A "Data Base Concept Diagram" is included in the report which is a free form graphical rendering of the proposed data base design.

A formal review of the phase is conducted where the results of the phase are discussed. Based on this review the project may proceed as proposed, discontinued, or revised as required.

After Phase 1, the project proceeds according to its scope. For ISEM related projects, the next step will be Phase 2. For EEM related projects, Phase 2 may be circumvented and the project will proceed to Phase 3.

PHASE 2 - "Application Logical Data Base Design"

Under the supervision of Data Engineering, this phase is performed by Systems Engineering for the application under development.

During this phase, the data used in the system is organized into logical files for use in the application's sub-systems. The application logical data base model is developed to express the objects included in the application, along with their relationships, e.g., one-to-one, one-to-many, many-to-many.

The formal deliverable resulting from this phase is the "Application Logical Data Base Design Manual" which includes an ALDBM diagram or matrix to express the relationships of the design. A technical review of the design is conducted prior to proceeding to the next phase.

PHASE 3 - "Enterprise Logical Data Base Design"

This phase is performed by Data Engineering who is fundamentally concerned with the design and control of the logical data base models.

During this phase, Data Engineering merges the application logical data base design into the enterprise design. Common application objects are consolidated into a single global object with the data needed to support all applications. Relationships between objects on the enterprise level are also reconciled.

The formal deliverable resulting from this phase is the "Enterprise Logical Data Base Design Manual" which is comparable to the manual produced in Phase 2 except it represents the total enterprise view of the data base. An ELDBM matrix is used to describe the extensive relationships within the model.

A technical review of the design is conducted prior to proceeding to the next phase.

PHASE 4 - "Enterprise Physical Data Base Design"

This phase represents the point where Data Base Administration (DBA) and Data Communications Administration (DCA) formally enter the methodology. Although they are consulted in Phase 1, Phases 4 and 5 represent their primary area of responsibility.

During this phase, the DBA and DCA develop an approach to physically implement the corporate data base. Various file management tools and techniques are evaluated at this time. To develop an effective design, they must review the requirements of all applications affected. Data volume, required processing speed, hit ratio, volatility, performance and security are issues considered at this time. This becomes the basis for developing the enterprise physical data base model.

The formal deliverable resulting from this phase is the "Enterprise Physical Data Base Design Manual" containing matrices that map the logical to the physical models. It also contains data definition language statements to implement a DBMS if one is selected. Any additional graphics produced during this phase follow the conventions as established by the file management tool or technique selected.

A technical review of the design is conducted prior to proceeding to the next phase.

PHASE 5 - "Application Physical Data Base Design"

This phase is comparable to the previous phase except it is concerned with only a single application and not the entire corporate data base. The application physical data base represents the subset of the EPDBM used to satisfy the data requirements of a single application.

The formal deliverable resulting from this phase is the "Application Physical Data Base Design Manual." Like the design manual resulting from the previous phase, it contains data definition language statements to implement a DBMS if one is used. Any additional graphics produced are native to the file management tool or technique selected. The file layouts produced from this phase are then delivered to Software Engineering in ISEM Phase 4-II, "Software Engineering."

A technical review of the design is conducted prior to proceeding to the next phase.

PHASE 6 - "DBEM Evaluation"

This phase is used to conduct an operational review of the data base models affected by the project. Based on the earlier design manuals, the data base is reviewed according to specifications. Also, actual project time is reviewed to determine how well estimates and schedules were met and to determine the total cost of the project.

The formal deliverable resulting from this phase is a "Project Evaluation Report" containing the findings from the phase.

DBEM METHODOLOGY STRUCTURE

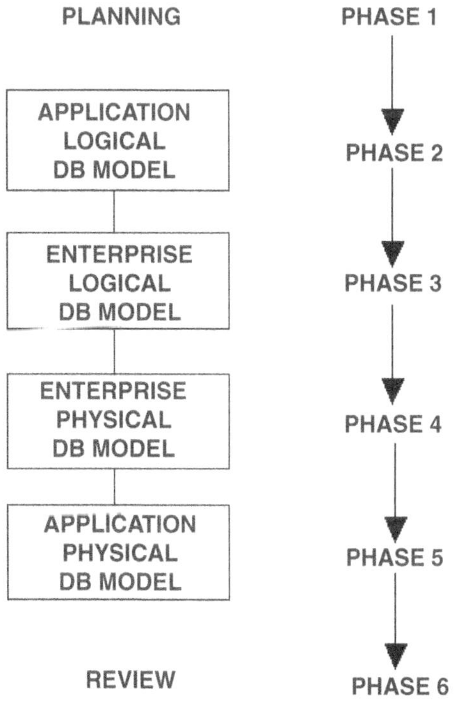

EEM/ISEM RELATIONSHIPS

The "PRIDE"-Enterprise Engineering Methodology (EEM) precedes both the Information Systems Engineering Methodology (ISEM) and the Data Base Engineering Methodology (DBEM). It is used to formulate an Enterprise Information Strategy that includes development projects for ISEM and DBEM to implement.

Because of the close relationship between systems and data, DBEM has a close working relationship with ISEM. In Phase 2 of ISEM, sub-systems and logical files are defined. This will then normally initiate a DBEM project to support the data requirements. In Phase 2 of DBEM the Application Logical Data Base Model (ALDBM) is prepared by Systems Engineers under the supervision of Data Engineering. Data Engineering will then incorporate the ALDBM into the Enterprise Logical Data Base Model (ELDBM). Data Base Administration will then design/update the Enterprise Physical Data Base Model (EPDBM) and Application Physical Data Base Model (APDBM). The result of the APDBM is the physical file structures for programming. In other words, the DBA must deliver the file layouts to the Software Engineer in Phase 4-II of ISEM.

WHO SHOULD PERFORM DBEM?

There are significant differences between Data Engineering and Data Base Administration (DBA). The Data Engineer tends to be more people and business oriented than the DBA, who is more concerned with the effective implementation of technology.

The Data Engineer is the chief data architect and Systems Engineering liaison. Essentially, they must be proficient in interpersonal relations/communications, possess good analytical/organizational skills, and knowledgeable about the business of the enterprise. They must also be able to interface with Data Base Administration and Data Communications Administration (DCA).

The DBA and DCA are more in tune with computer hardware/ software than the business. They typically lack the business and people skills as possessed by Data Engineering. This is not to demean their role, only to clarify it. Although Phases 4 and 5 are the principal phases where DBA/DCA works, this does not limit their involvement in the methodology. In fact, they actively participate in the early phases of DBEM in a consulting capacity to Data Engineering. During these phases, DBA and DCA analyzes the viability of designs as prepared by Data Engineering, as well as participate in project estimating/scheduling activities. This participation in the early phases promotes cooperation and communications between the two functions.

ISEM	EVENT	GOES TO DBEM
PHASE 2	Systems Engineering defines the Application Logical Files (Objects). These files ultimately represent the primary storage files.	PHASE 2
PHASE 3	Systems Engineering defines the need for Input/Output related files. This includes computer input transaction files and output data files. This may also include manual files to store input/output documents.	PHASE 5
PHASE 4-II	Software Engineering defines the need for computer working files to pass data between programs and modules.	PHASE 5

Data base design is separated from system design so that a neutral third party can accommodate the data requirements for the entire enterprise, not just a specific application.

OTHER FUNCTIONS PARTICIPATING IN DBEM

As in any other work effort, Project Managers are required to plan, estimate, schedule, and control a DBEM project. Further, Quality Assurance personnel review and verify deliverables resulting from the various phases of the methodology.

Data Processing Operations are interviewed and consulted during the course of a DBEM project. As a result, their participation is vital. End-users are also consulted in terms of physical file requirements (particularly manual files).

Systems Engineering and Software Engineering also play important roles in DBEM. As chief architect of the information system, Systems Engineering is responsible for defining the data elements needed to support information requirements and the objects (logical files) used in the system. Consequently, Systems Engineering must perform Phase 2 of DBEM to assure that object views and relationships are properly defined. Software Engineering is actually the benefactor of DBEM, receiving the necessary file layouts for use in programming. However, in the absence of a robust Data Resource Management organization, a Software Engineer may have to fulfill the Data Base Administration function.

When and where these other functions are involved during DBEM is defined by the methodology. Consult the DBEM Functional Matrices for a summary description.

BENEFIT$

The benefits of data base engineering are numerous:

- It assures that Data Resource Management personnel are "doing the right things," performing data base design that will accurately satisfy the needs of systems applications.
- Provides for improved control over the data base design process by simplifying the work effort.
- It demystifies the data base design process.
- It improves staff performance through a standard and consistent approach.
- Improves communications between systems development and data base personnel, as well as the data base staff internally.
- Improved consistency and quality of data bases.
- Data bases behave reliably and are easier to maintain and modify.
- More efficient use and control of physical file management tools and techniques.
- Improved control over project planning and execution.
- Improved corporate control over information resources.
- Competitive edge through accurate and timely information under changing conditions.

END SECTION

Section 20

"PRIDE"-DBEM PHASE 1 DATA BASE STUDY & EVALUATION

Data is stored; Information is produced.
- Bryce's Law

The purpose of this phase is to specify and analyze a data base related problem and/or opportunity, and to propose to management a viable solution. It is the most critical phase of the DBEM methodology. Several events occur during this phase:

- The scope of the DBEM project is defined.
- An analysis of the current data base is performed.
- Information requirements are reviewed and updated as required.
- A data base approach is defined for satisfying the requirements. The approach may include new development, modifications/improvements to existing data base, a packaged solution, or combinations of all three.
- An evaluation of alternatives is prepared, including estimates, costs, and schedules.
- A formal review with management is conducted to verify the phase findings.

METHODOLOGY NAVIGATION

A DBEM Phase 1 is normally initiated by an Objective as defined by the Enterprise Information Strategy (EIS). This strategy was developed during Phase 4 of the Enterprise Engineering Methodology (EEM) and is issued to Data Resource Management during EEM's Phase 5. The Objective defines the business problem/opportunity to be addressed by the project, which typically relates to new development, modification/improvements, or maintenance.

DBEM related projects are initiated primarily in support of ISEM related projects. However, special projects can be initiated to prepare tentative descriptions of Objects and Views in support of the Information Requirements and Functional Entities identified under EEM.

As an adjunct to an ISEM project, there are two ways of treating a DBEM project: either establish a separate DBEM specific project, or; include DBEM phases within an ISEM project. In most situations, the Data Resource Management organization tends to manage its own projects. However, from a project management perspective, it may be more desirable to define a separate parallel DBEM path within an ISEM project. Either scenario is plausible.

During Phase 1, Activity G a formal review of the phase findings is conducted with Data Resource Management. Consequently, management makes decisions regarding the project; they may elect to:

1. Approve the project for continuation as proposed.

2. Request revisions to the findings produced in the Phase.

3. Discontinue the project.

Assuming acceptance, the project will normally proceed to Phase 2 where the Application Logical Data Base Model (ALDBM) will be designed. The exception to this is when a DBEM project is initiated based on an EEM project. In this situation, the DBEM project proceeds to Phase 3 where a preliminary Enterprise Logical Data Base Model (ELDBM) can be formulated.

GENERAL DISCUSSION

THE PHILOSOPHY OF PHASE 1

In its simplest form, Phase 1 represents a definition of the problem, an analysis of the current mode of operation, a definition of requirements, an evaluation of alternatives, and an agreed upon course of action. Phase 1 is based on generic principles of product evaluation. As such, the activities of Phase 1 can be applied to any type of project. Because of the type of work involved, the Phase 1 in DBEM is compatible to the Phase 1 in ISEM.

PHASE 1/ACTIVITY A

Activity A, "Preliminary Project Scope," is critical to the start-up of the ISEM project. A superficial analysis during this activity can result in misunderstanding and severe project problems later. Activity A, therefore, is used to specify the project scope and to produce a Detail estimate and schedule for the remaining activities of Phase 1.

The Project Scope is a concise definition of the business problem/opportunity to be addressed by the project and the areas of the enterprise affected both directly and indirectly (user departments). This textual definition is important. It establishes the project boundaries which will determine estimates, schedules and work effort. Projects tend to lose direction without clear boundaries and tend to wander into areas unintentionally. This will result in performing more work (or less) than what is necessary to satisfy project objectives.

Work Requests/Objectives are documented in the IRM using the MI resource (Modification/Improvement). There is not necessarily a one-to-one relationship between an MI and a Project Description (PD). One MI may be implemented by many projects; one PD may implement many objectives. The Project Manager must have a clear understanding of these relationships.

The expression "Modification/Improvement" is based on the fact that most development work is of this nature. The "PRIDE" methodologies recognize that change is constant; that systems and data bases are built by evolution, not by revolution. As soon as a system or data base is installed, users will probably request changes to it.

DBEM was designed to accommodate change. Corporate information needs will fluctuate based on economic conditions, government regulation, competition, product changes, acquisitions, etc. The operating objective of the Information Resource Management organization, therefore, is to be sensitive and responsive to this changing environment.

Modifications and improvements are not to be confused with Maintenance. Maintenance is an unscheduled activity and occurs primarily as a result of operating problems; that is, the data base is not performing as intended or specified. Documentation changes are usually not involved and the action to be taken is corrective in nature. Modification/Improvements can be planned and scheduled, and should be in response to new or changing information needs.

IMPACT ANALYSIS

In order to accurately evaluate change, an "Impact Analysis" is performed to study the relationships between information resources and appraise what resources need to be modified. Some examples:

- If physical files require modification, an analysis is required to identify the logical files affected and the systems.

• For referential integrity purposes, if a logical view (record) is modified, an analysis is needed to evaluate how it will affect other objects and views. This is also required at the physical level when using a relational DBMS.

• If the length of a data element is to be modified, an analysis must be performed to identify the panels, maps, records, files, inputs, outputs, modules, programs, etc. affected.

This type of analysis reflects a "bill-of-material" philosophy. In other words, if one part is changed, what other parts will feel its effect? An IRM Repository is used to perform this type of analysis. An "Impact Analysis" is invaluable for scoping the size of a project and for developing project estimates and schedules.

Before proceeding with the project, the Project Scope and estimates/schedules are reviewed with the user sponsor(s). It is important that the Project Manager and the users have consensus in terms of the work to be performed, the approach, and anticipated estimated time/costs/schedule. This is true regardless of the type of work effort.

CURRENT DATA BASE ANALYSIS

The purpose of current data base analysis is to inventory data resources for control purposes. Such control eases maintenance and modification/improvement activity. It is also useful for reasons of back-up/recovery of key files. For development projects where a new system is being developed to replace an aging system, it is necessary to document the existing data base so that a strategy can be devised to transfer data from one file format to another.

Under Phase 1, Activity B of ISEM, Systems Engineering is concerned with documenting system resources. This same type of activity is performed under DBEM to document data resources. As such, both activities can be conducted in parallel and are complementary.

Documenting a data base can be a major undertaking. It is important to remember that **the intent here is to ONLY IDENTIFY AND DESCRIBE THE CURRENT DATA BASE, NOT TO CORRECT DEFICIENCIES.** Quite often, there is a temptation to try to correct an obvious problem at this stage. As a consequence, current data base analysis takes longer and longer. When errors or problems are spotted, they should be documented as future Modification/Improvements and not corrected as part of this effort.

How much definition is enough? DBEM recommends up to a Phase 5 level of detail (all four data base models):

APPLICATION LOGICAL DATA BASE MODEL (ALDBM) - represents the objects, views, and primary data used in the current system. The primary data used here will most likely be re-used, regardless of physical implementation. The logical views provide invaluable intelligence regarding relationships between objects (one-to-many, many-to-many). Consult Phase 2 for the required structures.

ENTERPRISE LOGICAL DATA BASE MODEL (ELDBM) - represents the global view of objects, views and primary data elements used in the enterprise. Consult Phase 3 for the required structures.

ENTERPRISE PHYSICAL DATA BASE MODEL (EPDBM) - represents all of the physical files used to store data in the enterprise. Consult Phase 4 for the required structures.

APPLICATION PHYSICAL DATA BASE MODEL (APDBM) - represents all of the physical files used to store data within a single information system. Consult Phase 5 for the required structures.

Current data base analysis normally begins from the application view, which is then merged into the enterprise view. The next question becomes which type of view to begin with: Application Logical or Application Physical? There are some trade-offs to consider. To most people, physical files will be the most visible and easiest to document. Although a replacement system with its own application physical files will probably supersede the need for the old files, documenting the existing files is necessary to transfer data from the old format to the new. However, if this in not a consideration, then documenting the APDBM is probably not necessary.

Documenting the ALDBM is somewhat more difficult to perform than its physical counterpart. Whereas tangible evidence exists for the physical (e.g., data description language statements, program source listings, etc.), the logical is defined based on deductive reasoning, normally by inference of inputs and outputs. The benefit of documenting the ALDBM is that the logical files, records, and data elements can be re-used in other applications, both existing and new. There are essentially two approaches for documenting data resources: Top/Down and Bottom/Up.

Top/Down Approach

This approach implies defining logical structures before the physical. To do so, Data Engineering analyzes data requirements from an application as defined by Systems Engineering. Inputs are analyzed in terms of the primary data elements collected. Outputs are analyzed in terms of both primary and generated data elements. Where a generated data element is encountered, it is traced backwards to the primary data elements needed to produce it. After all of the primary data elements have been defined, they can be organized into objects and views.

When the ALDBM has been defined, it can then be merged into the ELDBM as required. Following this, the physical files of the application are defined based on available documentation (e.g., source code, DDL, etc.). The APDBM is then related to both the EPDBM and the ALDBM. Relationships between the ELDBM and the EPDBM are established last.

Bottom/Up Approach

This approach begins with a definition of the APDBM based on available source code, DDL, etc. Following this, analysis proceeds backwards to the EPDBM, the ELDBM, and the ALDBM. Relationships between the ALDBM and the APDBM are defined at the end.

As to which approach is best, Top/Down or Bottom/Up, is actually irrelevant. In all likelihood, a company will probably use both simultaneously. Since Data Engineers and Systems Engineers primarily deal with logical structures, they will concentrate on the ALDBM and work back to the ELDBM. In contrast, Data Base Administrators and Software Engineers will concentrate on the physical structures, thereby defining the APDBM before backing into the EPDBM. The logical structures are then related to the physical to complete the models.

INFORMATION = DATA + PROCESSING

Specifying information requirement is one of the most important activities in all of "PRIDE." All designs that follow must ultimately satisfy the requirements. Therefore, they must be defined with a high degree of precision and accuracy. Instead of physical media requirements, "PRIDE" focuses on the types of business actions and decisions that need to be supported. Information and Data are not synonymous. Data is the raw material used to produce information.

During EEM and ISEM, end-users are interviewed in terms of the information they need to support their area of the business. This becomes the basis for a formal definition of the information requirements, including:

- The **BUSINESS PURPOSE** of the requirement.

- The **ACTIONS AND/OR BUSINESS DECISIONS** to be made, including WHO must make them and WHEN.

- The **BENEFITS** of having this information.

Data Engineering represents the function in DBEM responsible for analyzing and refining information requirements. From the requirements, Data Engineering can interpret the types of logical objects and views required to support the various requirements, along with indicative data elements that may need to be defined. To demonstrate how objects are used in relation to information requirements, Data Engineering prepares an Information Requirements/Objects Matrix.

PHASE 1/ACTIVITY D

Activity D represents the first major review with Data Resource Management. This is where the Information Requirements and Project Scope are reviewed and confirmed prior to seeking a solution. In particular, each requirement is evaluated in terms of the affected objects and required data elements (to assure they are accurately defined and related to the requirements). The net result is to reach consensus in terms of the requirements and project scope.

ROUGH DESIGN

Following confirmation of the information requirements, the Data Engineer is then charged with the task of devising a data base solution. To do so, the engineer prepares a complete rough design of the data base. This involves a tentative design of the four data base models. In this respect, the design process takes on the appearance of mini-Phases 2, 3, 4, and 5.

Participating in the rough design are representatives from Data Base Administration, Data Communications, Systems Engineering, Software Engineering, and DP Operations; all of which consult with Data

Engineering during design. The intent is to arrive at a viable solution for the company to implement. In terms of the physical data base, Data Base Administration considers various file management techniques. Data Communications considers data transmission/conversion requirements.

The data base design is considered "rough" in the sense that it represents a tentative design for evaluation purposes. Regardless, there is considerable attention to detail in terms of identifying the data resources needed to satisfy the information requirements. During this activity, the Data Engineer will consider alternatives. For example, the Data Engineer will identify...

- Reusable resources - data resources that have already been developed and are suitable for inclusion in the application.

- Existing resources requiring modification to accommodate requirements.

- New resources that will have to be designed and developed.

- Resources that can be implemented by commercial software packages.

- Data interface requirements between all of the above.

The point is, rough designs are used to propose practical cost-effective solutions for management to consider. However, there are additional benefits. The rough design will become the basis for formulating a project plan, estimate and schedule. This can then be used for "make versus buy" decisions. It will also improve communications and cooperation between the development staff due to early participation in the planning phase.

One final word of advise on rough designs: Do not attempt to perform the whole project in Phase 1. It must be remembered that the intent of Phase 1 is to propose a viable solution for management to consider. Although the rough design may be rigorous, it must not be considered the final design; this is what the succeeding phases of DBEM are for. In other words, do not try to build the perfect system in Phase 1; simply take it to the level where the project team has confidence in the design and that a realistic project plan/estimate/schedule can be prepared.

DATA BASE CONCEPT DIAGRAM

To graphically represent the data base solution, the Data Engineer prepares a Data Base Concept Diagram. This is a free-form schematic that illustrates to the user how the envisioned data base will operate upon completion. This is similar in intent to an artist's rendering as used in architecture.

Although there are no formal standards for this type of graphic, the Data Base Concept Diagram normally shows the logical objects involved and how they are to be implemented physically. See the "Examples" section under Phase 1, Activity E for a sample Data Base Concept Diagram.

EVALUATING A PACKAGED SOLUTION

There are two situations where a packaged solution can be considered:

1. APPLICATION SOFTWARE TO IMPLEMENT PARTS OF AN INFORMATION SYSTEM.

A commercial software package provides one alternative for implementing an information system. Normally, it is not a complete information system but only tested computer programs that is sometimes accompanied by user procedures for input preparation. Most packages tell the purchaser what input is required to produce a specified output without telling the purchaser who does what, when and how. Because of this, fundamental data structures may have to be documented to convert a package into an information system before installation can be accomplished. In other words, a package should be documented in the IRM Repository like any other resource.

One of the most common pitfalls in purchasing a commercial package is that little consideration is given to the information requirements they must satisfy. Quite often packages are purchased based on marketing hype rather than factual evidence. In other words, a superficial package evaluation is performed. Little investigation is made as how well the package satisfies requirements, how complete the documentation is, what modifications might be required to suit in-house needs, how it will interface with other systems, how long it may take to implement, etc., etc.

Despite this, experience often shows that the purchase decision is a viable alternative to 're-inventing the wheel'. This is especially true in those situations where resource and/or time constraints have necessitated the purchase consideration. When handled intelligently and adequately managed, packages can be successfully installed to the mutual satisfaction of both development personnel and users to solve pressing business problems.

How then is it possible to avoid the common pitfalls and evaluate packages intelligently? Fortunately, the answer is simple: Phase 1. It provides all of the background information necessary to intelligently investigate packages. If performed correctly, the Phase 1 will have produced a complete rough design of the data base. This becomes the focal point for preparing a comparative analysis. The Data Engineer should be able to illustrate:

• The types of file structures that accompany the package and if the files are accessible for additional customer specific processing requirements.

• How well the package's physical data characteristics conform to in-house requirements. For example, does the package express dates, times, names, monetary values, etc. in the same manner as is commonly used in the company? In other words, how does the vendor express data length, precision, scale, etc? Severe conflicts may be incompatible for in-house use.

• How the files will interface with other in-house file structures.

Whenever possible, package documentation should be incorporated into the IRM Repository for management and control. This means that narrative and specifications should be entered in the IRM Repository regarding Files, Records, and Data Elements. This does not necessarily mean entering the product's documentation in its entirety into the IRM Repository, although this would be ideal. It may be more desirable to simply establish cross-reference's to the package's own documentation. However, because the package will ultimately become a natural part of the company's systems, sufficient intelligence should be put into the IRM Repository so that the customer can merge the package's information resources with their own. This provides invaluable control to the customer, particularly when performing an "impact analysis." It also provides the customer with the ability to evaluate a product release prior to putting it into production.

The customer may elect to execute ensuing ISEM phases in order to properly document the package. This may be accomplished by either the 'in-house' development staff, by vendor personnel, or both working together.

The decision to purchase a package to reduce development time and costs can produce satisfying results for both the user and development staff. This does, however, require that the organization first develop the information requirements, evaluate and install the package that best meets the requirements, and finally, perform an audit of the project. The DBEM approach is a means for good communications among the participants and earns their commitment, thereby increasing the chances for a successful implementation.

2. FILE MANAGEMENT SOFTWARE TO ACCOMMODATE A NEW SYSTEM OR COMPUTER CONFIGURATION.

Although physical files are not finalized until Phase 4, Phase 1 begins to highlight the need for file management techniques. Although it may still be too early to make some conclusions, Data Base Administration can begin to consider performance and security issues. From this, some assumptions can be made regarding a suitable approach. In many cases, commercial file management software may be considered, particularly a Data Base Management System (DBMS). It is essential that packages be considered from a practical point-of-view, not necessarily what is technically fashionable. Remember, **AN ELEGANT SOLUTION TO THE WRONG PROBLEM SOLVES NOTHING**. To this end, Data Base Administrator considers such things as: transaction volume, processing speed, file volatility and hit ratio, backup/recovery, security, number of users, etc. The data base vendor should be able to provide benchmark tests as required to demonstrate how well their product works. Consult Phase 4 for additional considerations.

HARDWARE CONSIDERATIONS

During the rough design, consideration must be given to the physical implementation of the data base. In many instances, the existing operating environment is assumed. However, the rough design may suggest the need for additional equipment and/or enhancements to the current operating environment. This is one reason why DP Operations and Data Communications participate in the preparation of the rough design; to appraise the impact of the proposed data base on the current operating environment.

In studying the rough design, consideration is given to:

- The areas of the business served.
- Processing requirements: such as, Interactive, Batch, etc. Also, distributed versus central processing.
- Types of media (Inputs/Outputs): Graphical, Textual, Audio, Optical, etc.
- Data transmission/conversion requirements.
- Required processing speed and volume of transactions.
- Storage requirements (files).
- Back-up/file retention requirements.

From this, assumptions can be made regarding:

- Computer equipment and operating systems.
- Input/Output peripherals; e.g., terminals, printers, etc.
- Communication hardware/software, including networking.
- Data Base hardware and software.
- Other equipment and files (non-computer related).

From this evaluation, an estimate of anticipated expenditures can be prepared for inclusion in the overall project estimate. At this point, hardware decisions are only tentative. The final decision on these matters should be reserved for subsequent DBEM Phases as designs are finalized.

PROJECT PLANNING

Before a project estimate and schedule can be calculated, the path of the project must first be defined. Under DBEM, the project path is based on the four data base models.

PHASE	NAME	RELATED RESOURCE
1	DB Study & Eval.	(none)
2	ALDBM Design	System
3	ELDBM Design	FD (EL)
4	EPDBM Design	FD (EP)
5	APDBM Design	System
6	DBEM Evaluation	(none)

There are two situations where the project path may deviate:

1. It is possible, but not likely, that a single DBEM project can be used to implement the data requirements for more than one system. In this event, there will be multiple Phase 2's and Phase 5's (one for each system).

2. For an EEM related project, the project path may consist of nothing more than Phases 1, 3, and 6.

The point is, DBEM provides a good framework for executing a development project. However, Project Management must consider the practical implications for controlling the project and, as such, must make the final decision on project execution.

After the path of the project has been determined an Order-of-Magnitude (OOM) estimate and schedule can be calculated. This is an estimate and schedule of the amount of effort required to perform the remaining phases in the project. The estimate, thereby, is an expression of labor charges.

HUMAN RESOURCE REQUIREMENTS

Following the rough design and the project plan, consideration must be given to the types of human resources required to implement the project. Based on the type of application being developed, Project Management considers the types of Data Base personnel required to work on the project, including the skills and proficiencies needed to perform the work. Based on this analysis, Project Management has four options:

- Use in-house personnel.
- Recruit additional personnel.
- Use outside personnel (contractors).
- Combinations of the above.

In all instances, the use of human resources is based on their qualifications, their availability, and their cost. Project Management, therefore, must balance these variables when selecting personnel. There may be trade-offs to consider; for example:

- One person may be more expensive than others but can deliver the work faster.

- A senior person could perform the work in fewer hours than a junior person, but is committed to other project assignments. Consequently, the person cannot devote sufficient time to the project. Whereas the junior person may have fewer conflicts and is more available for project work.

The use of known human resources has a direct impact on the Order-of-Magnitude estimate and schedule. However, in the situation where resources are unknown, the Order-of-Magnitude estimate and schedule can still be calculated. Here, the Project Manager must make some assumptions as to the type of skills and proficiencies required and the number of Data Base personnel needed for the project. The estimate thereby becomes the basis for recruiting additional personnel or hiring contractors. It also becomes the basis for determining training requirements, the cost of which should be absorbed by the project.

COST/BENEFIT ANALYSIS

The purpose of a Cost/Benefit Analysis is to perform an economic impact analysis on the data base and the project. Prior to performing this, Project Management has prepared an Order-of-Magnitude estimate to complete the project, along with any additional planned expenditures (such as for additional hardware/software acquisitions).

"PRIDE" provides a worksheet to assist in the preparation of the Cost/Benefit Analysis. However, in-house standards should be observed when performing the analysis.

During the analysis Project Management must consider how the proposed data base solution will affect the company:

- Will it increase or reduce staffing?
- Will it increase or reduce equipment resources?
- What will be the annual savings and expenses of the data base as it compares to the existing environment?
- What will be the one-time savings and expenses of the project as it applies to the existing environment?
- What will be the intangible benefits of the data base?
- What is the investment evaluation? For example, what will be the Break Even Point of the project? What will be the Return On Investment (ROI)?

The Cost/Benefit Analysis becomes the basis for the preparation of the Cost and Evaluation Summary, a textual justification for the project.

PHASE 1 REVIEW

The formal deliverable resulting from Phase 1 is the "Data Base Study & Evaluation Report." This represents a packaging of the various elements produced during the phase (e.g., Project Scope, Information Requirements, Data Base Concept Diagram, Data Base Approach, Project Plan, Order-of-Magnitude, Cost and Evaluation Summary, etc.). A formal meeting with management is conducted to review the deliverable. At this time, management may elect to approve the project as proposed in the deliverable, request revisions before proceeding, or cancel the project completely.

There may be many reasons for discontinuing a project at this time, perhaps the project team cannot produce an adequate solution for satisfying the business problem, or maybe the economics of the project are not justifiable (the Return on Investment may be too low for example). There is little reason in expending human and financial resources towards an endeavor that cannot be proven worthwhile.

Assuming cancellation of the project, the company retains specifications regarding:

- The current data base.
- Information Requirements.
- A rough design of a proposed new data base.

These specifications are maintained in the IRM for use either at a later date, or by other applications. The knowledge and design decisions formulated during Phase 1 have been captured and will not be lost even with personnel turnover. In fact, this intelligence can be used by others at a later date. It is entirely conceivable for a project to be discontinued after Phase 1 and be resumed at a later date with minor changes to the specifications.

DESCRIPTION OF PHASE ACTIVITIES

Activity A - Preliminary Project Scope

Information Resource Management submits a Work Request/Objective to Data Resource Management who initiates a DBEM project. The request is normally prepared as part of the Enterprise Information Strategy (EIS) as prepared under the Enterprise Engineering Methodology (EEM).

Data Resource Management initiates a DBEM project which includes the assignment of appropriate Project Management and Data Engineering personnel.

Project Management prepares/revises a Project Scope for the project. Based on the scope, Project Management prepares an estimate/schedule for review and approval by Data Resource Management.

Activity B - Current Data Base Analysis

Data Engineering studies the current data base. Data Base and DP Operations provide assistance as required. Consequently, current data base documentation is updated as required and a Current Data Base Analysis is prepared highlighting the strengths and weaknesses of the current data base. This analysis is reviewed with Data Resource Management for accuracy.

Activity C - Analyze Requirements/Project Scope

Data Engineering prepares/revises formal information requirement descriptions and a final Project Scope. As part of this effort, Data Engineering defines skeletal objects needed to support the requirements.

Activity D - Review Requirements & Scope

Project Management and Data Engineering review the Information Requirements and the Project Scope with Data Resource Management. The outcome of this meeting is to either continue the project, revise the requirements and/or scope, or discontinue the project completely.

Activity E - Develop Data Base Approach

Data Engineering prepares a rough data base design (logical and physical) which will satisfy the overall information requirements within the Project Scope. The design is represented in the Data Base Concept Diagram, and the Data Base Approach Logic Narrative.

Activity F - Prepare Data Base Evaluation

Based on the rough data base design prepared in the previous activity, Project Management prepares a Project Plan which details the remaining phases in the project. From the plan an Order-of-Magnitude estimate and schedule is prepared to complete the project. This estimate combined with any other anticipated financial expenditures (for such things as hardware/software) is used in the preparation of a Cost/Benefit Analysis. This becomes the basis for developing the Cost and Evaluation Summary, a textual justification for the project.

Project Management and Data Engineering then assemble the deliverables produced in the phase and packages them into a single phase review document which is evaluated by Quality Assurance prior to conducting the formal phase review with management.

Activity G - Phase 1 Review

Project Management conducts a formal review of the Phase 1 deliverables with Information Resource Management, and Enterprise/Systems/Data Resource Management. The review is used to evaluate the work performed thus far and revise as required. At this time, management will review the formal Phase 1 **"Data Base Study & Evaluation Report"** consisting of:

- Phase Cover Page
- Project Scope
- Current Data Base Analysis
- Information Requirements
- Information Requirements/Objects Matrix
- Data Base Concept Diagram
- Data Base Approach
- Project Plan
- Project Estimate/Schedule Recap
- Cost and Evaluation Summary
- Phase Review Checklist

Based on this report and subsequent review meeting, management may elect to revise parts of the report, discontinue the project, or approve it for continuation.

The next few pages include sample deliverables resulting from this phase.

OD-90401	CURRENT DATA BASE ANALYSIS	"PRIDE"-DBEM
NOV 02, 2004	RELEASE 1.2.0	PAGE 2.1

AS DEFINED ON PD-00017 ORDER PROCESSING DATA BASE

The order processing data base has been extensively modified and fragmented over the years. This was primarily due to the lack of a formal Data Resource Management function to control data and promote sharing between applications.

STRENGTHS

* Considerable documentation was produced to create physical data base designs. This documentation was developed in accordance with the Data Base Management System (DBMS) in place.

WEAKNESSES

* Little (if any) documentation exists regarding logical data base designs.

* Rampant data redundancy exists within the DBMS. Although there was considerable work in terms of defining and controlling the physical data structures, there has been no emphasis on integrating applications through the data base. Consequently, the DBMS is misused; It is used as nothing more than an elegant access method, not as a tool for integrating systems.

- E.R. Object
Data Engineer
November 2, 2004

* * * END OF OUTPUT * * *

OD-90189	INFORMATION REQUIREMENTS	"PRIDE"-ISEM
SEP 01, 2004	RELEASE 1.2.0	PAGE 1

```
****************************************************************
*              IR-00101  CUSTOMER ORDER                        *
****************************************************************
```

ENTITY(S) REQUIRING INFORMATION:	NUMBER	INFO. FLOW
1. CUSTOMER SERVICE	FE-00118	ORIGINATOR
2. ORDER PROCESSING	FE-00127	RECEIVER
3. TMC STOCK CUSTOMER	FE-00326	ORIGINATOR
4. CUSTOMER SERVICE REP	OE-00453	RECEIVER
5. CUSTOMER SERVICE REP	OE-00453	ORIGINATOR

TYPE REQUIREMENT: INFORMATION INFORMATION TYPE: OPERATIONAL

FREQUENCY:	OFFSET:	RESPONSE TIME:
UPON REQUEST	N/A	15 SECOND(S)

BUSINESS PURPOSE

The purpose of this requirement is to interact with a customer to place or revise an order for corporate products.

ACTIONS AND/OR BUSINESS DECISIONS

Order Processing will...

1. Check product availability and advise the customer of anticipated delivery dates.

2. Create, revise or delete an order.

   NOTE: Deletion is only permitted if the order has not yet been processed by Customer Service.

3. Notify Customer Service to take necessary actions to fill the order.

4. Request changes, as required, to the customer's credit rating (i.e., increase, decrease) to Customer Credit.

BENEFITS

1. Improved communications/relations with the customer, thereby enhancing customer satisfaction.

2. Improved communications internally, particularly with Customer Service and Customer Credit. Such communications improves control over customer credit, thereby preventing unauthorized shipments.

OBJECTS/FILES AFFECTED:

* PRODUCT OBJECT	(FD-00100)
* CUSTOMER OBJECT	(FD-00101)
* ORDER OBJECT	(FD-00102)

REQUIRED DATA ELEMENTS:

* PRODUCT NUMBER	(DD-00003)
* NAME	(DD-00018)

(continued on next page)

SAMPLE PHASE 1 DELIVERABLES

(continued from previous page)

* UNIT PRICE	(DD-00111)
* ORDER NUMBER	(DD-00012)
* DATE	(DD-00016)
* QUANTITY ORDERED	(DD-00036)
* TERMS	(DD-00044)
* STATUS	(DD-00041)
* QUANTITY ON HAND	(DD-00508)
* CONTACT NUMBER	(DD-00042)
* DELIVERY PRIORITY	(DD-00050)
* CUSTOMER NUMBER	(DD-00001)

IMPLEMENTED BY INPUTS/OUTPUTS:

* PRODUCT AVAILABILITY	(OD-00010)
* CUSTOMER ORDER	(OD-00011)
* CUSTOMER PROFILE	(OD-00012)

SUPPORTED BY SYSTEM RESOURCES:

* TRUNDLE ORDER PROC. SYS (TOPS)	(OP)
* CUSTOMER ORDER INQUIRY (	OP-00)
* ORDER-PROC FILE MAINTENANCE	(OP-08)

*** END OF OUTPUT ***

OD-90214 I N F O R M A T I O N R E Q U I R E M E N T S / O B J E C T S M A T R I X

OCT 10, 2004 RELEASE 1.2.0

INFO RQMTS > OBJECTS V	IR-00110 SALES PERFORMANCE APPRAISAL	IR-00111 SALES REP INFO	IR-00112 MONTHLY SALES ANALYSIS	IR-00113 WEEKLY SALES COMMISSIONS				
FD-94002 EMPLOYEE OBJECT	X	X	X	X				
FD-94003 PRODUCT OBJECT			X	X				
FD-94004 ORDER OBJECT	X		X	X				
FD-94005 BILLING OBJECT	X		X	X				
FD-94006 SHIPMENT OBJECT	X		X	X				
FD-94007 BACK-ORDER OBJECT			X					

SAMPLE PHASE 1 DELIVERABLES

OD-90120	DATA BASE CONCEPT DIAGRAM		"PRIDE"-DBEM
RESOURCE: OP	NAME: TOPS SYSTEM	PREPARED BY: OBJECT	EFFECTIVE DATE: NOV 10, 2004
PAGE: 1	COPYRIGHT (C) TMC 2004	RELEASE 1.2.0	DATE PRINTED: NOV 10, 2004

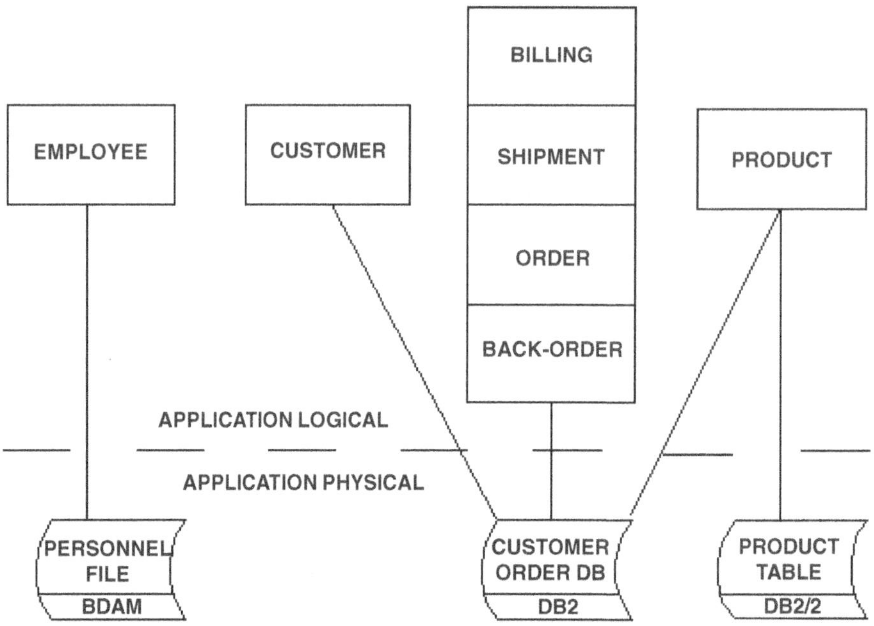

OD-90402	DATA BASE APPROACH	"PRIDE"-DBEM
NOV 15, 2004	RELEASE 1.2.0	PAGE 1

AS DEFINED ON PD-00017 ORDER PROCESSING DATA BASE

The proposed data base for the Order Processing System requires seven objects for the Application Logcial Data Base Model: Employee, Customer, Order, Billing, Shipment, Back Order, and Product. No new Enterprise Objects are required for this application.

PHYSICAL IMPLEMENTATION

The proposed design uses a combination of central and distributed files. It uses the communications network and computer hardware/ software that was developed for the existing Order Processing System.

The proposed files are:

* Personnel Master File - as currently implemented as a BDAM file on the mainframe.

* DB2 data base - to store and retrieve the bulk of the order processing data.

* DB2/2 file to keep track of products locally. *

** END OF OUTPUT ***

OD-90396	COST AND EVALUATION SUMMARY	"PRIDE"-DBEM
NOV 14, 2004	RELEASE 1.2.0	PAGE 9.1

AS DEFINED ON PD-00017 ORDER PROCESSING DATA BASE

The estimated cost for the project is $ 19,750.00 (consult the Project Estimate/Schedule Recap). No additional "out of pocket" expenditures are anticipated for data base hardware/software; existing resources can be effectively utilized.

Based on a technical Cost/Benefit Analysis, we conservatively estimate an annual savings of $40,000 in operating expenses. This is due to the fact that the new Order Processing data base will be integrated and eliminate data redundancy. Consequently, less time will be required tracking and correcting data errors.

INVESTMENT EVALUATION

Break Even Point = $19,750 / $40,000 = 6 months

Return on Investment = 202%

BENEFIT$

1. More accurate and less redundant data due to systems integration of orders with credit, billing, inventory and shipping.

2. Improved customer service with fewer clerks.

3. Improved corporate image and market share.

- Goh Touchen
Project Manager
November 14, 2004

* * * END OF OUTPUT * * *

END SECTION

Section 21

"PRIDE"-DBEM PHASE 2 APPLICATION LOGICAL DATA BASE DESIGN

A data element has only one logical definition, but may be represented physically in many different ways.
- Bryce's Law

The purpose of this phase is to design or modify an Application Logical Data Base Model (ALDBM) for an information system. Several events occur during this phase:

- Application Logical objects are defined, complete with views, and data elements.
- Relationships between objects are established via the views.
- A schematic of the Application Logical Data Base Model is prepared and reviewed for accuracy.
- A technical review of the Application Logical Data Base Model is performed to validate its integrity.

METHODOLOGY NAVIGATION

A Phase 2 is initiated following a Phase 1. There is a one-to-one relationship between Phase 2 and an information system. In other words, a Phase 2 will be executed for each system identified in Phase 1 (normally there is just one).

The formal deliverable resulting from Phase 2 is an "ALDBM Design Manual" consisting of a description of the Application Logical Data Base Model. This is reviewed with Data Resource Management who must decide whether to:

- Approve the project for continuation.
- Request revisions to the phase findings.
- Discontinue the project.

Following Phase 2, the DBEM project will proceed to Phase 3, "Enterprise Logical Data Base Design," where the Application Logical Data Base Models will be merged into the Enterprise model.

GENERAL DISCUSSION

The Application Logical Data Base Model (ALDBM) represents a logical view of the data resources required by a single system (application). It does not represent "global" data resources, but rather a "local" view. The "global" view is defined by the Enterprise Logical Data Base Model (ELDBM) as defined in Phase 3. The ALDBM is intended to represent the major facts and events pertaining to a specific application. There is one such model for each Information System.

As opposed to the ELDBM, the ALDBM does not have stability as a major objective, but rather is designed to satisfy the needs of one system at one particular time. As object views change, so do the corresponding ALDB Models. Each ALDBM is a consistent subset of the ELDBM, with redundancy and overlaps being allowed among the various ALDB Models.

Because the ALDBM is defined for a system, it is normally prepared by Systems Engineering, the architect of the system. By this time, the Systems Engineer should be the leading authority regarding the data requirements for the system and, thereby, is the in the best position to define the ALDBM. During this phase, Systems Engineering is supervised by Data Engineering who provides assistance in logical data base design related problems.

BASIC CONSTRUCTS

The ALDBM is composed of a set of interrelated Objects as used in a specific application. To briefly recap the Object Concept (as described in the Methodology Overview), Objects represent business facts and events. Facts are somewhat tangible in nature; e.g., Products, Parts, Employees, Customers, Vendors, etc. Events represent some form of interaction between facts; e.g., Shipments, Orders, Billings, Payments, Purchases, etc. Each object is uniquely identified by a single data element referred to as the "primary basic grouping." **AN OBJECT IS AN OBJECT ONLY WHEN A DATA ELEMENT HAS BEEN CREATED TO UNIQUELY IDENTIFY IT.**

Objects are decomposed into "views" to describe different aspects (perspectives) of the object. There are three types of views:

1. **Identification Views** - to uniquely identify the object. All objects have an identification view.

2. **Characteristic Views** - describe specific internal characteristics about an individual object, such as Types of Addresses, Customer Contacts, etc. Characteristic Views normally apply to factual objects and are used to uniquely identify a multiple-occurrence data element; for example, a customer may have many types of addresses (billing, shipping, mailing), and many contacts (people).

3. **Relationship Views** - are used to establish a relationship between two or more objects. This is typically used by event related objects to bind two factual objects; for example:

```
CUSTOMER ----------- ORDER ----------- PRODUCT
  (Fact)             (Event)            (Fact)
```

BASIC GROUPING

Each view consists of the data elements that are used to identify, describe or quantify the object. A "basic grouping" (logical key) is established for each view. The basic grouping uniquely identifies the view and the data elements within it. The "basic grouping" is composed of one or more of the following:

- **Primary Basic Grouping** - the data element used to uniquely identify the overall object. All "views" must include this data element in the basic grouping. It is the element that binds all common views together. An "Identification View" will only require the "Primary Basic Grouping" data element in its "basic grouping."

- **Secondary Keys** - two types are available:

- For "Characteristic Views" secondary keys are used to uniquely identify a single occurrence of data. For example, "Contact Number" is used to uniquely identify a single person. "Type Address" uniquely identifies a specific type of address. This approach is similar to a given "Line Number" to uniquely identify a single text line in a body of text.

- For "Relationship Views" the secondary keys represent "foreign keys" to other objects. For example, to qualify "Quantity Ordered," a logical record is established with "Order Number" to represent the primary basic grouping (the Order Object); however, the "basic grouping" will also include "Customer Number" and "Product Number" to explicitly define the relationship and to uniquely identify "Quantity Ordered." In other words, "Customer Number" is the primary basic grouping for the Customer Object, and "Product Number" is the primary basic grouping for the Product Object.

OBJECT RELATIONSHIPS

Objects relate to one and other through "Relationship Views," not "Characteristic Views." Under this approach, a "Relationship View" is used to link other objects; for example:

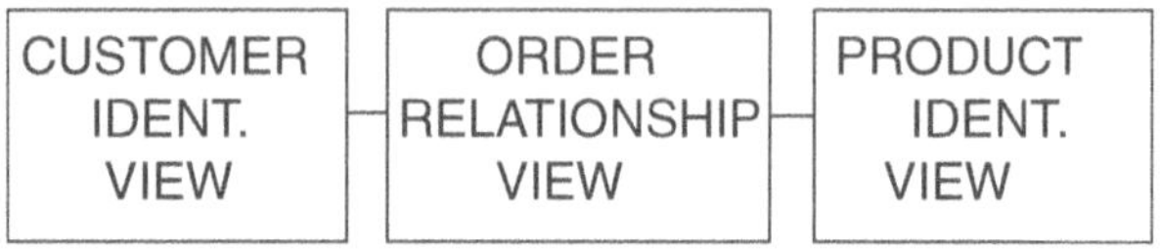

These relationships are implicitly defined by the "basic grouping" of the "Relationship View." In this example, the "basic grouping" for the "Order Relationship View" would be: "Order Number," "Customer Number," and "Product Number." In other words, "Customer Number" and "Product Number" are foreign keys to other objects.

To explicitly define the relationships between the objects pointers are defined from the "Relationship View" to the "Identification Views." Further, it is necessary to establish rules about the relationships. For example; Can a customer place many orders? Does an order relate to many products? This knowledge reflects important business rules and will be used as the integrity constraints when defining the physical data base. The typical types of relationship rules between objects include one-to-many (1:M), many-to-many (M:M), and many-to-one (M:1). For example:

The first relationship between Customer/Order represents a one-to-many relationship; a Customer can place many Orders, but an Order pertains to a single Customer. The Order/Product Relationship expresses a many-to-many relationship; a single Order may be for many Products, and one Product may be used by many Orders.

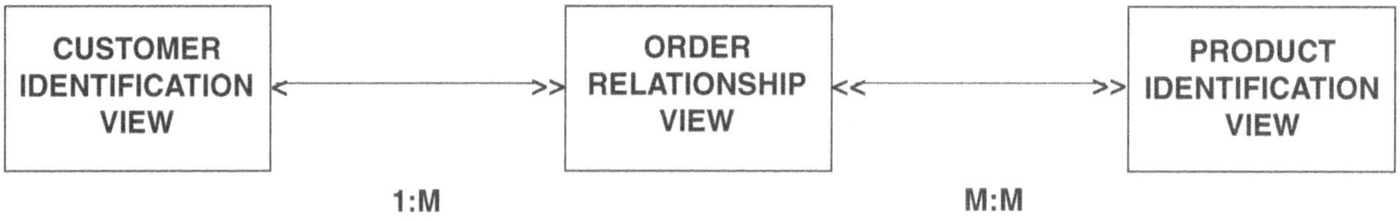

APPLICATION LOGICAL DATA BASE MODEL (ALDBM)
BASIC CONSTRUCTS

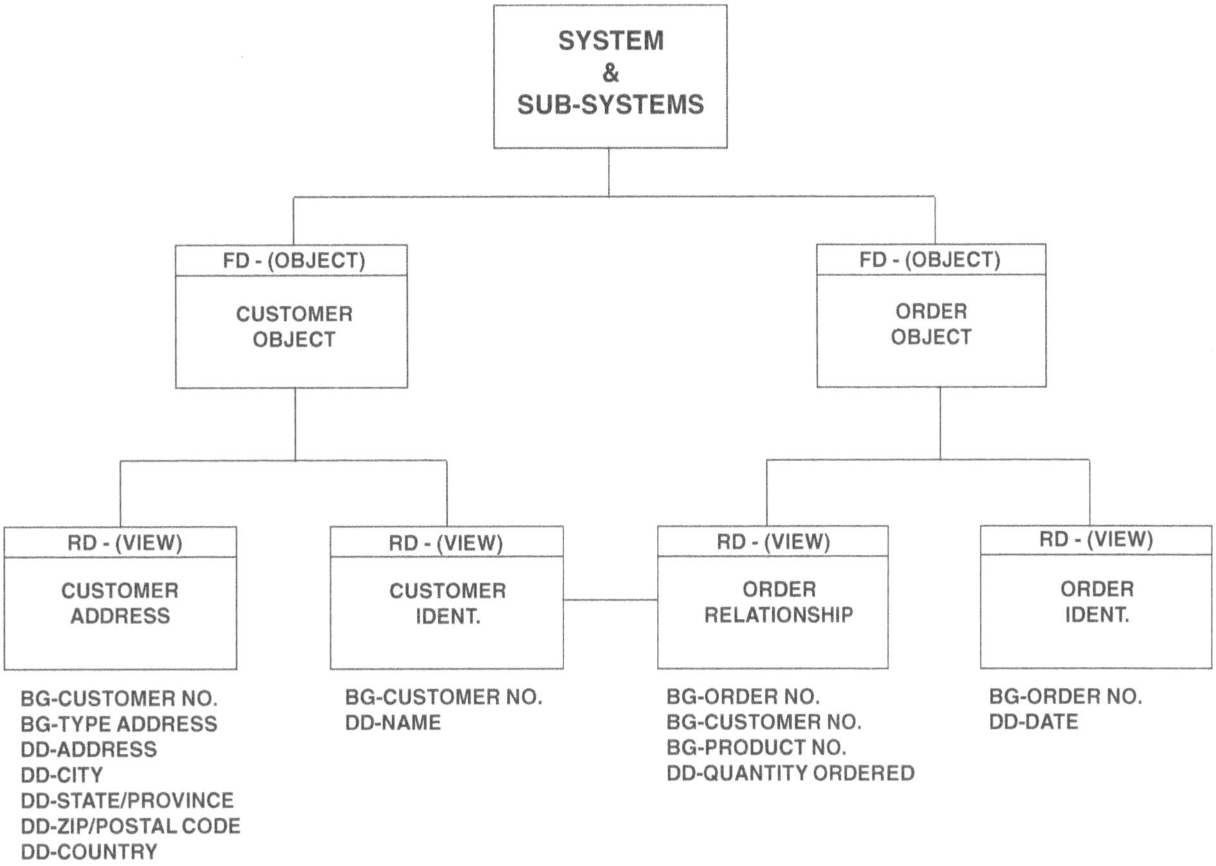

NOTES: The pointers represent relationships between the various resources as physically recorded in the IRM Repository. The relationship between the Order and the Customer would be used to indicate types of relationships between views (e.g., 1:M, M:M). Also, it implies a relationship to a Product Identification View.

- FD represents File Description
- RD represents Record Description
- BG represents the "Basic Grouping" data elements.
- DD represents Data Definitions

ALDBM/ELDBM RELATIONSHIP

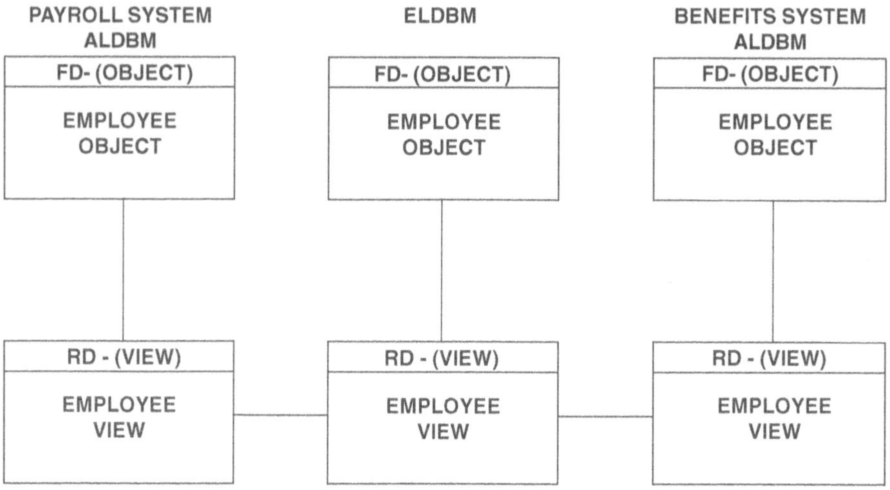

Each ALDBM Object corresponds to an ELDBM Object. As new Objects and/or data elements are uncovered by Systems Engineering, they are created first in an ALDBM and later integrated into the ELDBM by Data Engineering during Phase 3. In this way, each ALDBM Object will eventually become a subset of the corresponding ELDBM Object. (See exhibit above).

This means that there may be many different types of Employee Objects which contain subsets of data to describe the object for a particular system. However, there may be only one "global" Employee Object expressed by the ELDBM to accommodate all ALDB Models. This "application" versus "enterprise" view implies an evolutionary approach to data base design, which is natural.

The integration of ALDBM Objects into the ELDBM is developed through an iterative process of reviews and modifications to either model by Data Engineering. Since Objects are identified through the same criteria for both logical models, they should be compatible. In fact, in a hypothetical situation of a completely new enterprise, the first Application Logical Data Base Model becomes the Enterprise Model.

To satisfy the requirements of Data Base Engineering, the ALDBM must be:

1. Comprehensive enough to satisfy all of the data elements needed to support the information requirements of the system being developed/modified.

2. Easy to modify independently, that is: without causing changes to other Applications.

3. Independent from the particular hardware or software used and even from the DBMS type (hierarchical, network, relational) used for implementation.

4. Described in business terms, without excessive detail, to allow Users to clearly understand the meaning of data and also be able to check its accuracy.

5. Compatible with the ELDBM.

ALDBM TO APDBM RELATIONSHIPS

During Phase 5, Data Base Administration will design and develop the Application Physical Data Base Model (APDBM). This will consist of physical files and records that must store the data needed to satisfy the ALDBM. There is not always a one-to- one relationship between logical and physical structures. One logical file may be implemented by multiple physical files. Conversely, one physical file may implement many logical files. As we will see in the physical design phases (4 and 5), physical implementation is based on what is cost-effective and may take many forms (e.g., indexed files, indexed sequential, flat files, hierarchical/network/relational DBMS, etc.). The point is, the relationships between logical and physical must be mapped in the IRM to demonstrate that the physical implementation satisfies the logical. Again, this mapping is performed in Phase 5.

APPLICATION LOGICAL DATA BASE MODEL (ALDBM)

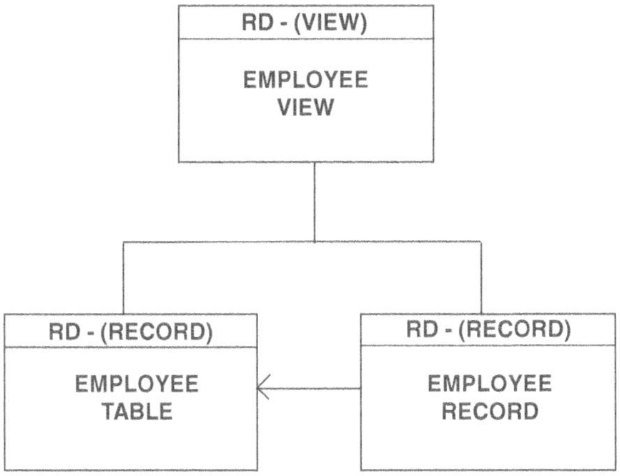

APPLICATION PHYSICAL DATA BASE MODEL (APDBM)

RECORD DESCRIPTION (RD)

As this discussion indicates, the Record Description (RD) resource in the IRM Repository is used for a variety of purposes. In a logical context, the RD represents a "View"; in a physical context, the RD represents a "Record." The RD is also used to represent other physical structures, such as: screen panels, print maps, input transactions, output data, messages, tables, etc. All of these structures exhibit the same characteristics; all require some form of unique key, and all contain data elements.

The File Description (FD) is used to represent logical files (Objects), physical files, and data bases. Files consist of records, thereby an FD/RD relationship is required. A data base represents a global view of data, therefore an FD/FD relationship is used to map all of the files within a data base.

The RD is also the primary resource used to map the relationships between all four data base models. Although the File Description (FD) provides for FD-to-FD relationships, it is the RD that is used to connect the models.

ALDBM REPRESENTATION

There are three complementary ways to graphically express the ALDBM:

• **ALDBM Diagram** - this is a simple graphic consisting of boxes to represent "objects" and arrows to represent relationships between objects. The types of arrows drawn on the graphic are used to represent relationship restraints. A single arrow represents "one" relationship, double-arrows represent "many" relationships; for example:

CUSTOMER<----->> ORDER<<----->> PRODUCT

1:M M:M

• The first relationship between Customer/Order represents a one-to-many relationship; a Customer can place many Orders, but an Order pertains to a single Customer.

The second relationship expresses a many-to-many relationship; an Order may be for many different Products, and a Product can relate to many Orders.

• **ALDBM Matrices** - the ALDBM Diagram may be fine for small applications but may be confusing for larger applications. To assist in this regard, matrices are used to express the same relationships; for example:

	CUSTOMER	ORDER	PRODUCT	SHIPMENT	BILLING
CUSTOMER		1:M		1:M	1:M
ORDER	M:1		M:M		
PRODUCT		M:M		M:M	M:M
SHIPMENT	M:1		M:M		
BILLING	M:1		M:M		

Two types of ALDBM Matrices are available. The ALDBM Views Matrix expresses the relationships between logical records. The ALDBM Objects Matrix expresses the relationships between logical files; this is deduced from subordinate record relationships.

• **Additional Matrices** - During Phase 5, Data Base Administration will define the Application Physical Data Base Model (APDBM). In order to substantiate that the APDBM correctly implements the ALDBM, additional matrices will be prepared in Phase 5 to represent:

ALDBM/APDBM Files Matrix
ALDBM/APDBM Records Matrix

These relations are recorded on the RD/RD relationship in the IRM Repository. They are used for defining the integrity constraints of the data base. Ultimately, they express business rules. It should be noted that comparable versions of these relationship graphics are available for the ELDBM.

• **Text reports** - a **"Data Structure Display"** is a structured list of FD/RD/DD relationships. It is used to express the basic structure of the ALDBM. The display is supported by an "Application Logical Files Discussion" which provides a textual explanation of each logical file (Object).

HOW DATA ELEMENTS ARE USED

The data element is the basic building block for all data resources. It is used to identify, describe and quantify the objects of the business. In this respect, it represents the facts about the business. The data element can be related to a variety of information resources, for different purposes. Data elements are documented in the IRM Repository using the Data Description (DD) resource and can be related to:

IR - INFORMATION REQUIREMENTS

The IR/DD relationship expresses all of the data elements, both primary and generated, needed to support the information requirement.

OD - OUTPUT DESCRIPTIONS

The OD/DD relationship is used to represent all of the data elements, both primary and generated, that will be physically contained on the output. Data will also be used as the Sort/Access Key for the output.

ID - INPUT DESCRIPTIONS

Data will be related to inputs to express its Sort/Access Key.

FD - FILE DESCRIPTIONS

Data will be related to files to express keys:

For logical files (Objects), a single indicative data element is selected as the "primary basic grouping."

For physical files, one or more data elements can be appointed the Sort/Access Key to the file.

RD - RECORD DESCRIPTIONS

In records, data will be used both as keys for the record, and as part of the record's content set (the collection of data elements contained in the record).

For logical records (Views), one or more indicative data elements are selected as the "basic grouping."

For physical files, one or more data elements can be appointed the Sort/Access Key to the record.

DD - DATA DESCRIPTIONS

Data to data relationships are used to express three different types of data dependencies:

1. The Source of a Generated Data Element - to express how a data element is calculated based on other data elements (primary as well as other generated data elements).

2. To express a Group Data Element (a concatenation of primary data elements). For example "Telephone Number" may be described as a group data element consisting of "Area Code," "Exchange," and "Account Number."

3. To express "Grouping" dependencies between data elements. This is used to express relationships between primary/indicative data elements and those that are primary/descriptive and quantitative. Such dependencies are used to express the meaning of the subordinate descriptive/quantitative items. For example, the descriptive data element "Name" can be related to "Customer Number" whereby it represents "Customer Name."

Of all of these resources, DBEM is primarily concerned with FD, RD, and DD resources.

DATA DEFINITION (DD)

Perhaps the most difficult task in all of DBEM is to correctly define a data element. This requires considerable skill. Historically, analysts and programmers have defined nothing more than a program label. Little insight was recorded in terms of the business facts and events being recorded. In order to properly define the ALDBM, it is necessary to supplant this physical perspective with a logical perspective.

Defining logical files and records is relatively easy in comparison to defining data elements. This is not said to intimidate the reader, but rather to issue a strong warning as to the ramifications of superficial work.

Systems Engineering will still be responsible for defining data requirements, but it is Data Engineering's responsibility to assure that data is defined correctly with a high degree of precision. If left unresolved in the ALDBM, problems will only cascade to the other data base models in subsequent phases.

Although a complete description of the characteristics of a data element are provided in the Methodology Introduction (see *"Anatomy of a Data Element"*), the following recaps the major points in defining data logically:

• A data element has one logical definition but can be physically implemented many different ways.

• The objective is to demonstrate that each data element is unique and not redundant with any other data element. Because of this, each data element must be classified according to its characteristics. A Data Taxonomy is used in this regard.

• Each data element is identified by a unique control number. A standard business name and definition is provided to describe the data element in standard business terminology (such as a Webster/Oxford-style definition).

• Data comes in two forms: primary and generated. Primary data refers to data in its virgin state; as introduced to a system from an external source (such as a person or department). Generated data refers to data that relies on other data elements in order to calculate a result (such as through some form of mathematics). A group data element is another form of generated data; it represents a concatenation of primary data elements. Generated data is perhaps the most important type of data to end-users. Primary data is important to the developers of the system.

Logical data base models only make use of primary data elements. Physical data base models can make use of both (primary and generated).

• There are three types of data elements: Indicative, Descriptive, and Quantitative.

Indicative data is used to identify objects, either in part or in full. This means that a data element may either be strong enough to uniquely identify a whole object (e.g., "Customer Number," "Order Number," "Territory Code"), or a view within an object, such as "Type Address," "Contact Number," etc. View-identifiers also uniquely identify a single occurrence of data such as where a line number is used to uniquely identify a single line of text.

• For primary data elements that are descriptive or quantitative, their dependency to indicative items are required to give them meaning (to put them in a specific context). For example, the data element "Name" should be linked to indicative data elements "Product Number," "Customer Number," etc.

APPROACH TO DESIGN

Ideally, File/Record/Data Definitions should have been created within ISEM Phase 2 (System Design) by Systems Engineering. If not, they are defined here in DBEM Phase 2. Subsequently, Systems Engineering is responsible for defining the relationships between objects using RD-to-RD Relationships.

During System Design, should Systems Engineering encounter FD/RD/DD structures in another ALDBM that satisfy the data requirements in the new application, the Systems Engineer may elect to incorporate these data structures into the new ALDBM. As part of this decision to re-use existing resources, Systems Engineering must consider the time frame by which the existing file is created/updated/referenced in other sub-systems. If compatible with the new application's requirements, it is perfectly acceptable to include the data structures in the new ALDBM. The Systems Engineer should, of course, check with Systems Resource Management to assure that this is acceptable.

During the preliminary design phases of system development/ modification projects, or as part of documentation projects, records and data elements are created and normalized and relationships are determined. Each such model is then integrated into the Enterprise Logical Data Base Model by Data Engineering during Phase 3.

DESCRIPTION OF PHASE ACTIVITIES

Activity A - Define Application Objects

Systems Engineering reviews the data requirements for all of the Information Requirements within an application. From them, an ALDBM is created consisting of File, Record and Data Descriptions. The files are attached accordingly to sub-systems.

Data Engineering supervises resource definitions.

Activity B - Design Application Logical DB Model

Systems Engineering defines the relationships between Objects using RD-to-RD relationships. Data Engineering supervises these relationships and referees discrepancies in resource relationships.

Systems Engineering then assembles the deliverables produced in the phase and packages them into a single phase review document which is evaluated by Quality Assurance prior to conducting the formal phase review with management.

Activity C - Phase 2 Review

Systems Engineering conducts a formal review of the Phase 2 deliverables with Data Resource Management, Systems Resource Management, Project Management, and Data Engineering. The review is used to evaluate the work performed thus far and revise as required. At this time, management will review the formal Phase 2 "ALDBM Design Manual" consisting of:

- Phase Cover Page
- Application Logical Files Discussion
- ALDBM Objects Matrix
- ALDBM Views Matrix
- ALDBM Diagram
- Data Structure Display
- Project Plan
- Project Estimate/Schedule Recap
- Phase Review Checklist

Based on this report and subsequent review meeting, management may elect to revise parts of the report, discontinue the project, or approve it for continuation.

The next few pages include sample deliverables resulting from this phase.

SAMPLE PHASE 2 DELIVERABLES

OD-90160	APPLICATION LOGICAL FILES DISCUSSION	"PRIDE"-ISEM
OCT 24, 2004	RELEASE 1.2.0	PAGE 1

```
*********************************************************
* FD-00101          CUSTOMER OBJECT     *
*********************************************************
```

BUSINESS PURPOSE

This logical file represents a Customer within the TMC Order Processing System (TOPS).

PRIMARY BASIC GROUPING-

DD-00001 CUSTOMER NUMBER

LOGICAL RECORDS IN FILE-

RD-00100	CUSTOMER IDENTIFICATION VIEW
RD-00101	CUSTOMER ADDRESS VIEW
RD-00102	CUSTOMER CONTACT VIEW

WHERE FILE IS USED AND HOW (CREATE/UPDATE/REFERENCE)-

OP00	CUSTOMER ORDER INQUIRY	R
OP06	CREDIT CHANGE REQUEST	R
OP07	CUSTOMER RECORD MANAGEMENT	R
OP08	ORDER-PROC FILE MAINTENANCE	C U R

ENTERPRISE LOGICAL FILES IMPLEMENTING THIS FILE-

NUMBER	NAME
FD-10101	CUSTOMER OBJECT (EL)

APPLICATION PHYSICAL FILES IMPLEMENTING THIS FILE-

NUMBER	NAME	TYPE FILE	RESIDENCE	MEDIUM
FD-00055	ORDER PROCESSING DATA BASE	PRIMARY	COMPUTER	DA
FD-00053	E-MAIL CREDIT REQUEST	WORKING	COMPUTER	DA

* * * END OF OUTPUT * * *

SAMPLE PHASE 2 DELIVERABLES

OD-90248 A L D B M O B J E C T S M A T R I X

NOV 23, 2004 RELEASE 1.2.0

"AL" OBJECTS > / "AL" OBJECTS V	FD-00100 PRODUCT OBJECT	FD-00101 CUSTOMER OBJECT	FD-00102 ORDER OBJECT	FD-00103 TERRITORY OBJECT	FD-00104 ASSIGNMENT OBJECT	FD-00105 SHIPMENT OBJECT	FD-00106 BACK ORDER OBJECT	FD-00107 PAYMENT OBJECT	FD-00108 GOVERNMENT AGENCY OBJECT
FD-00100 PRODUCT OBJECT			M:M			M:M	M:1		
FD-00101 CUSTOMER OBJECT			M:1		1:M	M:1			
FD-00102 ORDER OBJECT							M:M		
FD-00103 TERRITORY OBJECT					M:M				
FD-00108 GOVERNMENT AGENCY OBJECT									
FD-00109 ACCOUNT OBJECT								M:1	
FD-00110 BILLING OBJECT								M:1	

OD-90403 APPLICATION LOGICAL DATA BASE MODEL (ALDBM) DIAGRAM "PRIDE"-DBEM

NOV 23, 2004 RELEASE 1.2.0 PAGE 4.2

OP TRUNDLE ORDER PROC. SYS (TOPS)

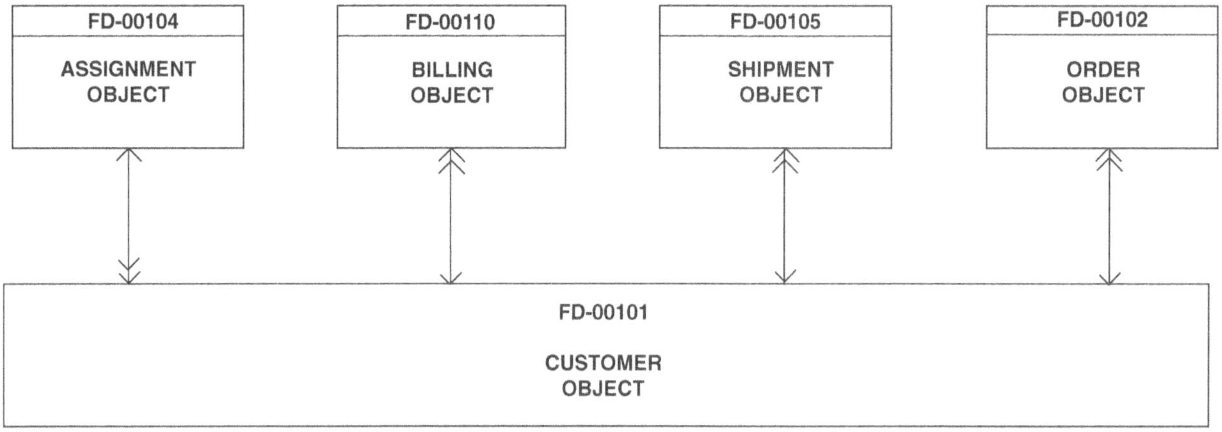

SAMPLE PHASE 2 DELIVERABLES

OD-90119	DATA STRUCTURE DISPLAY		"PRIDE"-DBEM
NOV 18, 2004	RELEASE 1.2.0		PAGE 5.1
RESOURCE	NUMBERS/NAMES:		TYPE:
FD-00049	PRODUCT FILE		APPLICATION LOGICAL
	DD-00096	PRODUCT NUMBER	PRIMARY BASIC GROUPING
RD-00049	PRODUCT IDENTIFICATION		APPLICATION LOGICAL
	DD-00096	PRODUCT NUMBER	PRIMARY BASIC GROUPING
	DD-00096	PRODUCT NUMBER	CONTENT SET
	DD-00515	NAME	CONTENT SET
RD-00050	PRODUCT CHARACTERISTICS		APPLICATION LOGICAL
	DD-00096	PRODUCT NUMBER	PRIMARY BASIC GROUPING
	DD-00177	PRODUCT TYPE	SECONDARY BASIC GROUPING
	DD-00177	PRODUCT TYPE	CONTENT SET
	DD-00178	PRODUCT STRENGTH	CONTENT SET
	DD-00179	PRODUCT SIZE	CONTENT SET
	DD-00180	PRODUCT COLOR	CONTENT SET
	DD-00111	UNIT PRICE	CONTENT SET
	DD-00156	PRODUCT UNIT OF MEASURE	CONTENT SET
	DD-00175	PRODUCT UNIT SHIP WEIGHT	CONTENT SET
	DD-00176	PRODUCT STANDARD COST	CONTENT SET

*** END OF OUTPUT ***

END SECTION

Section 22

"PRIDE"-DBEM PHASE 3 ENTERPRISE LOGICAL DATA BASE DESIGN

Data Resource Management is a neutral third party who represents the enterprise's overall interests, not just a single application.
- Bryce's Law

The purpose of this phase is to design or modify the Enterprise Logical Data Base Model (ELDBM). Several events occur during this phase:

- Enterprise Logical objects are defined, complete with views, and data elements.
- Relationships between objects are established via the views.
- Relationships between the ELDBM and pertinent ALDB Models are established.
- A schematic of the Enterprise Logical Data Base Model is prepared and reviewed for accuracy.
- A technical review of the Enterprise Logical Data Base Model is performed to validate its integrity.

METHODOLOGY NAVIGATION

For normal application development projects, a Phase 3 is performed following Phase 2. The one exception to this is when a special project is subsequently performed from an EEM related project. The phase 3, therefore, is used to define a preliminary set of enterprise level objects.

There is a one-to-one relationship between Phase 3 and the Enterprise Logical Data Base Model as represented by an FD resource in the IRM Repository.

The formal deliverable resulting from Phase 3 is an "ELDBM Design Manual" consisting of a description of the Enterprise Logical Data Base Model. This is reviewed with Data Resource Management who must decide whether to:

- Approve the project for continuation.
- Request revisions to the phase findings.
- Discontinue the project.

Following Phase 3, the DBEM project will normally proceed to Phase 4, "Enterprise Physical Data Base Design," where physical file structures will be created to implement the logical data base design. If EEM related, the project will then proceed to the Phase 6 DBEM Evaluation.

GENERAL DISCUSSION

The Enterprise Logical Data Base Model represents a global view of all Objects currently of interest to the enterprise, as well as the relationships between them. It is intended to present a unified, agreed upon, view of the major facts and events that the enterprise must keep track of in conducting business. As such, there is only one ELDBM associated with a single enterprise.

The model is designed in such a way that it can be built through an evolutionary process and when completed, becomes very stable, in the sense that new Objects need be created only as the enterprise changes its areas of business operation.

The ELDBM is composed of a set of interrelated Objects, each of which represents a certain type of enterprise fact or event, which may be as concrete or abstract as one desires, and is uniquely identifiable by a single indicative data element. Objects consist of logical records (views), which in turn are composed of data elements. A detailed discussion on Objects, Views and Data Elements is presented in both the Methodology Overview and Phase 2.

To satisfy the requirements of Data Base Engineering, the Enterprise Logical Data Base Model must be:

1. Comprehensive enough to be able to satisfy the Information Requirements of the entire enterprise.

2. Stable in the sense that after the basic Objects and relationships are identified, changes will only occur as the enterprise expands into new business areas, and not as the result of technological changes or new policies.

3. Independent from hardware or software used and even from the physical file management approach (hierarchical, network, relational) used for implementation.

4. Described in business terms, without excessive detail, to allow Users to clearly understand the meaning of data and also be able to check its accuracy.

5. Flexible to allow an evolutionary approach to design.

When completed, the ELDBM contains all Objects of interest to the Enterprise along with their relationships, records and data elements. When supporting an EEM project, such detail will not necessarily be possible. In this situation, the ELDBM will consist of nothing more than skeletal file descriptions, with their primary basic grouping, and identification view. Details about the objects, including their relationships, will be established later as a result of normal DBEM Phase 2 activity.

ENTERPRISE LOGICAL DATA BASE MODEL (ALDBM)
BASIC CONSTRUCTS

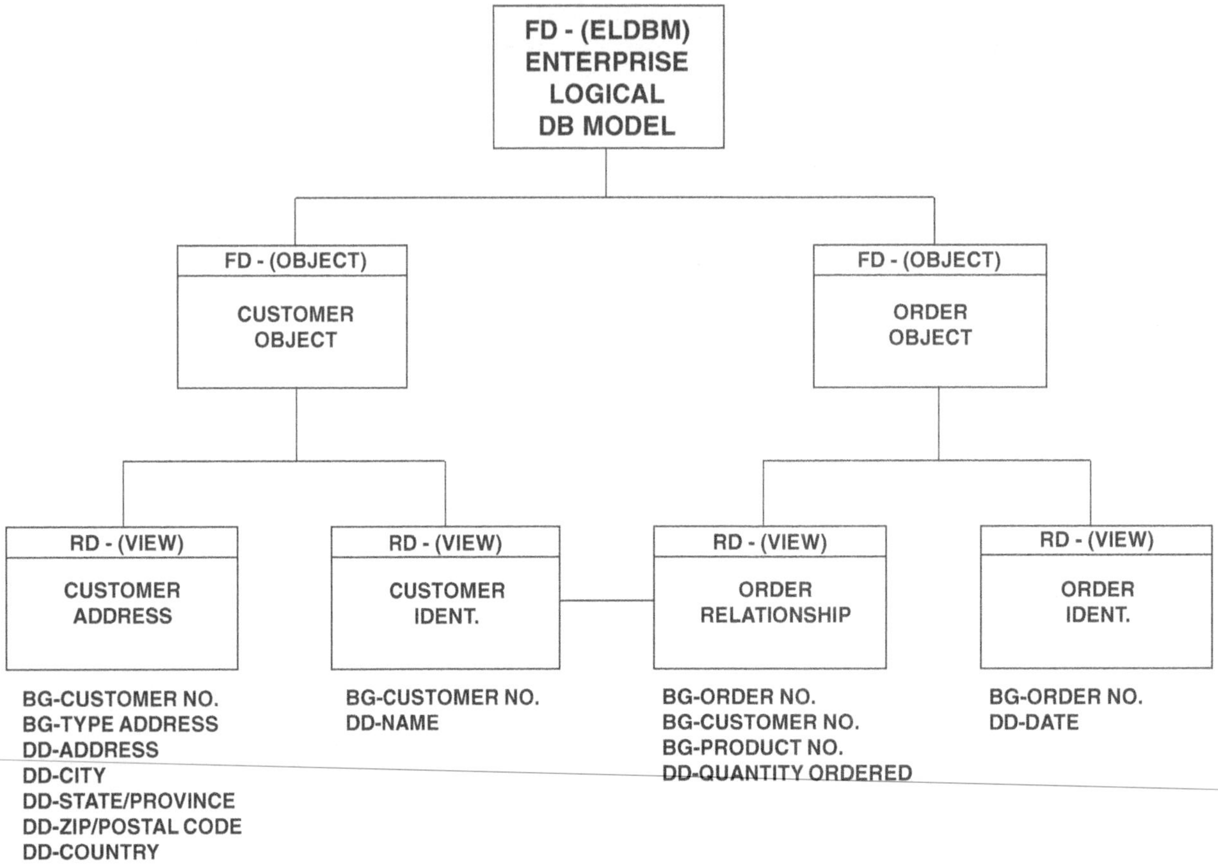

NOTES: The pointers represent relationships between the various resources as physically recorded in the IRM Repository. The relationship between the Order and the Customer would be used to indicate types of relationships between views (e.g., 1:M, M:M). Also, it implies a relationship to a Product Identification View.

- FD represents File Description
- RD represents Record Description
- BG represents the "Basic Grouping" data elements.
- DD represents Data Definitions

ALDBM/ELDBM RELATIONSHIP

Normally, Phase 3 represents a merging of the various ALDB Models. This means, for example, that there may be many "Customer" objects defined in the various ALDB Models, yet only one global "Customer" object in the ELDB Model.

As part of this merging of the various objects, Data Engineering reviews and referees discrepancies in object relationships. Although this is a normal part of Phase 2, the Data Engineer must eliminate such inconsistencies during Phase 3.

If DBEM is performed correctly, Data Engineering should not have to define any additional data elements in Phase 3 (other than those defined by Systems Engineering in Phase 2). The only exception to this is when a DBEM project follows an EEM project, instead of ISEM. To satisfy the tentative model, Data Engineering may have to define indicative data elements for use as "primary basic grouping" items in objects.

Each ALDBM Object corresponds to an ELDBM Object. In this way, each ALDBM Object can be considered a subset of a corresponding ELDBM Object. To illustrate:

This means that there may be many different types of Employee Objects which contain subsets of data to describe the object for a particular system. However, there may be only one "global" Employee Object expressed by the ELDBM to accommodate all ALDB Models. This "application" versus "enterprise" view implies an evolutionary approach to data base design, which is natural.

The integration of ALDBM Objects into the ELDBM is developed through an iterative process of reviews and modifications to either model by Data Engineering. Since Objects are identified through the same criteria for both logical models, they should be compatible. In fact, in a hypothetical situation of a completely new enterprise, the first Application Logical Data Base Model becomes the Enterprise Model.

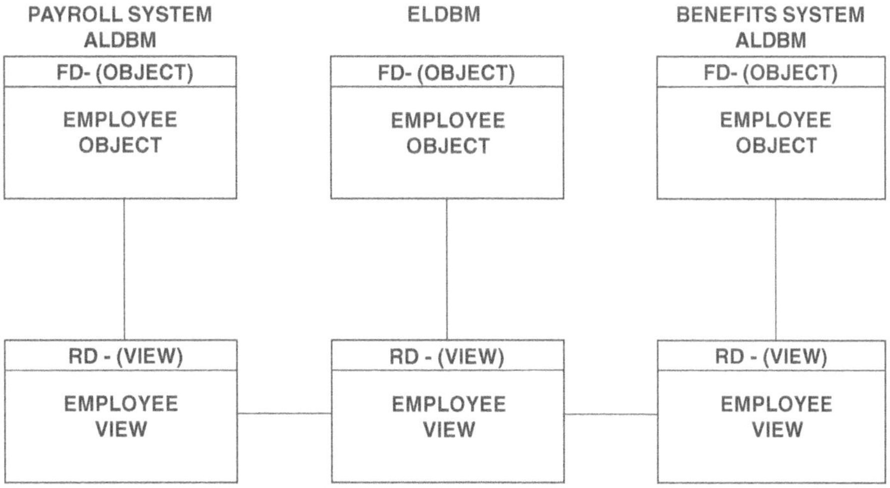

ELDBM TO EPDBM RELATIONSHIPS

During Phase 4, Data Base Administration will design and develop the Enterprise Physical Data Base Model (EPDBM). This will consist of physical files and records that must store the data needed to satisfy the ELDBM. There is not always a one-to- one relationship between logical and physical structures. One logical file may be implemented by multiple physical files. Conversely, one physical file may implement many logical files. As we will see in the physical design phases (4 and 5), physical implementation is based on what is cost-effective and may take many forms (e.g., indexed files, indexed sequential, flat files, hierarchical/network/relational DBMS, etc.). The point is, the relationships between logical and physical must be mapped in the IRM Repository to demonstrate that the physical implementation satisfies the logical. Again, this mapping is performed in Phase 4.

ENTERPRISE LOGICAL DATA BASE MODEL (ELDBM)

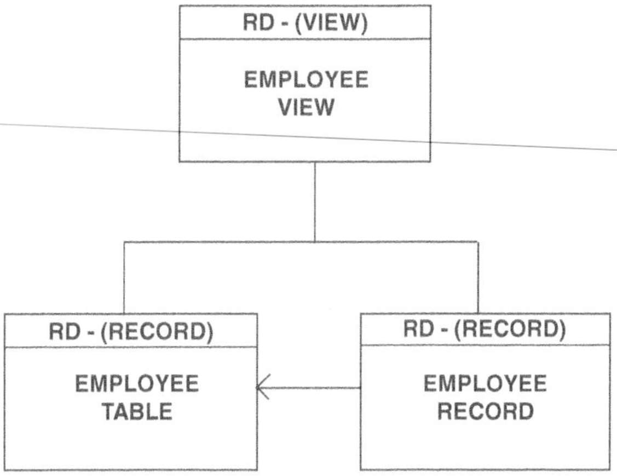

ENTERPRISE PHYSICAL DATA BASE MODEL (EPDBM)

RECORD DESCRIPTION (RD)

As this discussion indicates, the Record Description (RD) resource in the IRM is used for a variety of purposes. In a logical context, the RD represents a "View"; in a physical context, the RD represents a "Record." The RD is also used to represent other physical structures, such as: screen panels, print maps, input transactions, output data, messages, tables, etc. All of these structures exhibit the same characteristics; all require some form of unique key, and all contain data elements.

The File Description (FD) is used to represent logical files (Objects), physical files, and data bases. Files consist of records, thereby an FD/RD relationship is required. A data base represents a global view of data, therefore an FD/FD relationship is used to map all of the files within a data base.

The RD is also the primary resource used to map the relationships between all four data base models. Although the File Description (FD) provides for FD-to-FD relationships, it is the RD that is used to connect the models.

The following diagram illustrates all of the BASIC relationships needed to map the four data base models in the IRM. Additional pointers are available for different purposes, particularly for complicated physical file structures.

IRM RESOURCE RELATIONSHIPS - MAPPING THE FOUR DATA BASE MODELS

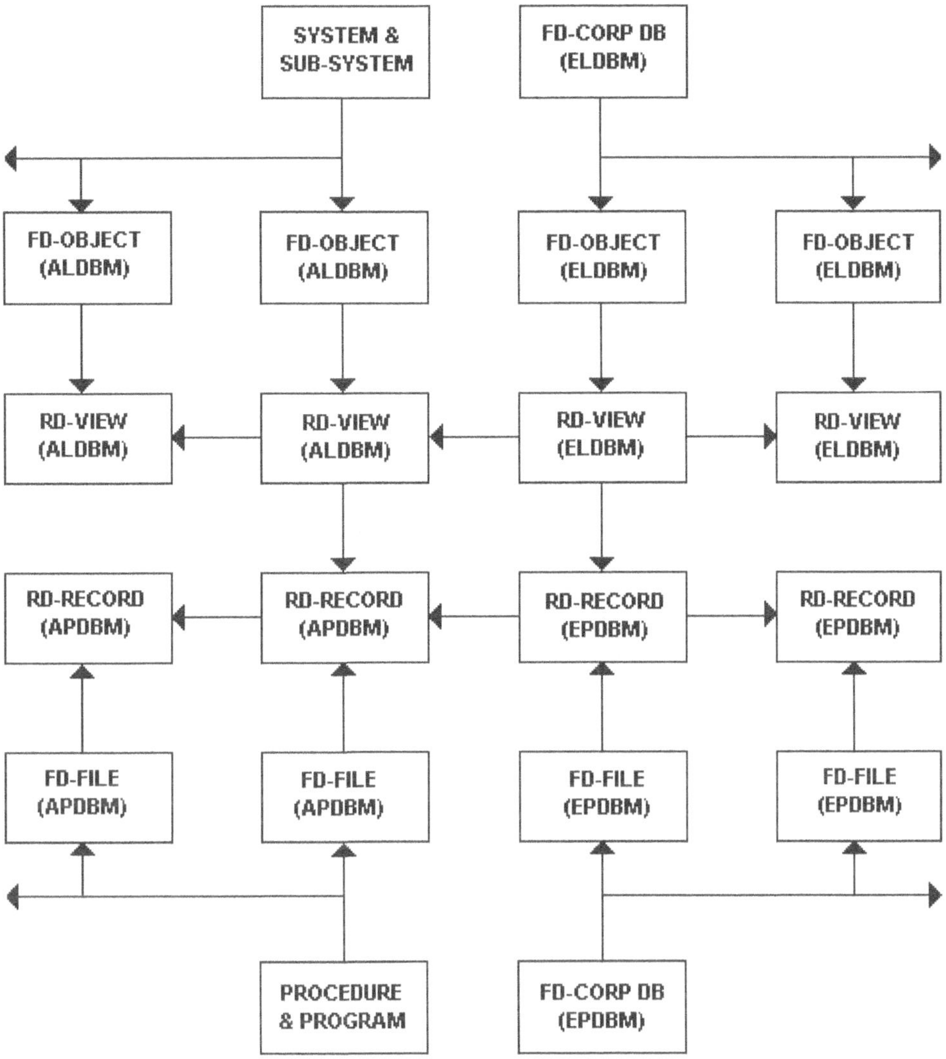

NOTES:
ALDBM - Application Logical Data Base Model FD - File Description
ELDBM - Enterprise Logical Data Base Model RD - Record Description
EPDBM - Enterprise Physical Data Base Model
APDBM - Application Physical Data Base Model

This diagram illustrates all of the BASIC relationships needed to map the four data base models. Additional pointers are available for different purposes, particularly for complicated physical file structures.

ELDBM REPRESENTATION

There are two complementary ways to graphically express the ELDBM:

• **ELDBM Diagram** - this is a simple graphic consisting of boxes to represent "objects" and arrows to represent relationships between objects. The types of arrows drawn on the graphic are used to represent relationship constraints. A single arrow represents "one" relationship, double-arrows represent "many" relationships; for example:

CUSTOMER<----->> ORDER<<----->> PRODUCT

1:M M:M

The first relationship between Customer/Order represents a one-to-many relationship; a Customer can place many Orders, but an Order pertains to a single Customer.

The second relationship expresses a many-to-many relationship; an Order may be for many different Products, and a Product can relate to many Orders.

• **ELDBM Matrices** - the ELDBM Diagram may be fine to express simple models, but is not conducive for expressing the overall data model. To assist in this regard, matrices are used to express the same relationships; for example:

	CUSTOMER	ORDER	PRODUCT	SHIPMENT	BILLING
CUSTOMER		1:M		1:M	1:M
ORDER	M:1		M:M		
PRODUCT		M:M		M:M	M:M
SHIPMENT	M:1		M:M		
BILLING	M:1		M:M		

Two types of ELDBM Matrices are available. The ELDBM Views Matrix expresses the relationships between logical records. The ELDBM Objects Matrix expresses the relationships between logical files; this is deduced from subordinate record relationships.

• **Additional Matrices** - To substantiate that the ELDBM correctly implements the various ALDB Models, comparable matrices are used to show relationships. This includes the ALDBM/ELDBM Objects Matrix and the ALDBM/ELDBM Views Matrix.

During Phase 4, when Data Base Administration designs the Enterprise Physical Data Base Model (EPDBM), additional matrices will be developed to express ELDBM/EPDBM relationships.

These relations are recorded on the RD/RD relationship in the IRM Repository. They are used for defining the integrity constraints of the data base. Ultimately, they express business rules. It should be noted that comparable versions of these relationship graphics are available for the other data base models.

In addition to the data base diagrams and matrices, a Data Structure Display is prepared during this phase which provides a concise and organized breakdown of the data resources used.

DESCRIPTION OF PHASE ACTIVITIES

Activity A - Define Enterprise Objects

Data Engineering reviews either:

• ALDBM Models as prepared by Systems Engineering in Phase 2 and, from this, creates or modifies the ELDB Model consisting of File, Record and Data Descriptions. Also, relationships between ALDBM models and the ELDBM model is established. This is based on RD-to-RD relationships. NOTE: ALDBM/ELDBM Relationships are not required when implementing an EEM related project.

• High level Information Requirements resulting from EEM and, from them, creates or modifies high level descriptions of objects. This includes an FD to represent each object; an RD to represent the "Identification View" for each object, and primary/ indicative data elements for use as the "primary basic grouping" of the objects and identification views.

Activity B - Design Enterprise Logical DB Model

Data Engineering defines the relationships between Objects using RD-to-RD relationships.

Data Engineering then assembles the deliverables produced in the phase and packages them into a single phase review document which is evaluated by Quality Assurance prior to conducting the formal phase review with management.

Activity C - Phase 3 Review

Data Engineering conducts a formal review of the Phase 3 deliverables with Data Resource Management, Project Management, and Data Base Administration. The review is used to evaluate the work performed thus far and revise as required. At this time, management will review the formal Phase 3 **"ELDBM Design Manual"** consisting of:

- Phase Cover Page
- ELDBM Narrative
- ELDBM Objects Matrix
- ELDBM Views Matrix
- ELDBM Diagram **
- Data Structure Display
- ALDBM/ELDBM Objects Matrix **
- ALDBM/ELDBM Views Matrix **
- Project Plan
- Project Estimate/Schedule Recap
- Phase Review Checklist

** May be considered optional due to the nature of the project.

Based on this report and subsequent review meeting, management may elect to revise parts of the report, discontinue the project, or approve it for continuation.

The next few pages include sample deliverables resulting from this phase.

```
--------------------------------------------------------------------------------------------
OD-90405        ENTERPRISE LOGICAL DATA BASE MODEL (ELDBM) NARRATIVE        "PRIDE"-DBEM
--------------------------------------------------------------------------------------------
DEC 5, 2004                        RELEASE 1.2.0                            PAGE 1.1

              ******************************************************************
              *        FD-00001           ENTERPRISE LOGICAL DB MODEL          *
              ******************************************************************

        BUSINESS PURPOSE

        This represents the collection of all Enterprise Logical "Objects"
        (Logical Files) in the enterprise. It specifically does not include
        Application Logical objects or physical files which should be maintained
        elsewhere.

        OBJECTS WITHIN DATA BASE:

        NUMBER          NAME

        FD 00309        ACCOUNT OBJECT
        FD-00304        ASSIGNMENT OBJECT
        FD-00306        BACK ORDER OBJECT
        FD-00310        BILLING OBJECT
        FD-00301        CUSTOMER OBJECT
        FD-00308        GOVERNMENT AGENCY OBJECT
        FD-00302        ORDER OBJECT
        FD-00307        PAYMENT OBJECT
        FD-00300        PRODUCT OBJECT
        FD-00305        SHIPMENT OBJECT
        FD-00303        TERRITORY OBJECT

VVVVVVVVVVVVVVVVVVVVVVVVVVVVVVVVVVVVVVVVVVVVVVVVVVVVVVVVVVVVVV

                          * * * END OF OUTPUT * * *
```

SAMPLE PHASE 3 DELIVERABLES

OD-90247 E L D B M O B J E C T S M A T R I X

DEC 5, 2004 RELEASE 1.2.0

"EL" OBJECTS > / "EL" OBJECTS V	FD-00300 PRODUCT OBJECT	FD-00301 CUSTOMER OBJECT	FD-00302 ORDER OBJECT	FD-00303 TERRITORY OBJECT	FD-00304 ASSIGNMENT OBJECT	FD-00305 SHIPMENT OBJECT	FD-00306 BACK ORDER OBJECT	FD-00307 PAYMENT OBJECT	FD-00308 GOVERNMENT AGENCY OBJECT	F A O
FD-00300 PRODUCT OBJECT			M:M			M:M	M:1			
FD-00301 CUSTOMER OBJECT			M:1		1:M	M:1				
FD-00302 ORDER OBJECT							M:M			
FD-00303 TERRITORY OBJECT					M:M					
FD-00308 GOVERNMENT AGENCY OBJECT										
FD-00309 ACCOUNT OBJECT								M:1		
FD-00310 BILLING OBJECT								M:1		

OD-90274 E L D B M V I E W S M A T R I X "]

DEC 5, 2004 RELEASE 1.2.0 P.

"EL" VIEWS > / "EL" VIEWS V	RD-00300 CUSTOMER IDENTIFICA TION VIEW	RD-00301 CUSTOMER ADDRESS VIEW	RD-00302 CUSTOMER CONTACT VIEW	RD-00303 PRODUCT IDENTIFICA TION VIEW	RD-00304 GOV'T IDENTIFICA TION VIEW	RD-00305 GOVERNMENT TAX RATE VIEW	RD-00306 ORDER IDENTIFICA TION VIEW	RD-00307 CUSTOMER/ PRODUCT REL. VIEW	RD-00308 BACK-ORDER IDENTIFICA TION VIEW	RD-(ORD] PRO] REL
RD-00300 CUSTOMER IDENT IFICATION VIEW								M:1		
RD-00303 PRODUCT IDENTIF ICATION VIEW								M:M		]
RD-00304 GOV'T INDENTIFI CATION VIEW										
RD-00306 ORDER IDENTIFIC CATION VIEW										]
RD-00312 ACCOUNT IDENTIF ICATION VIEW										
RD-00313 BILLING IDENTIF ICATION VIEW										
RD-00318 TERRITORY IDENT IFICATION VIEW										

SAMPLE PHASE 3 DELIVERABLES

OD-90404	ENTERPRISE LOGICAL DATA BASE MODEL (ELDBM) DIAGRAM	"PRIDE"-DBEM
DEC 5, 2004	RELEASE 1.2.0	PAGE 4.2

FD-00001 ENTERPRISE LOGICAL DB MODEL

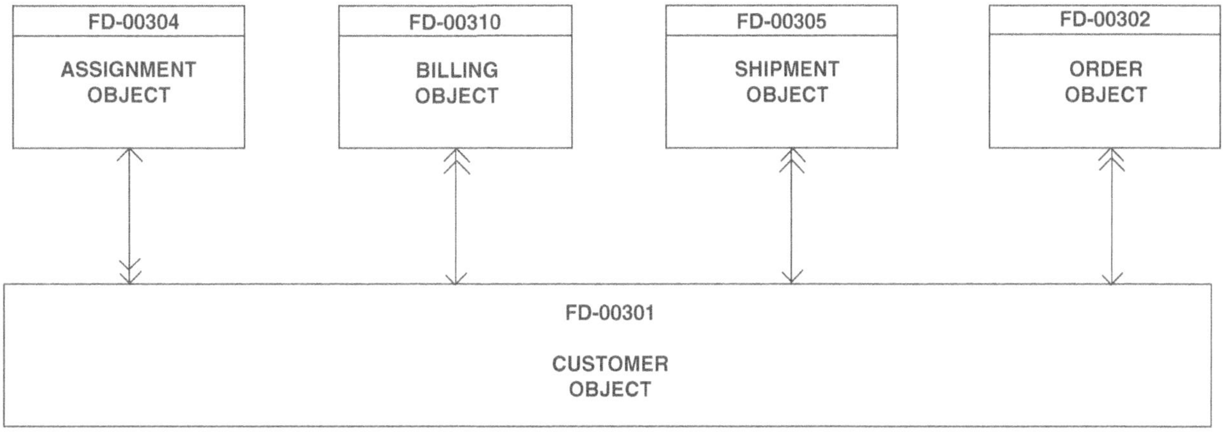

OD-90119	DATA STRUCTURE DISPLAY	"PRIDE"-DBEM
NOV 18, 2004	RELEASE 1.2.0	PAGE 5.1

RESOURCE	NUMBERS/NAMES:		TYPE:
FD-00049	PRODUCT FILE		APPLICATION LOGICAL
	DD-00096	PRODUCT NUMBER	PRIMARY BASIC GROUPING
RD-00049	PRODUCT IDENTIFICATION		APPLICATION LOGICAL
	DD-00096	PRODUCT NUMBER	PRIMARY BASIC GROUPING
	DD-00096	PRODUCT NUMBER	CONTENT SET
	DD-00515	NAME	CONTENT SET
RD-00050	PRODUCT CHARACTERISTICS		APPLICATION LOGICAL
	DD-00096	PRODUCT NUMBER	PRIMARY BASIC GROUPING
	DD-00177	PRODUCT TYPE	SECONDARY BASIC GROUPING
	DD-00177	PRODUCT TYPE	CONTENT SET
	DD-00178	PRODUCT STRENGTH	CONTENT SET
	DD-00179	PRODUCT SIZE	CONTENT SET
	DD-00180	PRODUCT COLOR	CONTENT SET
	DD-00111	UNIT PRICE	CONTENT SET
	DD-00156	PRODUCT UNIT OF MEASURE	CONTENT SET
	DD-00175	PRODUCT UNIT SHIP WEIGHT	CONTENT SET
	DD-00176	PRODUCT STANDARD COST	CONTENT SET

* * * END OF OUTPUT * * *

SAMPLE PHASE 3 DELIVERABLES

OD-90285 A L D B M / E L D B M O B J E C T S M A T R I X

DEC 5, 2004 RELEASE 1.2.0

"AL" OBJECTS > / "EL" OBJECTS V	FD-00100 PRODUCT OBJECT	FD-00101 CUSTOMER OBJECT	FD-00102 ORDER OBJECT	FD-00103 TERRITORY OBJECT	FD-00104 ASSIGNMENT OBJECT	FD-00105 SHIPMENT OBJECT	FD-00106 BACK ORDER OBJECT	FD-00107 PAYMENT OBJECT	FD-00108 GOVERNMENT AGENCY OBJECT	F A C
FD-00300 PRODUCT OBJECT	X									
FD-00301 CUSTOMER OBJECT		X								
FD-00302 ORDER OBJECT			X							
FD-00303 TERRITORY OBJECT				X						
FD-00304 ASSIGNMENT OBJECT					X					
FD-00305 SHIPMENT OBJECT						X				
FD-00306 BACK ORDER OBJECT							X			

OD-90286 A L D B M / E L D B M V I E W S M A T R I X

DEC 5, 2004 RELEASE 1.2.0

"AL" VIEWS > / "EL" VIEWS V	RD-00100 CUSTOMER IDENTIFICA TION VIEW	RD-00101 CUSTOMER ADDRESS VIEW	RD-00102 CUSTOMER CONTACT VIEW	RD-00103 PRODUCT IDENTIFICA TION VIEW	RD-00104 GOV'T IDENTIFICA TION VIEW	RD-00105 GOVERNMENT TAX RATE VIEW	RD-00106 ORDER IDENTIFICA TION VIEW	RD-00107 CUSTOMER/ PRODUCT REL. VIEW	RD-00108 BACK-ORDER IDENTIFICA TION VIEW	R O P R
RD-00300 CUSTOMER IDENT IFICATION VIEW	X									
RD-00301 CUSTOMER ADDRESS VIEW		X								
RD-00302 CUSTOMER CONTACT VIEW			X							
RD-00303 PRODUCT IDENTIF ICATION VIEW				X						
RD-00304 GOV'T IDENTIFIC ATION VIEW					X					
RD-00305 GOVERNMENT TAX RATE VIEW						X				
RD-00306 ORDER IDENTIFIC ATION VIEW							X			
RD-00307 CUSTOMER/PRODUCT								X		

END SECTION

Section 23

"PRIDE"-DBEM PHASE 4 ENTERPRISE PHYSICAL DATA BASE DESIGN

Whereas the logical Data Base will remain relatively static, the physical data base changes dynamically.
- Bryce's Law

The purpose of this phase is to design or modify the Enterprise Physical Data Base Model (EPDBM). Several events occur during this phase:

- A method of implementation is selected to physi cally store data to satisfy the logical data base models.
- Enterprise Physical files are defined, complete with records and data elements.
- Relationships between physical files and records are established.
- Relationships between the EPDBM and ELDBM are established.
- A schematic of the Enterprise Physical Data Base Model is prepared and reviewed for accuracy.

METHODOLOGY NAVIGATION

A Phase 4 is initiated following completion of a normal Phase 3 (ISEM related). There is a one-to-one relationship between Phase 4 and the Enterprise Physical Data Base Model as represented by an FD resource in the IRM Repository.

The formal deliverable resulting from Phase 4 is an "EPDBM Design Manual" consisting of a description of the Enterprise Physical Data Base Model. This is reviewed with Data Resource Management who must decide whether to:

- Approve the project for continuation.
- Request revisions to the phase findings.
- Discontinue the project.

Following Phase 4, the DBEM project will proceed to Phase 5, "Application Physical Data Base Design," where physical file structures will be created to implement the logical data base design for a single application.

GENERAL DISCUSSION

Phase 4 represents a major transitional stage in the methodology: going from logical data base design to physical. Whereas Data Engineering had primary responsibility for executing the early phases of DBEM, responsibility now shifts to Data Base Administration and Data Communications Administration who are responsible for the design and development of physical hardware/software related to the collection, storage and retrieval of data.

The Enterprise Physical Data Base Model (EPDBM) is a global view of all of the files used to store data, this includes manual files, computer related files, and any other storage media. As such, there is only one EPDBM associated with a single enterprise.

Whereas the ELDBM is a relatively static view of all corporate data, the EPDBM will be more volatile. It reflects the technology used for implementation at any given time. To create the EPDBM, Objects of the ELDBM may be regrouped or split in different ways, according to performance, security and administrative considerations. In other words physical files and records will not necessary be designed in the same manner as their logical counterparts.

The model consists of multi-level aggregations of data within one or more structures (Network, Hierarchical, Relational, flat files), which will depend on the type of technology and particular product(s) being used. This narrative refers to these aggregates as data bases, files, records and data elements, but relates them to the terminology used in a particular product. Some products refer to this level of modeling as the "Schema" or "Logical Data Bases" and reserves the term Physical Model to descriptions at a lower level of abstraction, sometimes involving even hardware descriptions. DBEM considers this level as the Physical Model and reserves the term Logical for the independent levels of abstraction.

The EPDBM provides data base designers with the freedom to choose the best technology for each new application, either a DBMS, disk, tape or even manual files, keeping it independent from the logical models. Logical files and records in the ELDBM are ultimately defined to reflect the rules of the business, independent of physical considerations. A straightforward implementation with the same logical and physical structures could possibly produce poor performance or place unreasonable physical constraints on the data base. Even more important, the physical implementation may be changed in response to performance tuning, new technology, or even varying usage loads without impacting the logical models known to Users and Systems Engineering. In particular, distributed data bases and data communications procedures are implemented and documented at this level only, and not in the logical models.

The Enterprise Physical Data Base Model must satisfy the following requirements:

1. Be comprehensive enough to represent all data stored in the enterprise's data base, regardless if physical files are centralized or distributed.

2. Be updated constantly to reflect any changes made to existing files as a result of new developments, modifications, improvements or even performance related reorganizations.

3. Contain an exact image of the physical implementation, including access paths or any other information necessary for the particular DBMS or file management technique being used.

4. Uses the structure and terminology of the specific product used to implement each part of the EPDBM.

The number of aggregation levels, their names and contents vary widely with each file management technique, which makes it impossible to refer to all of them in this narrative. When describing the procedures to convert the ELDBM to either flat files or each of the major DBMS types (Hierarchical, Network and Relational) standard terminology is used within each one. It is quite possible to have several DBMS packages as well as any other types of files within the EPDBM. Each product's terminology and constructs is used in the parts of the EPDBM it implements.

LOGICAL VERSUS PHYSICAL RESOURCE STRUCTURES

The differences between logical and physical data structures can be considerable. Aside from using the same type of IRM resources to model them in the IRM Repository (FD, RD, DD resources), there is little else that the two models share in common.

The logical model is governed based on some very specific rules. For example, "primary basic groupings," the three types of "views," fact/event relationships, use of indicative data, etc. These rules were created to develop simple data abstractions about the various objects used in the operation of the business. Ultimately, they promote data independence from the physical environment.

The physical model is totally different. It was created to model the physical environment, and, as such, is dependent on it. This is why it will evolve as technology evolves. Unlike the logical model where the "basic grouping" requires indicative data, the physical model can use whatever sort/access key is convenient for the storage and processing of data. Further, records can be merged and separated as required. This reflects the fact that the physical model is dependent on the physical file management tools and techniques in use.

FORMS OF PHYSICAL FILES

The media used to implement physical files can take many forms: disk, tape, diskettes, cards, manual files, even memory. Defining a file in memory is somewhat controversial, yet important. If programs are going to write working files to memory as part of their normal processing, understanding the layout of the file is just as important as if it were written to disk. Data Base Administration needs to know the capabilities of physical files and processing requirements in order to advise DP Operations for hardware capacity planning.

Computer files may use a variety of file management techniques: Sequential, Index Sequential, Random, or DBMS specific. The decision to use a particular file management technique should be based on what is effective and economical, not on what is technically fashionable. For example, not everything requires a DBMS. There is little point to using a DBMS if it is going to be used as nothing more than an elegant access method.

When selecting a suitable file management technique, Data Base Administration must consider the following:

- **PERFORMANCE** - anticipated volume of transactions versus speed. Typically, the greater the volume of transactions, the slower the computer processing speed. The fewer the transactions, the greater the responsiveness. Systems Engineering should be able to estimate the required transaction volumes, processing speed, file volatility and hit ratio. This will become the basis for establishing benchmark tests.

- **SECURITY & SHARING** - this will influence the file access method to be selected, distribution, and use in multiple applications.

- **BACKUP/RECOVERY REQUIREMENTS** - will dictate file retention and other maintenance considerations.

This highlights the fact that Data Base Administration requires information about the various applications in order to arrive at a suitable physical file design. This includes data and processing specifications as recorded in the IRM by Systems Engineering. Data Communications will also use these specifications to devise a data transmission/conversion strategy.

SELECTING A DATA BASE MANAGEMENT SYSTEM (DBMS)

When should you consider a DBMS? Any time when data is going to be shared between applications on the same computer. As with any file management technique, selection should be based on the criteria mentioned above. Because there are significant differences between various file management techniques, it is unlikely that a single DBMS will be able to serve all of the needs of a company. In fact, DBMS packages will proliferate in companies as data bases become more distributed.

Currently, there are essentially three different types of DBMS products commercially available:

- **Hierarchical** - Featuring a parent/child organization structure (a tree structure).

- **Network** - Featuring user defined chains and pointers to relate the different sections in a data base.

- **Relational** - Featuring user defined tables based on keys.

NOTE: As of this writing, Object-Oriented DBMS packages are starting to emerge. However, the technology is still in its infancy and, as such, has not been included in this narrative.

Although the intent of the three DBMS' is to promote the sharing of data and simplify data access, their implementations are significantly different. To illustrate:

DBMS PROS AND CONS

Feature	HIERARCHICAL	NETWORK	RELATIONAL
ACCESS PATH	Tree Processing.	User defined chains; can be extensive.	Based on selected keys.
REFERENTIAL INTEGRITY	Excellent.	Good if chains are well maintained.	Good; shows promise.
DATA SHARING	Hindered due to complicated global DB. Multi data bases normally occur.	Good. Maintaining global DB	Excellent. Encourages sharing.
PROCESSING CONSIDERATIONS	Excellent for heavy transaction volume.	Excellent for heavy transaction volume.	Slower for heavy transaction volume. Consumes resources.
DEVELOPMENT CONSIDERATIONS	Harder to develop. Mostly supports procedural languages.	Harder to develop. Mostly supports procedural languages.	Easier to develop. Supports both procedural & non-proc lang.
MISCELLANEOUS	No standard. Predominantly IBM's IMS.	CODASYL standard. Many packages.	SQL Standard-ANSI. Trend is to relational.

Obviously there are trade-offs between the various approaches. One is not necessarily better than the other. Bottom-line: Data Base Administration must resolve how well the DBMS fulfills the data and processing requirements of the various applications.

In order to represent the different structures found in the major DBMS products within a standard framework, it is necessary to draw a correspondence among the various levels of aggregation they support. The table below shows such a correspondence and the IRM resources used to represent them:

IRM Resources	Flat File	Hierarchical (IMS)	Network (CODASYL)	Relational (DB2)
DD	Field	Field	Field	Attribute
RD	Record	Segment	Record	Table
FD	File	Physical DB	Area	Tablespace
FD		Logical DB	Schema	Database

While the first two levels (field and record) are quite similar for most DBMS packages, the clustering provided by an IMS Physical Data Base, a CODASYL Area or a DB2 Tablespace are quite different. However, these and higher level aggregations are still represented as FD resources.

ENTERPRISE PHYSICAL DATA BASE MODEL (EPDBM)
BASIC CONSTRUCTS

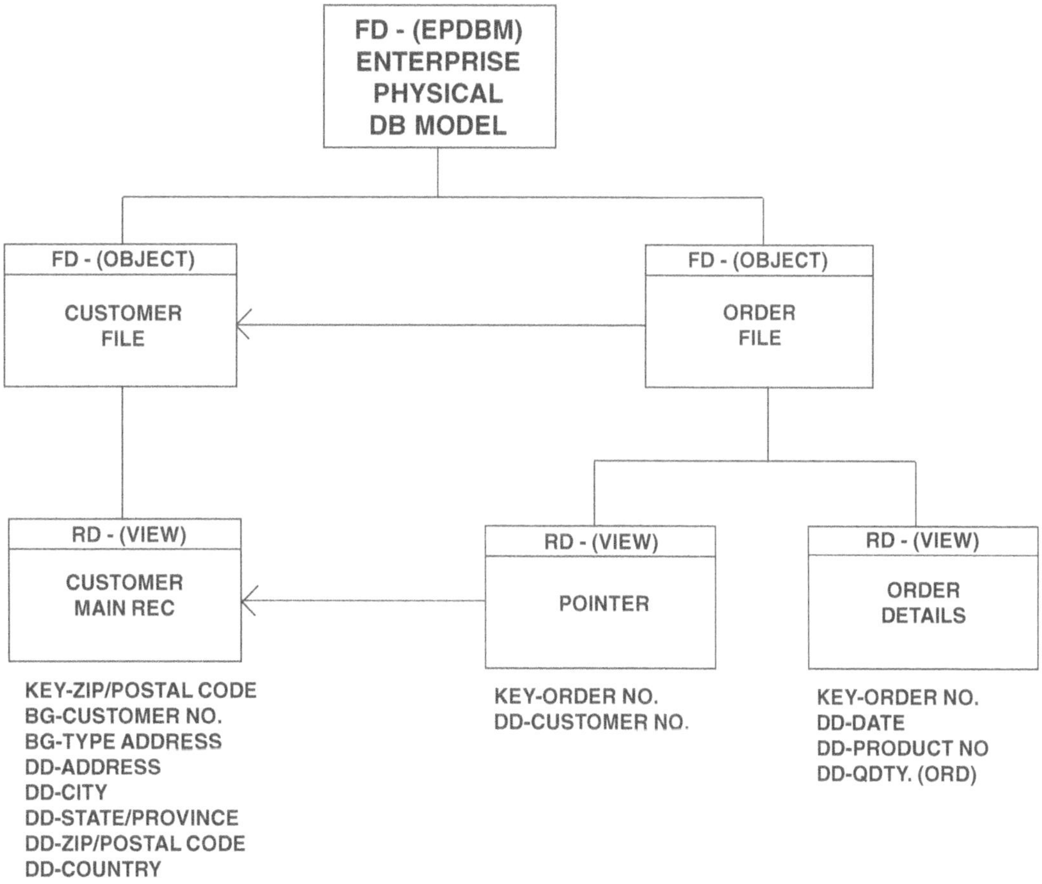

NOTES: The pointers represent relationships between the various resources as physically recorded in the IRM. RD-to-RD and FD-to-FD relationships depend on the selected file management technique.

- FD represents File Description
- RD represents Record Description
- KEY represents the "Sort/Access Key" data elements.
- DD represents Data Definitions

MANUAL FILES

Manual files, which are used to store documents, follow a slightly different relationship:

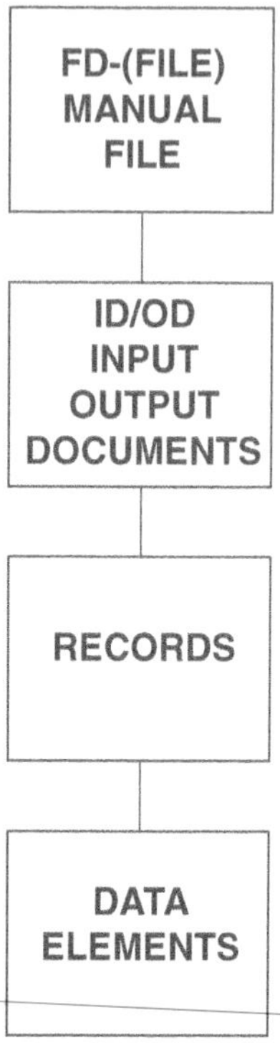

ELDBM TO EPDBM RELATIONSHIPS

Data Base Administration has the responsibility of mapping the relationships between the ELDBM and the EPDBM. As mentioned, there is not always a one-to-one relationship between logical and physical structures. One logical file may be implemented by multiple physical files. The point is, the relationships between logical and physical must be mapped in the IRM Repository to demonstrate that the physical implementation satisfies the logical.

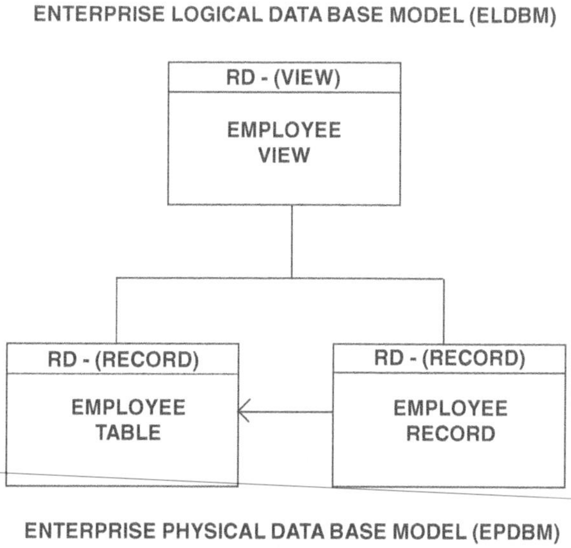

EPDBM TO APDBM

The EPDBM represents the global view of data storage requirements. During Phase 5, Data Base Administration must consider physical requirements for individual systems. This is where the Application Physical Data Base Model (APDBM) is defined. Ultimately, the APDBM represents subsets of the EPDBM. Relationships between the EPDBM and the APDBM can be expressed using RD-to-RD and FD-to-FD relationships.

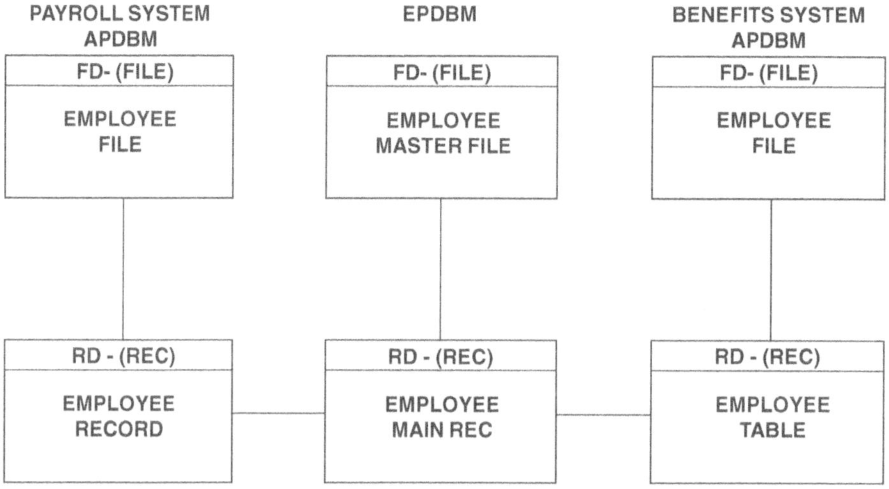

RECORD DESCRIPTION (RD)

As this discussion indicates, the Record Description (RD) resource in the IRM is used for a variety of purposes. In a logical context, the RD represents a "View"; in a physical context, the RD represents a "Record." The RD is also used to represent other physical structures, such as: screen panels, print maps, input transactions, output data, messages, tables, etc. All of these structures exhibit the same characteristics; all require some form of unique key, and all contain data elements.

The File Description (FD) is used to represent logical files (Objects), physical files, and data bases. Files consist of records, thereby an FD/RD relationship is required. A data base represents a global view of data, therefore an FD/FD relationship is used to map all of the files within a data base.

The RD is also the primary resource used to map the relationships between all four data base models. Although the File Description (FD) provides for FD-to-FD relationships, it is the RD that is used to connect the models.

The following diagram illustrates all of the BASIC relationships needed to map the four data base models in the IRM Repository. Additional pointers are available for different purposes, particularly for complicated physical file structures.

IRM RESOURCE RELATIONSHIPS - MAPPING THE FOUR DATA BASE MODELS

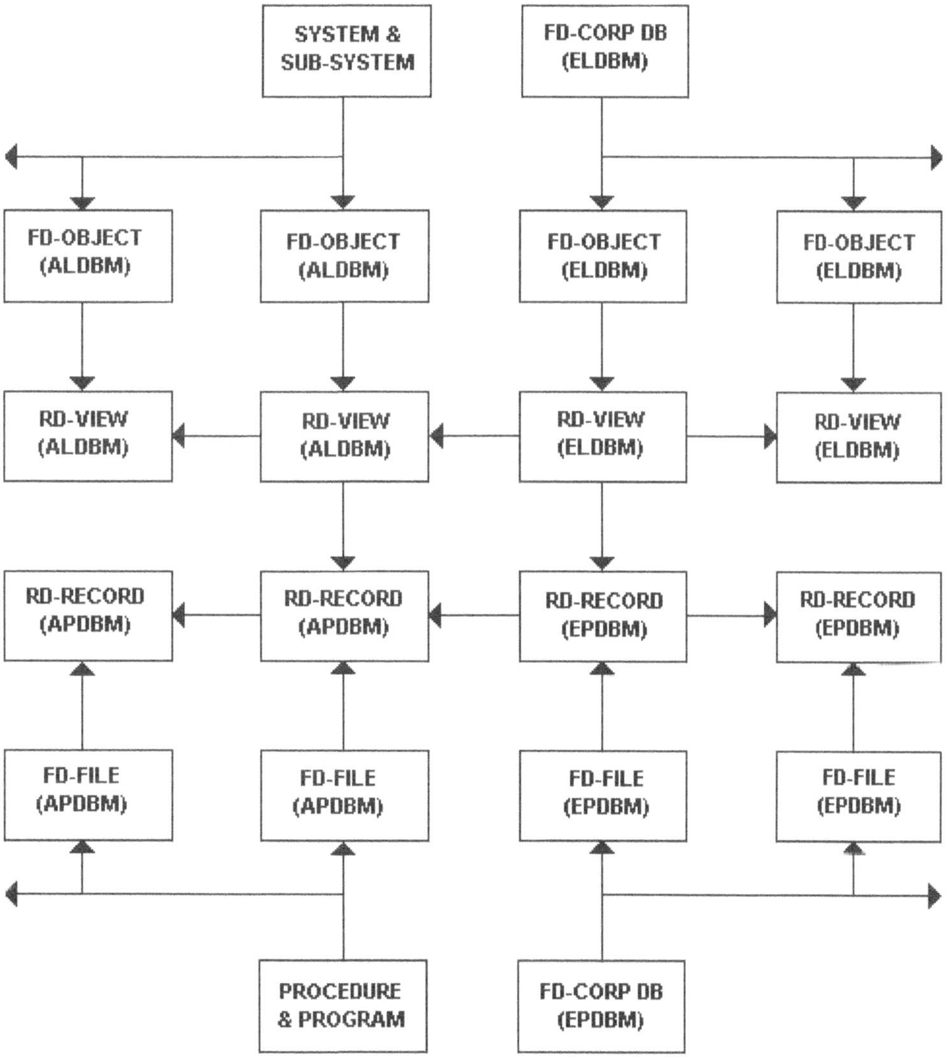

NOTES:
ALDBM - Application Logical Data Base Model FD - File Description
ELDBM - Enterprise Logical Data Base Model RD - Record Description
EPDBM - Enterprise Physical Data Base Model
APDBM - Application Physical Data Base Model

This diagram illustrates all of the BASIC relationships needed to map the four data base models. Additional pointers are available for different purposes, particularly for complicated physical file structures.

EPDBM REPRESENTATION

Graphical techniques to represent physical file management techniques may vary from vendor to vendor. Data Base Administration should use whatever technique is available. To represent FD/RD/DD structures in the IRM, a variety of matrices are available:

- **EPDBM Files Matrix** - expresses relationships between enterprise physical files using FD-to-FD Relationships.

- **EPDBM Records Matrix** - expresses relationships between enterprise physical records using RD-to-RD Relationships.

- **Additional Matrices** - To substantiate that the EPDBM correctly implements the ELDBM and various ALDB Models, additional matrices are used to show:

  - ELDBM/EPDBM Files Matrix
  - ELDBM/EPDBM Records Matrix
  - ALDBM/EPDBM Files Matrix
  - ALDBM/EPDBM Records Matrix

Aside from the matrices, the other major deliverable resulting from this phase are file layouts which detail the physical data structures.

DESCRIPTION OF PHASE ACTIVITIES

Activity A - Develop Hardware/Software Approach

Data Base Administration (DBA) and Data Communications Administration (DCA) reviews the current ELDBM and ALDBM designs as created in the previous phases. From this, they determine the most suitable method of implementation for physically collecting, storing, and retrieving data. This includes both hardware and software.

Activity B - Review Approach

DBA/DCA review the Data Base Approach with management to obtain approval. Hardware/Software acquisitions are decided here.

Activity C - Define Enterprise Physical DB Model

Data Base Administration designs/modifies/documents the Physical Data Base either according to the specifications of the previous activity or existing operational statements. Data Communications Administration designs/ modifies network software as required.

Activity D - Create D.D.L. Statements

Data Base Administration creates the necessary Data Definition Statements (SCHEMA, DBD, CREATE, etc) to implement the design of the previous activity. Data Communications Administration creates the corresponding network statements.

Also, relationships between ELDBM models and the EPDBM model are established. This is based on RD-to-RD relationships.

DBA/DCA then assembles the deliverables produced in the phase and packages them into a single phase review document which is evaluated by Quality Assurance prior to conducting the formal phase review with management.

Activity E - Phase 4 Review

DBA/DCA conducts a formal review of the Phase 4 deliverables with Data Resource Management, Data Engineering, and Project Management. The review is used to evaluate the work performed thus far and revise as required. At this time, management will review the formal Phase 4 "EPDBM Design Manual" consisting of:

- Phase Cover Page
- EPDBM Narrative
- EPDBM File Matrix
- EPDBM Records Matrix
- Data Resource Layouts
- ELDBM/EPDBM File Matrix
- ELDBM/EPDBM Records Matrix
- ALDBM/EPDBM File Matrix
- ALDBM/EPDBM Records Matrix
- Phase Review Checklist

Based on this report and subsequent review meeting, management may elect to revise parts of the report, discontinue the project, or approve it for continuation.

The next few pages include sample deliverables resulting from this phase.

OD-90406 ENTERPRISE PHYSICAL DATA BASE MODEL (EPDBM) NARRATIVE "PRIDE"-DBEM

DEC 13, 2004 RELEASE 1.2.0 PAGE 1.1

****
* FD-00002 ENTERPRISE PHYSICAL DB MODEL *
****

BUSINESS PURPOSE

This represents the collection of all Enterprise Physical files in the enterprise. It specifically does not include Application Physical files or logical objects which should be maintained elsewhere.

FILES WITHIN DATA BASE:.

NUMBER	NAME	RESIDENCE	ORGANIZATION
FD-00054	E-MAIL MESSAGES	COMPUTER	SEQUENTIAL
FD-00052	ORDER PROC STATUS HISTORY	COMPUTER	INDEXED SEQUENT
FD-00050	RELATIONAL DATA BASE	COMPUTER	DBMS

/\

* * * END OF OUTPUT * * *

SAMPLE PHASE 4 DELIVERABLES

OD-90287 E P D B M F I L E S M A T R I X

DEC 19, 2004 RELEASE 1.2.0

"EP" FILES > / "EP" FILES V	FD-00002 ENTERPRISE PHYSICAL DB MODEL	FD-00050 RELATIONAL DATA BASE	FD-00052 ORDER PROC STATUS HISTORY	FD-00054 E-MAIL MESSAGES	FD-00055 PERSONNEL MASTER FILE	FD-00056 ORDER HISTORY	FD-00057 DATA BASE BACKUP	FD-00058 PURCHASE ORDER SCHEMA	FD-00059 ORDER VIEWS	FD IM
FD-00050 RELATIONAL DATA BASE	X									
FD-00052 ORDER PROC STAT US HISTORY	X									
FD-00054 E-MAIL MESSAGES	X									
FD-00055 PERSONNEL MASTER FILE	X	X								
FD-00056 ORDER HISTORY	X									
FD-00057 DATA BASE BACKUP	X	X								
FD-00058 PURCHASE ORDER SCHEMA	X									
FD-00059										

OD-90288 E P D B M R E C O R D S M A T R I X "PRIDI

DEC 19, 2004 RELEASE 1.2.0 PAGE

"EP" RECORDS > / "EP" RECORDS V	RD-00406 CUSTOMER IDENTIFICA TION TABLE	RD-00407 CUSTOMER/ PRODUCT RELATION	RD-00408 BACK-ORDER ID TABLE	RD-00409 ORDER/PROD UCT RELATI ONSHIP	RD-00410 SHIPMENT IDENTIFICA TION TABLE	RD-00411 CUSTOMER/ PRODUCT RELATION	RD-00412 ACCOUNT ID ENTIFICATI ON TABLE	RD-00413 BILLING ID IDENTIFICA TION TABLE	RD-00414 CUSTOMER/ PRODUCT RELATION	RD-004 ACCOUN GOVERN RELATI
RD-00400 CUSTOMER IDENT IFICATION TABLE		M:1				M:1			M:1	
RD-00403 PRODUCT IDENTIF ICATION TABLE		M:N		M:M		M:M			M:M	
RD-00404 GOV'T INDENTIFI CATION TABLE										M:1
RD-00406 ORDER IDENTIFIC CATION TABLE				M:M						
RD-00412 ACCOUNT IDENTIF ICATION TABLE										1:M
RD-00413 BILLING IDENTIF ICATION TABLE										
RD-00418 TERRITORY IDENT IFICATION TABLE										

NOTE: THE TOP ROW REPRESENTS THE SUPERIOR COMPONENT. THE LEFT COLUMN REPRESENTS THE SUBORDINATE COMPONEN

A CELL COORDINATE CONTAINS THE RD "MEMBERSHIP," "LEVEL," AND "QUALIFIER" FIELDS.
IF ALL THREE FIELDS ARE BLANK, THEN AN "X" IS DISPLAYED IN THE CELL ON THE MATRIX.

SAMPLE PHASE 4 DELIVERABLES

```
------------------------------------------------------------------------------------------------
OD-90119                       DATA STRUCTURE DISPLAY                       "PRIDE"-DBEM
------------------------------------------------------------------------------------------------
NOV 18, 2004                        RELEASE 1.2.0                           PAGE 5.1

RESOURCE    NUMBERS/NAMES:                              TYPE:

FD-00049    PRODUCT FILE                                APPLICATION LOGICAL

            DD-00096    PRODUCT NUMBER                  PRIMARY BASIC GROUPING

RD-00049    PRODUCT IDENTIFICATION                      APPLICATION LOGICAL

            DD-00096    PRODUCT NUMBER                  PRIMARY BASIC GROUPING

            DD-00096    PRODUCT NUMBER                  CONTENT SET
            DD-00515    NAME                            CONTENT SET

RD-00050    PRODUCT CHARACTERISTICS                     APPLICATION LOGICAL

            DD-00096    PRODUCT NUMBER                  PRIMARY BASIC GROUPING
            DD-00177    PRODUCT TYPE                    SECONDARY BASIC GROUPING

            DD-00177    PRODUCT TYPE                    CONTENT SET
            DD-00178    PRODUCT STRENGTH                CONTENT SET
            DD-00179    PRODUCT SIZE                    CONTENT SET
            DD-00180    PRODUCT COLOR                   CONTENT SET
            DD-00111    UNIT PRICE                      CONTENT SET
            DD-00156    PRODUCT UNIT OF MEASURE         CONTENT SET
            DD-00175    PRODUCT UNIT SHIP WEIGHT        CONTENT SET
            DD-00176    PRODUCT STANDARD COST           CONTENT SET

                              * * * END OF OUTPUT * * *
```

SAMPLE D.D.L. STATEMENTS

Physical file structures are unique to the selected computer hardware/software. Consult vendor literature. Enclosed are some random samplings:

SAMPLE SQL STATEMENTS

```
CREATE TABLE USERID.CUSTOMER_ID
( CUSTOMER CHAR(8) NOT NULL,
  NAME CHAR(38),
  DUNS_NO CHAR(1) NOT NULL,
  INDUSTRY CHAR(4) NOT NULL,
  PRIMARY KEY
    ( CUSTOMER ) )
```

(continued on next page)

SAMPLE PHASE 4 DELIVERABLES

(continued from previous page)

SAMPLE COBOL COPYBOOK

```
*****************************************************************
*FD-00040  CUSTOMER ORDER FILE
*
*This file contains data that describes customer orders.
*
*****************************************************************
 FD   ORDER-FILE                                                     FD-00040
      RECORDING MODE IS F
      BLOCK CONTAINS   10 RECORDS
      RECORD CONTAINS   72 CHARACTERS
      LABEL RECORDS ARE STANDARD.
*****************************************************************
*RD-00047  ORDER HEADER
*
*CONTAINS GENERAL CUSTOMER ORDER INFORMATION.
*
*****************************************************************
 01   ORDER-HEADER.                                                  RD-00047
      05   CUST-NUMBER                                               DD-00001
                             PIC 9(8)

      05   PO-NUMBER                                                 DD-00502
                             PIC 9(14)

      05   DATE                                                      DD-00016
                             PIC 9(8)

      05   ORD-TOT-AMOUNT                                            DD-00510
                             PIC S9(7)V99

      05   VENDOR-TYPE                                               DD-00231
                             PIC XXX

      05   FILLER
                             PIC X(30).

*****************************************************************
*RD-00048  ORDER DETAIL
*
```

SAMPLE PHASE 4 DELIVERABLES

OD-90289 E L D B M / E P D B M F I L E S M A T R I X "PRII

DEC 19, 2004 RELEASE 1.2.0 PAGE

"EL" OBJECTS > "EP" OBJECTS V	FD-00300 PRODUCT OBJECT	FD-00301 CUSTOMER OBJECT	FD-00302 ORDER OBJECT	FD-00303 TERRITORY OBJECT	FD-00304 ASSIGNMENT OBJECT	FD-00305 SHIPMENT OBJECT	FD-00306 BACK ORDER OBJECT	FD-00307 PAYMENT OBJECT	FD-00308 GOVERNMENT AGENCY OBJECT	FD-003(ACCOUN' OBJECT
FD-00050 RELATIONAL DATA BASE	X	X	X	X	X	X	X	X	X	X
FD-00052 ORDER PROC STAT US HISTORY			X			X	X			
FD-00054 E-MAIL MESSAGES		X						X	X	X

OD-90290 E L D B M / E P D B M R E C O R D S M A T R I X "PRII

DEC 19, 2004 RELEASE 1.2.0 PAGE

"EL" VIEWS > "EP" RECORDS V	RD-00300 CUSTOMER IDENTIFICA TION VIEW	RD-00301 CUSTOMER ADDRESS VIEW	RD-00302 CUSTOMER CONTACT VIEW	RD-00303 PRODUCT IDENTIFICA TION VIEW	RD-00304 GOV'T IDENTIFICA TION VIEW	RD-00305 GOVERNMENT TAX RATE VIEW	RD-00306 ORDER IDENTIFICA TION VIEW	RD-00307 CUSTOMER/ PRODUCT REL. VIEW	RD-00308 BACK-ORDER IDENTIFICA TION VIEW	RD-003(ORDER/ PRODUC' REL. V:
RD-00124 ORDER DATA							X	X		
RD-00125 BACK-ORDER DATA								X		X
RD-00126 SHIPMENT DATA										
RD-00128 E-MAIL MESSAGES	X				X					
RD-00400 CUSTOMER IDENTI FICATION TABLE	X									
RD-00401 CUSTOMER ADDRESS TABLE		X								
RD-00402 CUSTOMER CONTACT TABLE			X							
RD-00403 PRODUCT IDENTIF ICATION TABLE				X						

SAMPLE PHASE 4 DELIVERABLES

OD-90291 A L D B M / E P D B M F I L E S M A T R I X "PRID

DEC 19, 2004 RELEASE 1.2.0 PAGE

"AL" OBJECTS > / "EP" FILES V	FD-00100 PRODUCT OBJECT	FD-00101 CUSTOMER OBJECT	FD-00102 ORDER OBJECT	FD-00103 TERRITORY OBJECT	FD-00104 ASSIGNMENT OBJECT	FD-00105 SHIPMENT OBJECT	FD-00106 BACK ORDER OBJECT	FD-00107 PAYMENT OBJECT	FD-00108 GOVERNMENT AGENCY OBJECT	FD-001 ACCOUN OBJECT
FD-00050 RELATIONAL DATA BASE	X	X	X	X	X	X	X	X	X	X
FD-00052 ORDER PROC STAT US HISTORY			X			X	X			
FD-00054 E-MAIL MESSAGES		X						X	X	X

OD-90292 A L D B M / E P D B M R E C O R D S M A T R I X "PRI

DEC 19, 2004 RELEASE 1.2.0 PAGE

"AL" VIEWS > / "EP" RECORDS V	RD-00100 CUSTOMER IDENTIFICA TION VIEW	RD-00101 CUSTOMER ADDRESS VIEW	RD-00102 CUSTOMER CONTACT VIEW	RD-00103 PRODUCT IDENTIFICA TION VIEW	RD-00104 GOV'T IDENTIFICA TION VIEW	RD-00105 GOVERNMENT TAX RATE VIEW	RD-00106 ORDER IDENTIFICA TION VIEW	RD-00107 CUSTOMER/ PRODUCT REL. VIEW	RD-00108 BACK-ORDER IDENTIFICA TION VIEW	RD-001 ORDER/ PRODUC REL. V
RD-00124 ORDER DATA							X	X		
RD-00125 BACK-ORDER DATA									X	X
RD-00126 SHIPMENT DATA										
RD-00128 E-MAIL MESSAGE	X				X					
RD-00400 CUSTOMER IDENTI FICATION VIEW	X									
RD-00401 CUSTOMER ADDRESS TABLE		X								
RD-00402 CUSTOMER CONTACT TABLE			X							
RD-00403 PRODUCT IDENTIF ICATION TABLE				X						

END SECTION

Section 24

"PRIDE"-DBEM PHASE 5 APPLICATION PHYSICAL DATA BASE DESIGN

A Data Base should naturally evolve over time and synchronize with all Information Systems.
- Bryce's Law

The purpose of this phase is to design or modify an Application Physical Data Base Model (APDBM) for an information system. Several events occur during this phase:

- Application Physical files are defined, complete with records, and data elements.
- Relationships between physical files and records are established.
- Data Resource Layouts, and program file structures are provided to Software Engineering for inclusion in programs.
- Data structures in the APDBM are related to those in the EPDBM and ALDBM.
- A technical review meeting is performed to validate the integrity of the APDBM and to issue file layouts to Software Engineering.

METHODOLOGY NAVIGATION

A Phase 5 is initiated following Phase 4. There is a one-to-one relationship between Phase 5 and an information system. In other words, a Phase 5 will be executed for each system in the project (normally there is just one).

Even if the scope of the project is to only modify a small portion of a system (such as a single sub-system, procedure, program or module), the whole system identifier should be used.

The formal deliverable resulting from Phase 5 is an "APDBM Design Manual" consisting of a description of the Application Physical Data Base Model. This is reviewed with Data Resource Management, Project Management, Data Communications Administration and Software Engineering.

Following Phase 5, the DBEM project will proceed to

Phase 6, "DBEM Evaluation."

GENERAL DISCUSSION

Phase 5 closes the loop between the four data base models. The objective is to define all of the physical file structures needed to support a single information system, not the enterprise as a whole. As such, the APDBM represents a subset of the EPDBM.

During Phase 2 of ISEM, Systems Engineering identified the primary logical files needed for the system. From these logical files, decisions were made as to how these files should be physically implemented. Ultimately, these files represent the primary storage files required by the system. Later, during Phase 3 of ISEM, Systems Engineering may have identified the need for manual files to store input/output documents, and computer files for input transactions and to capture output data. During Phase 4-II of ISEM, Software Engineering identified the need for working files to pass data between programs and modules. Whereas the thrust in DBEM has primarily been concerned with the ISEM Phase 2 files, Data Base Administration must also accommodate those files identified in Phases 3 and 4 of ISEM.

As Data Base Administration completes the APDBM, the physical file structures are delivered to Software Engineering as part of Phase 4-II/Activity C of ISEM. These data structures will then be included as part of the programming specifications. In other words, Software Engineering is the ultimate beneficiary of DBEM.

The purpose of the APDBM is to provide data structures in a way that is convenient for the specific application, avoiding the complexities of the global picture. In the absence of a DBMS, the APDBM is just a subset of the EPDBM which refers to the particular application, that is: the physical files used by the application. If a DBMS is used, it can be different from the enterprise model to an extent that varies with the DBMS used. Differences between the two physical models that cannot be accommodated by the DBMS, (or in the absence of one), must be implemented by the application programs. The View mechanism of IBM's DB2 is one of the most powerful in this respect. The Subschema mechanism is used for CODASYL type DBMS package, while IMS uses the Logical Data Base.

APPLICATION PHYSICAL DATA BASE MODEL (APDBM)
BASIC CONSTRUCTS

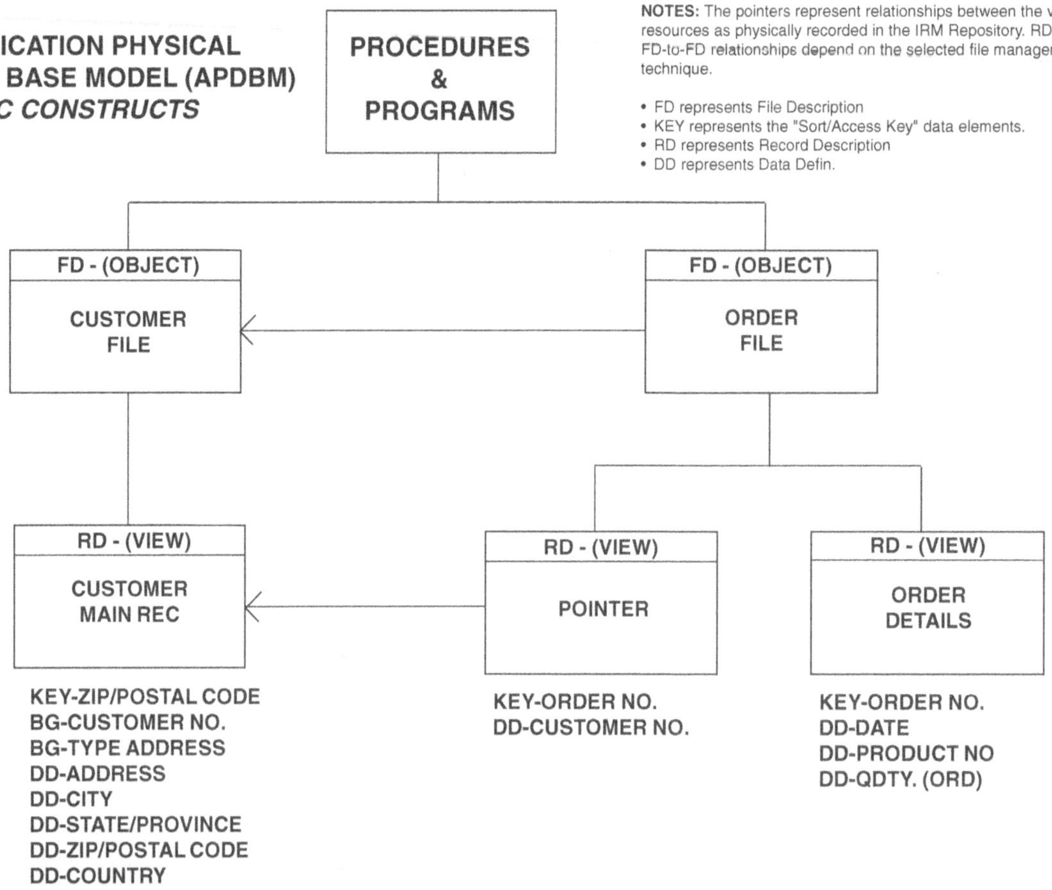

NOTES: The pointers represent relationships between the various resources as physically recorded in the IRM Repository. RD-to-RD and FD-to-FD relationships depend on the selected file management technique.

- FD represents File Description
- KEY represents the "Sort/Access Key" data elements.
- RD represents Record Description
- DD represents Data Defin.

Ideally, if the DBMS could provide 100% data independence, the APDBM should have exactly the same structure as the ALDBM, or else match it as closely as possible. In this way, Software Engineering would have no difficulty in translating the logical data structures designed by Systems Engineering into physical ones. In practice, one has to adhere to the constraints of the DBMS and provide additional differentiation through program code.

Security and integrity constraints can also be implemented through this model by hiding certain records, data elements and relationships present in the Enterprise Model from a particular application.

For a more complete description of physical file considerations, consult the Phase 4 narrative.

ALDBM TO APDBM RELATIONSHIPS

The Application Physical Data Base Model (APDBM) will consist of physical files and records that must store the data needed to satisfy the ALDBM. There is not always a one-to-one relationship between logical and physical structures. One logical file may be implemented by multiple physical files. Conversely, one physical file may implement many logical files. Physical implementation is based on what is cost-effective and may take many forms (e.g., indexed files, indexed sequential, flat files, hierarchical/network/relational DBMS, etc.). The point is, the relationships between logical and physical must be mapped in the IRM to demonstrate that the physical satisfies the logical.

APPLICATION LOGICAL DATA BASE MODEL (ALDBM)

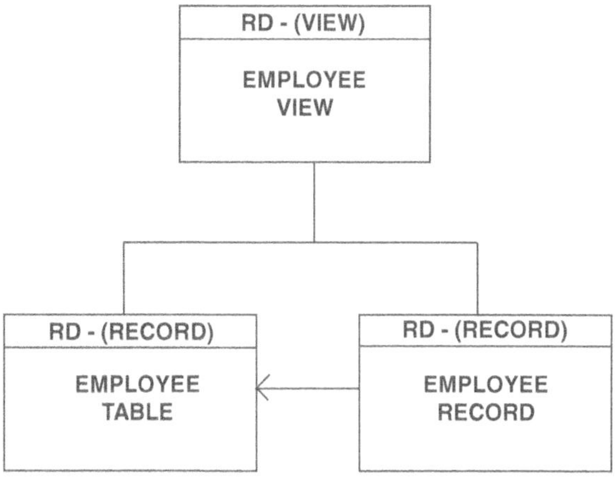

APPLICATION PHYSICAL DATA BASE MODEL (APDBM)

EPDBM TO APDBM RELATIONSHIPS

The EPDBM represents the global view of data storage requirements. During Phase 5, Data Base Administration must consider physical requirements for individual systems. Ultimately, the APDBM represents subsets of the EPDBM. Relationships between the EPDBM and the APDBM can be expressed using RD-to-RD and FD-to-FD relationships.

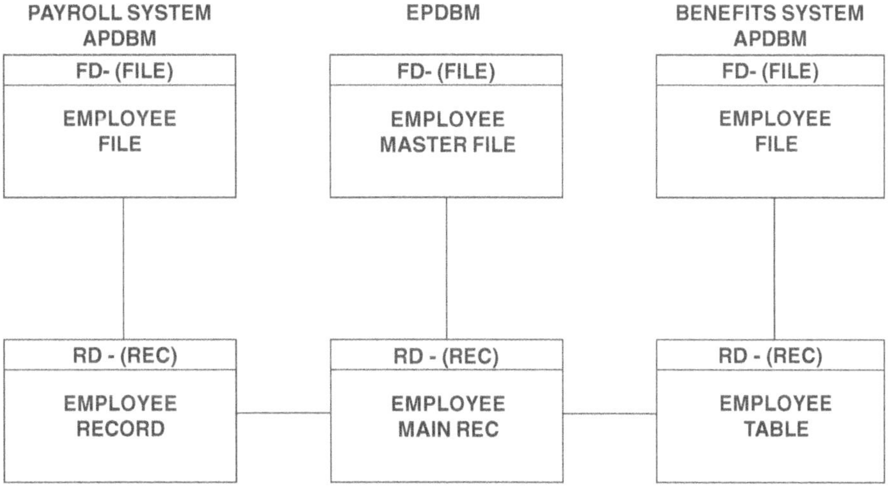

RECORD DESCRIPTION (RD)

As this discussion indicates, the Record Description (RD) resource in the IRM is used for a variety of purposes. In a logical context, the RD represents a "View"; in a physical context, the RD represents a "Record." The RD is also used to represent other physical structures, such as: screen panels, print maps, input transactions, output data, messages, tables, etc. All of these structures exhibit the same characteristics; all require some form of unique key, and all contain data elements.

The File Description (FD) is used to represent logical files (Objects), physical files, and data bases. Files consist of records, thereby an FD/RD relationship is required. A data base represents a global view of data, therefore an FD/FD relationship is used to map all of the files within a data base.

The RD is also the primary resource used to map the relationships between all four data base models. Although the File Description (FD) provides for FD-to-FD relationships, it is the RD that is used to connect the models.

The following diagram illustrates all of the BASIC relationships needed to map the four data base models in the IRM. Additional pointers are available for different purposes, particularly for complicated physical file structures.

IRM RESOURCE RELATIONSHIPS - MAPPING THE FOUR DATA BASE MODELS

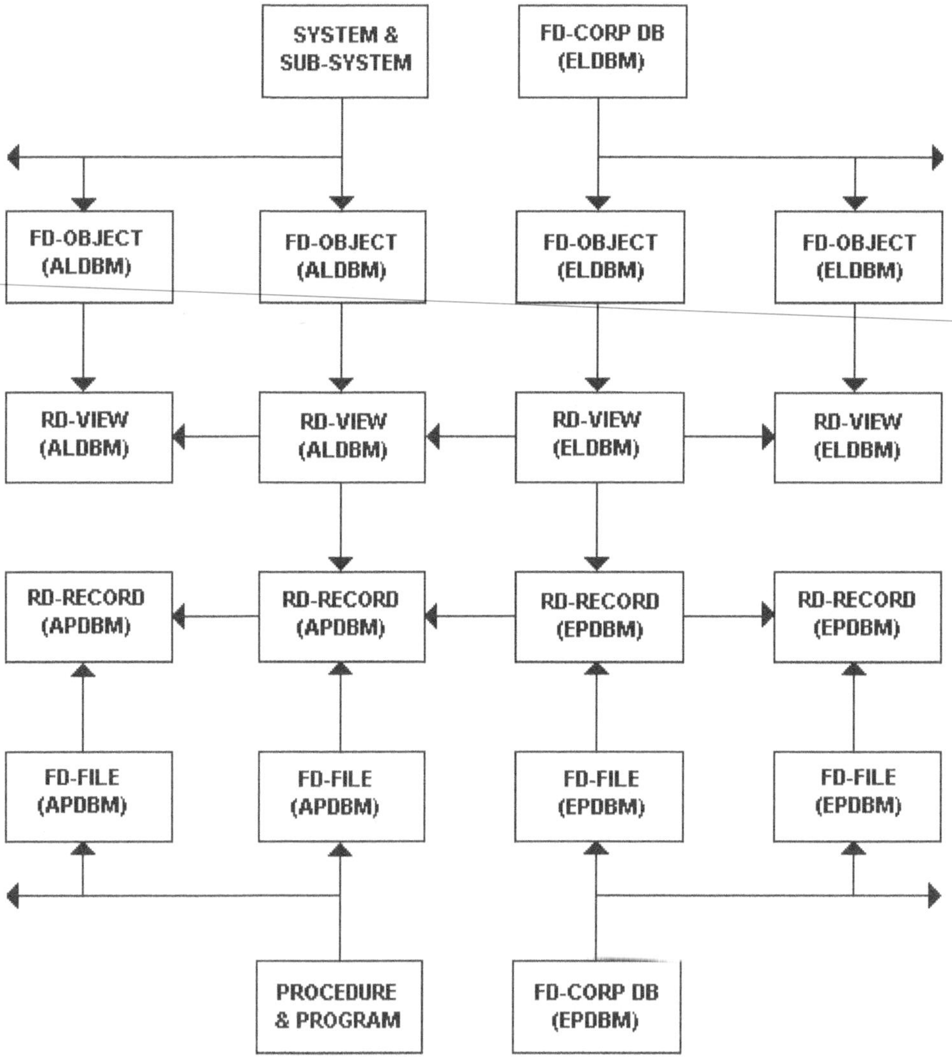

NOTES:
ALDBM - Application Logical Data Base Model
ELDBM - Enterprise Logical Data Base Model
EPDBM - Enterprise Physical Data Base Model
APDBM - Application Physical Data Base Model

FD - File Description
RD - Record Description

APDBM REPRESENTATION

Graphical techniques to represent physical file management techniques may vary from vendor to vendor. Data Base Administration should use whatever technique is available. To represent FD/RD/DD structures in the IRM, a variety of matrices are available:

• **APDBM Files Matrix** - expresses relationships between application physical files using FD-to-FD Relationships.

• **APDBM Records Matrix** - expresses relationships between application physical records using RD-to-RD Relationships.

• **Additional Matrices** - To substantiate that the APDBM correctly implements the ALDBM and represents a subset of the EPDBM, additional matrices are used to show:

- ALDBM/APDBM Files Matrix
- ALDBM/APDBM Records Matrix
- EPDBM/APDBM Files Matrix
- EPDBM/APDBM Records Matrix

All of these are based on FD/RD relationships as recorded in the IRM Repository.

Aside from the matrices, the other major deliverable resulting from this phase are file layouts which detail the physical data structures. These are provided to Software Engineering as part of their programming specifications.

DESCRIPTION OF PHASE ACTIVITIES

Activity A - Analyze ALDBM & EPDBM

Data Base Administration reviews the ALDBM, the EPDBM and performance, security and integrity requirements as reflected in the phase documentation.

Activity B - Define Application Physical DB Model

Data Base Administration designs the Application Physical Data Base Model to match the ALDBM and EPDBM. Physical file/record/data structures are entered in the IRM and mapped between the three models.

Activity C - Create D.D.L. Statements

Data Base Administration finalizes the design by preparing program file structure statements (such as Data Definition Language statements, COBOL copybooks, etc.).

DBA then assembles the deliverables produced in the phase and packages them into a single phase review document which is evaluated by Quality Assurance prior to conducting the formal phase review.

Activity D - Phase 5 Review

Data Base Administration conducts a formal review of the Phase 5 deliverables with Data Resource Management, Data Communications, Project Management, and Software Engineering. The review is used to evaluate the work performed thus far and revise as required. At this time, management will review the formal Phase 5 "APDBM Design Manual" consisting of:

- Phase Cover Page
- APDBM File Matrix
- APDBM Records Matrix
- Data Resource Layouts
- ALDBM/APDBM File Matrix
- ALDBM/APDBM Records Matrix
- EPDBM/APDBM File Matrix
- EPDBM/APDBM Records Matrix
- Phase Review Checklist

Based on this report and subsequent review meeting, management may elect to revise parts of the report. If acceptable, the designs are turned over to Software Engineering for inclusion in programming.

The next few pages include sample deliverables resulting from this phase.

SAMPLE PHASE 5 DELIVERABLES

OD-90293 — A P D B M   F I L E S   M A T R I X — "PRIDE

JAN 6, 2005 — RELEASE 1.2.0 — PAGE

"AP" FILES > / "AP" FILES V	FD-00051 ORDER PROC STATUS HISTORY	FD-00053 E-MAIL CREDIT REQUEST	FD-00055 ORDER PROC ESSING DATA BASE	FD-00150 WEEKLY ORD ER STATUS HOLD FILE	FD-00151 ORDER STAT US INPUT T RANSACTION	FD-00152 SALES MANAGEMENT FILE	FD-00153 ORDER REQU EST HOLD FILE	FD-00154 DAILY ORDE R INPUT TR ANSACTIONS	FD-00160 DAILY WORK ORDERS	FD-0016 DATA CONVERS HOLD FI
FD-00051 ORDER PROC STATUS HISTORY			X							
FD-00150 WEEKLY ORDER STATUS HOLD FILE			X							
FD-00160 DAILY WORK ORDERS			X							

OD-90294 — A P D B M   R E C O R D S   M A T R I X — "PRID

JAN 6, 2005 — RELEASE 1.2.0 — PAGE

"AP" RECORDS > / "AP" RECORDS V	RD-00506 CUSTOMER IDENTIFICA TION TABLE	RD-00507 CUSTOMER/ PRODUCT RELATION	RD-00508 BACK-ORDER ID TABLE	RD-00509 ORDER/PROD UCT RELATI ONSHIP	RD-00510 SHIPMENT IDENTIFICA TION TABLE	RD-00511 CUSTOMER/ PRODUCT RELATION	RD-00512 ACCOUNT ID ENTIFICATI ON TABLE	RD-00513 BILLING ID IDENTIFICA TION TABLE	RD-00514 CUSTOMER/ PRODUCT RELATION	RD-0051 ACCOUNT GOVERNM RELATIO
RD-00500 CUSTOMER IDENT IFICATION TABLE		M:1				M:1			M:1	
RD-00503 PRODUCT IDENTIF ICATION TABLE		M:N		M:M		M:M			M:M	
RD-00504 GOV'T INDENTIFI CATION TABLE										M:1
RD-00506 ORDER IDENTIFIC CATION TABLE				M:M						
RD-00512 ACCOUNT IDENTIF ICATION TABLE										1:M
RD-00513 BILLING IDENTIF ICATION TABLE										
RD-00518 TERRITORY IDENT IFICATION TABLE										

NOTE: THE TOP ROW REPRESENTS THE SUPERIOR COMPONENT. THE LEFT COLUMN REPRESENTS THE SUBORDINATE COMPONENT

A CELL COORDINATE CONTAINS THE RD "MEMBERSHIP," "LEVEL," AND "QUALIFIER" FIELDS.
IF ALL THREE FIELDS ARE BLANK, THEN AN "X" IS DISPLAYED IN THE CELL ON THE MATRIX.

SAMPLE PHASE 5 DELIVERABLES

OD-90295 A L D B M / A P D B M F I L E S M A T R I X "PRIDE

JAN 6, 2005 RELEASE 1.2.0 PAGE

"AL" OBJECTS > / "AP" FILES V	FD-00100 PRODUCT OBJECT	FD-00101 CUSTOMER OBJECT	FD-00102 ORDER OBJECT	FD-00103 TERRITORY OBJECT	FD-00104 ASSIGNMENT OBJECT	FD-00105 SHIPMENT OBJECT	FD-00106 BACK ORDER OBJECT	FD-00107 PAYMENT OBJECT	FD-00108 GOVERNMENT AGENCY OBJECT	FD-0010 ACCOUNT OBJECT
FD-00051 ORDER PROC STATUS HISTORY			X			X	X			
FD-00053 E-MAIL CREDIT REQUEST		X						X	X	X
FD-00055 ORDER PROCESSING DATA BASE	X	X	X	X	X	X	X	X	X	X

OD-90296 A L D B M / A P D B M R E C O R D S M A T R I X "PRII

JAN 6, 2005 RELEASE 1.2.0 PAGE

"AL" VIEWS > / "AP" RECORDS V	RD-00100 CUSTOMER IDENTIFICA TION VIEW	RD-00101 CUSTOMER ADDRESS VIEW	RD-00102 CUSTOMER CONTACT VIEW	RD-00103 PRODUCT IDENTIFICA TION VIEW	RD-00104 GOV'T IDENTIFICA TION VIEW	RD-00105 GOVERNMENT TAX RATE VIEW	RD-00106 ORDER IDENTIFICA TION VIEW	RD-00107 CUSTOMER/ PRODUCT REL. VIEW	RD-00108 BACK-ORDER IDENTIFICA TION VIEW	RD-001C ORDER/ PRODUCT REL. VI
RD-00121 ORDER DATA							X	X		
RD-00122 BACK-ORDER DATA								X		X
RD-00123 SHIPMENT DATA										
RD-00127 E-MAIL CONTENT	X				X					
RD-00500 O-P CUSTOMER ID ENT TABLE	X									
RD-00501 O-P CUSTOMER ADDRESS TABLE		X								
RD-00502 O-P CUSTOMER CONTACT TABLE			X							
RD-00503 O-P PRODUCT IDENT TABLE				X						

SAMPLE PHASE 5 DELIVERABLES

OD-90297 E P D B M / A P D B M F I L E S M A T R I "PRI[

JAN 5, 2005 RELEASE 1.2.0 PAGE

"EP" FILES > "AP" FILES V	FD-00050 RELATIONAL DATA BASE	FD-00052 ORDER PROC STATUS HISTORY								
FD-00051 ORDER PROC STATUS HISTORY		X								
FD-00053 E-MAIL CREDIT REQUEST										
FD-00055 ORDER PROCESSING DATA BASE	X									
FD-00150 WEEKLY ORDER STATUS HOLD FILE										
FD-00151 ORDER STATUS IN PUT TRANSACTION										
FD-00152 SALES MANAGEMENT FILE										
FD-00153 ORDER REQUEST HOLD FILE										
FD-00154 DAILY ORDER INPUT TRANSACT.										

OD-90298 E P D B M / A P D B M R E C O R D S M A T R I "PRI[

JAN 6, 2005 RELEASE 1.2.0 PAGE

"EP" RECORDS > "AP" RECORDS V	RD-00124 ORDER DATA	RD-00125 BACK-ORDER DATA	RD-00126 SHIPMENT DATA	RD-00400 CUSTOMER IDENT. TABLE	RD-00401 CUSTOMER ADDRESS TABLE	RD-00402 CUSTOMER CONTACT TABLE	RD-00403 PRODUCT IDENT. TABLE	RD-00404 GOV'T IDENT. TABLE	RD-00405 GOVERNMENT TAX RATE TABLE	RD-0040[ORDER IDENT. TABLE
RD-00121 ORDER DATA	X									
RD-00122 BACK-ORDER DATA		X								
RD-00123 SHIPMENT DATA			X							
RD-00127 E-MAIL CONTENT										
RD-00500 O-P CUSTOMER ID ENT TABLE				X						
RD-00501 O-P CUSTOMER ADDRESS TABLE					X					
RD-00502 O-P CUSTOMER CONTACT TABLE						X				

SAMPLE PHASE 5 DELIVERABLES

For sample D.D.L. Statements & Data Structure Displays, see the Sample Deliverables in Section 23.

END SECTION

Section 25

"PRIDE"-DBEM PHASE 6 DBEM EVALUATION

Most organizations use a DBMS as nothing more than an elegant file access method. Consequently, the opportunity to share data and integrate systems is lost.
- Bryce's Law

The purpose of this phase is to evaluate the data base and project in terms of how well they were delivered as compared to what was proposed in the early phases of the methodology. It is similar in intent to Phase 5 of EEM "EEM Evaluation" and Phase 9 of ISEM "ISEM Evaluation." Several events occur during this phase:

- The data base is analyzed in terms of how well it satisfies its requirements. Also, performance issues are evaluated. From this, strengths and weaknesses are evaluated.
- Project performance is evaluated in terms of how well the project met estimates, schedules and proposed costs. Further, an analysis of the project is prepared.
- A data base evaluation report is prepared summarizing the phase findings and reviewed by management.

METHODOLOGY NAVIGATION

This represents the concluding phase of the DBEM project. It is performed either after Phase 5, or if a project has been discontinued earlier on (such as after Phase 1, 2 or 3).

The formal deliverables resulting from Phase 6 will be the "DBEM Project Evaluation Report" which is reviewed by Quality Assurance prior to being issued to management.

GENERAL DISCUSSION

During Phase 1 Data Engineering and Project Management prepared the "Data Base Study and Evaluation Report." On the basis of these reports, management either approved or disapproved the project for design and implementation of the data base.

If the project was disapproved, the reasons are outlined during this phase and placed with the original documentation for future reference. If the project was approved, this approval was given as a result of certain expectations outlined within the reports. The purpose of this evaluation is to determine if these expectations were realized.

THE BENEFITS OF PHASE 6

"Unless we learn from history, we are doomed to repeat it." This may be an old adage, but it is certainly true. Many organizations tend to regard project reviews as a waste of precious time and resources. Consequently, data bases are installed without any type of follow-up. Because of this, many of the benefits were never realized.

The data base organization can realize many benefits from such a review. It is not just a platitudinous statement to say that "we learn from our mistakes." It is a clear and established fact. A detailed audit provides Data Engineering and Data Base Administration with an opportunity to review their estimating and design skills. This knowledge, along with historical data, can be gainfully applied to new data base designs and projects. As such, estimating guidelines can be updated.

A carefully executed audit can also add to the credibility of the data base organization by showing how well they performed and that they can account for their actions. Further, project data can be charged back to user sponsors for services rendered. For example, if a DBEM project is performed in support of an ISEM project, the charged incurred for the DBEM project could conceivably be charged back to the ISEM project.

Finally, as a result of an audit, the data base organization may find that the new data base is simply not functioning in spite of maintenance and revisions. In this event, the audit may indicate an entirely new approach should be taken as opposed to continually fighting the problem.

"WHEN" SHOULD YOU CONDUCT PHASE 6?

The review should be scheduled for execution after the data base and system has operated for an adequate period of time. Typically, this will be between thirty (30) and ninety (90) days after installation. It may also be advantageous to execute this phase on occasion, just to review the performance of an existing data base. This might be relevant if significant changes in transaction volumes or file sizes have occurred.

"WHO" SHOULD PERFORM THE REVIEW?

Ideally, the audit should not be performed by the same individuals who developed the data base. A neutral third party should be involved who can audit objectively the data base and the project.

Project Management, Data Base Administration, and Data Communications Administration are normally charged with the task of evaluating a DBEM project. Project Management is concerned with evaluating project data and the others are concerned with performance considerations pertaining to the data base.

AUDIT CONSIDERATIONS

During the audit the DBA and DCA may identify errors, omissions and severe weaknesses in the new data base. In this event, they may initiate a Work Request/Objective to document the modification/improvement. This will then go through the normal process of evaluation and priority calculation as provided for under the Enterprise Engineering Methodology (EEM).

If you have been collecting Project Management data through computer software (e.g., time, dates, money), this data can be exported to other financial packages and spreadsheets for further analysis as required.

In addition to controlling information resources, the IRM Repository provides auditors with the "roadmap" for locating the control points in a system. Further, the "impact analysis" feature is useful for considering enhancements.

BENCHMARKING

Benchmarking a data base typically involves analyzing the volume of transactions processed in a prescribed time frame. During Phase 6 benchmark tests are performed to assure that the data base can perform according to specifications and to look for new ways to "tune" the data base.

DESCRIPTION OF PHASE ACTIVITIES

Activity A - Perform Data Base Audit

Data Base Administration (DBA) and Data Communications Administration (DCA) reviews the original data base design documentation and/or Project Scope to become familiar with the project objectives, if a new or modified data base has been implemented. Next, a survey is conducted to identify how it is actually operating. Any differences with respect to original objectives, as well as operational problems are identified and noted as part of the survey. If any problem areas are identified, Data Base Administration works with Systems Engineering to determine the problem and possible solutions. If necessary, a Work Request/Modification-Improvement is prepared to resolve the issue.

Activity B - Prepare Project Evaluation

DBA/DCA prepares a "**Project Evaluation Report**" which identifies the actual hours, costs and schedules during project execution. This evaluation is compared to the original evaluations prepared during Phase 1. Any discrepancies beyond a reasonable tolerance are highlighted with explanation. The report includes:

- Phase Cover Page
- Data Base Evaluation - a textual analysis of the completed data base.
- Project Evaluation - a textual analysis of the completed project. This should include sections describing "Lessons Learned" and the final "Cost/Benefit Analysis."
- O-O-M Estimate/Actual - detailing all project estimates, schedules and expenditures.
- Phase Review Checklist - featuring criteria to evaluate the report.

Quality Assurance checks this report to assure conformance with standards. Information/Data/System/Enterprise Resource Management is then provided a copy of the report for review.

The next few pages include sample deliverables resulting from this phase.

SAMPLE PHASE 6 DELIVERABLES

OD-90407	DATA BASE EVALUATION	"PRIDE"-DBEM
MAR 23, 2005	RELEASE 1.2.0	PAGE 1.1

AS DEFINED ON PD-00017 ORDER PROCESSING DATA BASE

A review of the Data Base structures for the Order Processing System was performed on 3/22/95. The basis for the analysis was the Phase reports as produced under "PRIDE"-DBEM.

Based on interviews and a review of the physical data base, it appears that the data base satisfies the requirements of System OP. However, concern has been expressed by the DBA regarding speed in processing transactions, particularly for general queries from the Customer Services department. Although no immediate action is required at this time, the application should be reviewed again in one or two years.

- P.T. Nelson
Data Base Administrator
March 23, 2005

* * * END OF OUTPUT * * *

OD-90392	PROJECT EVALUATION	"PRIDE"-DBEM
MAR 23, 2005	RELEASE 1.2.0	PAGE 2.1

AS DEFINED ON PD-00017 ORDER PROCESSING DATA BASE

The Order Processing Data Base project was recently concluded. The new system was installed on time with a minimal overrun of costs.

Although the project was successfully completed, the learning curve used to develop new types of D.D.L. resulted in a modest 4% overrun ($950) from the Phase 1 Order-of-Magnitude Estimate. The project took approximately five (5) elapsed months to perform at a total cost of $20,600.00. Consult the attached "PRIDE"-PM reports for details on estimated versus actual figures.

We did experience a two week delay in delivering the Phase 5 Application Physical Data Base Model (APDBM), but this did not impede progress on project PD-00132 (TOPS project).

LESSONS LEARNED

1. Data Base design standards need to be enhanced, particularly in the areas of physical design and D.D.L.

2. Cooperation from Systems Engineering and DP Operations significantly expedited project development for which we give our thanks.

3. Estimating guidelines need to be updated to reflect the results produced from this project.

- Goh Touchen
Project Manager
March 23, 2005

* * * END OF OUTPUT * * *

END SECTION

Section 26

"PRIDE" PROJECT MANAGEMENT

Manage from the bottom-up, not just from the top-down; this creates personal commitment and accountability.
- Bryce's Law

CONCEPTS & PHILOSOPHIES

If we lived in a perfect world, there would not be a need for managers. Everyone would know precisely what their assignments were and would successfully accomplish them on time and within budget. However, the reality is we live in an imperfect world. We as human beings make mistakes; we work on multiple assignments concurrently, and require guidance. It must be recognized from the outset that project management does not come free. Nor does it come naturally to people.

Traditionally, the typical approach to project management has most often been to find a project manager, provide resources, and then give them an assignment with no direction as to how the project will be conducted or controlled. With this approach, the success or failure of the project is dependent on the abilities and experience of the project manager and how well the manager can organize and train the project team, plan the project, estimate, etc. Consequently, there is significant trial and error in the process. This approach usually results in a unique method for the particular project because it reflects the thinking of the project manager. Different managers use different techniques and ideas.

Another common approach used was the "brute-force approach." Simply stated, *"I don't care how you get the job done; just have it completed by (date)."* This approach shows a lack of sensitivity to the complexity of project management.

There is more to project management than maintaining costs and time schedules. It is the process of applying resources to a defined goal and attaining this goal within time and cost objectives. Fundamentally, it is a people oriented function as opposed to an administrative or clerical function. Project management, therefore, is not a tool or technique, but rather a philosophy of management.

To put it into proper perspective, project management is analogous to the production control function in manufacturing, which is concerned with controlling the orderly assembly of products. This involves monitoring the assembly lines and assuring that materials management delivers the necessary parts to the assembly lines on time and within reasonable cost. Project management is to a methodology, what production control is to an assembly line. Without the assembly line, production control is a useless exercise. Conversely, without a methodology, project management is useless.

The ultimate test of a methodology is if it can operate independent of project management. The two are not synonymous. Although they work in concert, there are distinct differences. Whereas a methodology dictates what work is required, project management controls the application of work. Just as an assembly line can produce a product without production control, a methodology can produce a product without project management. Therefore, a methodology is independent of project management, but project management is totally dependent upon a methodology.

A PROJECT REQUIRES A METHODOLOGY

BUT A METHODOLOGY DOES NOT REQUIRE A PROJECT

A project is an application of effort towards prescribed objectives through the execution of a defined sequence of events. All projects have a life cycle; a beginning for planning, a middle for execution, and an end for review. Each project has a unique scope, set of objectives and defined sequence of events. These events are defined by a methodology, whether it is to develop enterprise resources (as in EEM), system resources (as in ISEM), or data resources (DBEM). The methodology thereby is the "roadmap" for a project. It provides organization and direction for the project.

Project managers must recognize that they are in the business of solving problems, not creating problems. They do not wait for things to happen, they make things happen. This "active" versus "reactive" management philosophy distinguishes management from supervision. Whereas management assigns and reviews work to assure that goals are met, supervision is responsible for implementing the assigned work and directing the daily activities of others. Problems arise when there is more time spent on supervision than on managing and actually performing the work. This phenomenon occurs when people are improperly trained and/or when they are not held accountable for their actions.

MINI-PROJECT MANAGER CONCEPT

At the root of the problem is the issue of accountability and commitment. Historically, companies have taken a unidirectional approach to management. Assignments, estimates and schedules are dictated by management with little concern for the opinion of the people who have to perform the work. This typically results in unrealistic estimates and schedules that are seldom achieved. Why? Because the individual human resources were not consulted and made accountable for achieving the goals. Workers may even go so far as to undermine project plans simply because they were not consulted.

Instead, companies must learn to manage from the bottom-up, not just top-down. We refer to this approach as the "Mini-Project Manager" concept. Under this approach, people are treated like professionals and are expected to act as such in return. It emphasizes less supervision and more personal management by the individual worker. Employees must

PRODUCTION CONTROL ANALOGY

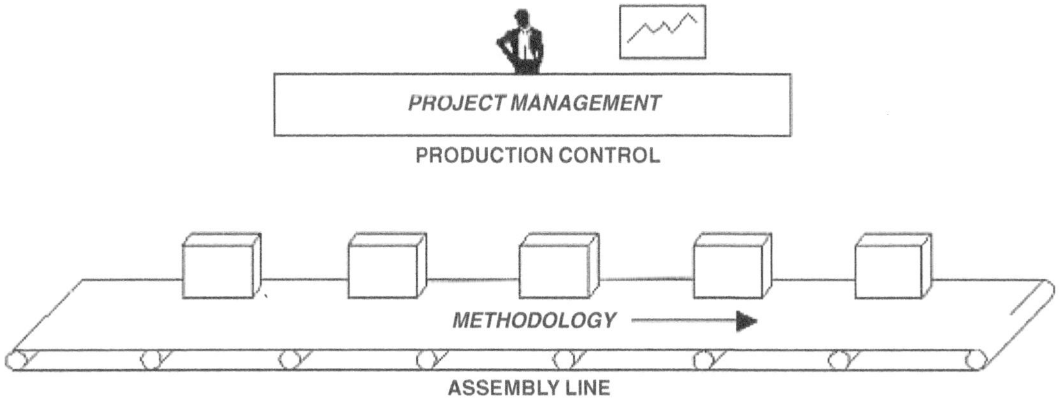

learn to supervise themselves. Although, management still provides leadership and direction over assignments, employees participate in the estimating, scheduling and control process. This produces far more realistic results which can be more effectively managed. It also creates greater personal commitment to project activities since the individual worker's voice is heard.

MINI-PROJECT MANAGER CONCEPT

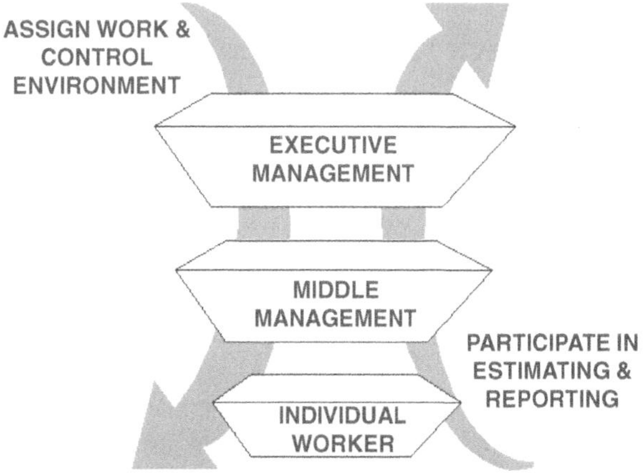

"Manage from the bottom-up, not just top-down. Create personal commitment and accountability."

TIME DEFINITIONS

The "Mini-Project Manager" concept begins with a different perspective of time. Fundamentally, there are two types of time, available and unavailable. Available time is time used for productive work as in a normal work day. Unavailable time is time that is not available to perform productive work, such as holidays, vacations, shutdowns, etc. Although both available and unavailable are schedulable, we are primarily concerned with the utilization of available time.

Available time is divided into two categories: direct and indirect time. Direct time is that amount of time expended on productive work, such as following the assigned phases/activities/steps of a methodology. Indirect time is time that does not contribute to productive work, such as non-project related meetings, training, breaks, etc.

Under the "Mini-Project Manager" concept, direct time is the responsibility of the individual employee to manage. In other words, they are personally responsible for managing their work effort. Conversely, indirect time is the responsibility of management to control the work environment.

WHAT IS TIME?

PRINCIPLE: Time is a resource that must be managed like any other resource.

AVAILABLE TIME	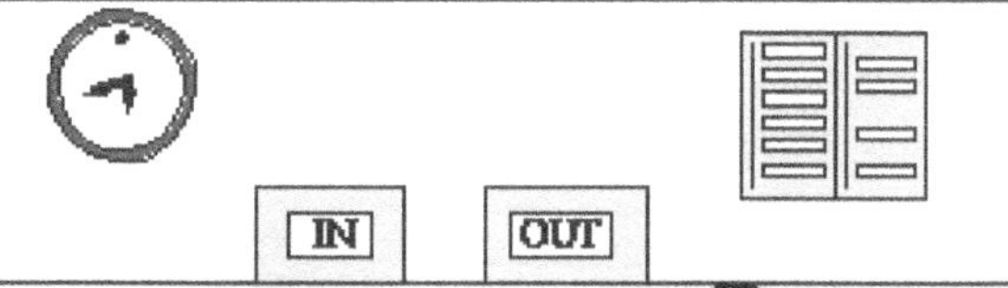
DIRECT TIME	
INDIRECT TIME	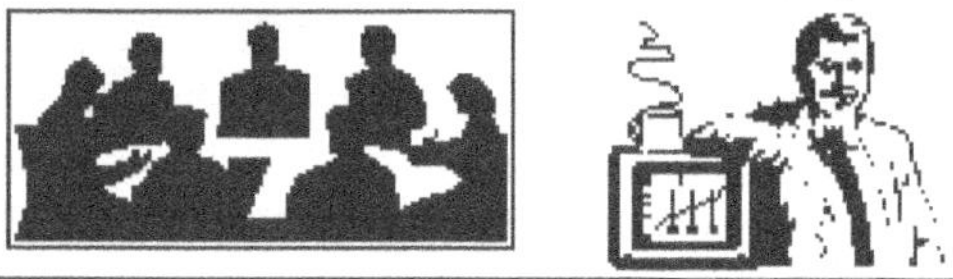
	UNAVAILABLE TIME

EFFECTIVENESS RATE = DIRECT ÷ (DIRECT + INDIRECT)

EFFECTIVENESS RATE

This view of time highlights the fact there are interferences, whether planned or unplanned, that prevent workers from expending 100% of the available time on assignments. The relationship of direct to indirect time is called "effectiveness rate" and is expressed as the percent of time spent on direct work. If someone has a low effectiveness rate, this may highlight to the manager that an employee is experiencing too many interferences, such as meetings or breaks. The manager can then exercise authority and minimize or eliminate the indirect activities if it is deemed unproductive or incorrect. In this way, the manager is controlling the work environment. On the other hand, a low effectiveness rate is not always considered undesirable. For example, a manager may send employees to a training class to sharpen their skills and become more productive. The employee's effectiveness rate for this period would be very low.

As should be apparent, "effectiveness rate" is noticeably different than an efficiency rate. It is simply an expression of the availability of an individual to do the "right" work. A worker can have a high effectiveness rate, yet may be inefficient at their assignments. Conversely, someone with a low effectiveness rate may be the most efficient worker in the company.

To calculate an individual's effectiveness rate, divide the total direct hours in a day by the total available hours in a day. For example, in an eight hour business day, if the employee spends six hours on direct assignments and two on indirect activities, this means the person is 75% effective for that day. Although effectiveness rates will vary, 60% to 70% is a typical rate. This ratio can apply to an individual, a department, or for an entire organization. Ultimately, the effectiveness rate provides a measure of how well time is being used in the work environment.

As we will see, this definition of time will have other beneficial uses, particularly in the areas of estimating, scheduling and control. This view is more realistic than the "man hour" concept which assumes a "one-to-one" relationship of the work to elapsed time and is not concerned with the control of indirects. According to the "man hour" approach, a person is 100% effective. When something should be completed is quite different than how much work is involved. For example, it could take three elapsed weeks to accomplish ten hours of direct work. It all depends on how much time is available; what other activities are involved; and what interferences exist.

The underlying philosophy behind all of this is that time is a resource that must be managed like any other resource. Before we can manage it, we must first understand it.

PROJECT MANAGEMENT ACTIVITIES

There are five basic activities associated with project management; each directly affects the others:

- Project Planning (Section 27)
- Project Estimating (Section 28)
- Project Scheduling (Section 29)
- Project Reporting (Section 30)
- Project Control (Section 31)

PROJECT MANAGEMENT SYSTEM

A Project Management system is an information system used to support the basic activities of project management. There are a variety of project management related software packages attacking the subject piecemeal. Some are used to develop project networks, others are used to calculate schedules or estimates, or simply to perform project bookkeeping. However, very few take an integrated approach to project management. This is a serious problem since dependencies exist between project management activities; to illustrate:

• Planning is the basis for developing estimates, schedules, and for project control.

• Estimates are used as the basis for key project decisions (approve, reject, revise), and are used to calculate schedules.

• Scheduling requires a project plan, involving a network and human resources with skills and proficiencies.

• Reporting requires assignment of tasks, estimates and schedules.

• Control requires effective planning, estimating, scheduling, and reporting.

A superficial approach to project management will produce shallow results. Since a project management system is used as a communications vehicle for management, the more people involved in the program, the greater the communications and coordinated use of resources.

Can the philosophies of project management be adopted and implemented by a single person? Yes. A group of people for a single project? Yes. A department or division? Yes. The entire company? Yes. In fact, as the scope grows, communications improves and the philosophy is more consistently applied.

REQUIREMENTS FOR A PROJECT MANAGEMENT SYSTEM

GENERAL SUPPORTS ANY TYPE OF PROJECT. DISTINGUISHES D/I/U ACTIVITIES. PROMOTES MINI-PM CONCEPT. INTEGRATED APPROACH. IMPORT/EXPORT CAPABILITIES.	**SCHEDULING SUPPORT** PROVIDES FOR EFFECTIVENESS RATE. MULTI-VERSIONING. RESOURCE ALLOCATIONS.
PLANNING SUPPORT VARIOUS WBS. INTERNAL & EXTERNAL DEPENDENCIES. MULTIPLE RESOURCES/MULTIPLE PROJECTS/MULTIPLE ASSIGNMENTS. SKILLS INVENTORY.	**REPORTING SUPPORT** RECORDS & VALIDATES TIME. ESITMATE TO DO. MAINTAINS HISTORY. RECORDS OUT-OF-POCKET EXPENDITURES.
ESTIMATING SUPPORT DETAIL & O-O-M. MULTI-VERSIONING. GUIDELINES (EST. GENERATOR)	**CONTROL SUPPORT** POSTS TIME TO PROJECTS. VARIOUS STATUS REPORTS TO ANALYZE PROJECTS & RESOURCES. BILLING/CHARGEBACK.

The scope of project management affects many people:

• The individual worker will prepare estimates and schedules, perform project work, and report on activities.

• The project manager will plan and direct the use of resources on projects.

• Department managers will administer resources and control projects within an area.

• Executive management will establish project priorities and monitor project progress.

Obviously, project management should not be restricted to a handful of people or projects. Dozens of projects may be active at any one time, involving hundreds of workers, across departmental boundaries. Synchronization of the work effort is required to maximize effect and minimize confusion. Project management, therefore, should be viewed as a corporate philosophy as opposed to a technique used by various people.

It must be remembered that a project management system is not a substitute for management. It does not make decisions; people do, based on information produced from the system. Once convinced of the action or decision to make, make it. There is little point to any information system if you do not take advantage of the information it provides. What this highlights is the fact that a project management system will only be as effective as the people who use it.

PROJECT ADMINISTRATION

This function should not be confused with Project Management. The primary responsibility for Project Control rests with Project Management. Project Administration is concerned with the collection, recording and maintenance of Project data and, as such, is the chief administrator of the Project Management system. At any one time, many projects may be in progress within the systems department and therefore, coordination is required for the processing of project data. The Project Administrator can be viewed as the administrative aid to both the Project Managers and Departmental Management.

The Project Administrator is concerned with the following "PRIDE" forms (see "Forms" section):

- Time Distribution Worksheet
- Project Control Sheet
- Planning Chart (Gantt)

TIME DISTRIBUTION WORKSHEET

A Time Distribution Worksheet is prepared weekly by each professional involved with direct project activities including systems maintenance. Before instituting the Weekly Time Distribution, all indirect activities should be identified. Generally, any activity not identified in the Phases of "PRIDE", or in Data and Project Management, is an indirect activity. Indirect activities include such things as: meetings, training, employee personal matters, travel, review, etc. Scheduled holidays or vacations should not be used when computing effectiveness since these days are "unavailable" for scheduling and will be identified accordingly in the Planning Charts. A list of the approved indirect activities should be issued to all personnel to use when preparing time sheets.

It is important for estimating and scheduling purposes to establish the Effectiveness Rate as quickly as possible. The Time Distribution Worksheet provides the ability to separate and identify both direct and indirect hours. As an aside, development organizations are typically 60% - 70% effective.

Time Reporting data is collected on a weekly cycle. Two decisions should be made in reference to this data collection:

1. What is the scheduled work week in hours? (e.g., 40 hours)

2. When does the work week start and end? (e.g., Monday - Friday)

The Project Administrator collects all Time Distributions at the end of the weekly cycle. They are checked for arithmetic correctness and completeness of hours in terms of the scheduled work week. Lack of discipline concerning the submission and correctness of the time distributions on the part of departmental personnel can create serious delays in processing. The Project Managers are responsible for ensuring this does not become a problem and resolving any problems encountered by the administrator. An automated Project Management system can materially help in validating project/time data.

After the Weekly Time Distributions have been audited by Project Administration, they are recapped by the administrator by Project, System Structure and Phase Activity numbers for Direct Hours and by type activity for Indirect Hours on to a Time Distribution Worksheet. Using the completed recap, the Effectiveness Rate is computed and entered on the recap. This data should be plotted on a spreadsheet for review and analysis.

After completing the Effectiveness Rate, the recap is forwarded to departmental management review. Since this rate is used for project planning, any variations can have significant effect on project schedules such as schedule overruns or underruns. Management should investigate the reasons for these variations and take corrective action where necessary.

PROJECT CONTROL SHEET

The primary responsibility for the preparation and maintenance of these sheets rests with the person assigned to the phase. A copy is retained by the project manager and Project Administrator.

After the weekly time sheet has been prepared by the individual, he posts the hours appropriate to each activity. The new hours are added to those previously recorded. The actual fields on the control sheets should be updated in pencil to permit the use of the same control sheet each successive week.

The Project Manager reviews the control sheet with the analyst involved. Jointly, they establish by review the completion status of the activities being performed. They enter on the Control Sheet the hours to complete and any start or end dates. This review is the most critical in Project Management. At this point, it can be established if the phase and activity are on schedule and within estimated hours. If there are any significant changes required at this juncture, the Project Manager and the individual prepare a new revised Control Sheet outlining on the reverse side the reasons for the revision the new course of action. The Project Manager also will review the effect of this change on the overall project and submit and/or make any appropriate changes. From the updated control sheets, the Project Manager updates the various Project Planning Charts. A copy of the updated overall project planning chart and completed control sheets are returned to the Project Administrator who in turn revises original Control Sheets.

PLANNING CHART (GANTT CHART)

Based on the updated project data from the control sheet, the Administrator updates the individual Resource Allocation Charts (also using the Planning Chart form) both as to planned additions, revisions and actual status.

Since the problem of resource allocation is an overall department problem, it is recommended that the Administrator maintain the individual resource allocations. This is an excellent check on the various Project Managers' execution of Project Management since these allocations cannot be updated without all steps being accomplished.

GENERAL COMMENTS - PROJECT ADMINISTRATION

At first blush, it may appear that the Project Administration function involves considerable clerical and Project Management effort. Experience will show that this function saves considerable time that was previously spent both by the Project Manager and others in the establishment and dissemination of project status data. The discipline once established provides an orderly and effective way of managing projects. The important aspect of this control system is that the individual is responsible for his own activities and reporting.

The foregoing general steps do not cover all the possibilities or actions that should be taken as a result of processing project data. Each week, Project Management and departmental management should review the project data for any conflicts between projects and take any additional steps necessary.

Again, an integrated and automated Project Management system can greatly simplify the life of a Project Administrator, but it does not eliminate the need for the function. Project Administration is the "Traffic Cop" that monitors all projects in progress and enforces policy in terms of time reporting and when to revise estimates and schedules. In many organizations the Project Administrator and the Technical Librarian are one and the same person.

BENEFIT$

The benefits of project management are numerous:

- It assures that project personnel are "doing the right things" to satisfy project objectives and plans.
- It creates personal commitment and accountability.
- It demystifies the activities of project management, such as estimating and scheduling.
- Improved control over the organization.
- Improved communications within the organization.
- Improved accuracy of estimates and schedules.
- Improved use of resources.
- Improved productivity.
- Improved control over cash flow.

END SECTION

Section 27

"PRIDE" PROJECT PLANNING

You cannot put two quarts of liquid into a one quart bottle. If you try, you will lose a quart.
- Bryce's Law

INTRODUCTION

We as human beings have a natural aversion to planning of any kind. Some of us cannot plan a day of activities, let alone a week, month, year, etc. This is primarily due to an inherent resistance to organizing and structuring our activities. Planning requires discipline, something that is sorely lacking in most information systems organizations. To obtain the results of planning requires an expenditure of time and some serious thinking. There is no "free lunch."

Projects are often preordained to failure before they even start, simply due to the lack of effective planning. This may be caused by our temptation to *"leap before we look."* Emotional desire must be overcome by logical fact in order for planning to succeed.

In order to perform project planning, we must resolve the following questions:

• What is the scope of the project? - The scope must state the project's objectives and the parts of the organization involved, both directly and indirectly.

• What are the steps required to meet the project's objectives? The old adage is true, *"If you don't know where you are going, any road will take you there."* Performing work in a logical sequence gives direction to the project. The inability to do so results in lost time and effort. Therefore, not only do the required steps in a project need to be defined, but the precedent relationships within a project must also be defined. The methodology thereby becomes the "roadmap" for the execution of the project.

• What are the deliverables and benchmarks of the project? In order to verify that a particular project task has been completed, it is necessary to substantiate that all aspects of the task have successfully been executed. An impartial and objective mechanism that checks the completeness of tasks is necessary. It is important to demonstrate tangible results from our project efforts in the form of accomplishments and deliverables. **Any task that does not result in a reviewable or tangible result is an unnecessary step that should be eliminated.**

• What resources are required to perform the work? Assigning the correct resources to the appropriate work steps is a critical factor in every project. By properly defining the work steps and the benchmarks, it is possible to clearly identify the skills required to execute the steps. Resources with the appropriate skills and availability can then be assigned to the project tasks.

GENERAL DISCUSSION

WORK BREAKDOWN STRUCTURE (WBS)

All projects have a structure depending on the methodology used. The methodology defines what is going to be produced. It can be as simple as one step or as extensive as several phases involving multiple activities and tasks. The methodology represents the selected approach for implementing a project. It is structured into a hierarchy consisting of one or more phases of work. A phase represents a major "key event" or milestone in the project. Each phase consists of one or more activities representing "sub-events" required to meet the milestone. Each activity consists of one or more operational steps or tasks representing the individual actions to be taken in the project.

The level of detail required to perform a project is ultimately left to the discretion of the Project Manager. If a simple project, perhaps the manager will only define a phase with a few activities. However, if a project is large and complex, the manager may wish to define and manage at the operation level.

PRECEDENT RELATIONSHIPS (DEPENDENCIES)

Up to this point we have only defined WHAT work is involved, not its sequencing. A methodology defines not only the various units of work, but also dependencies between the worksteps. Such dependencies are referred to as "precedent relationships."

Project worksteps may be conducted either sequentially, in parallel, or combinations of the two. Precedent relationships define what worksteps precede and succeed a single workstep.

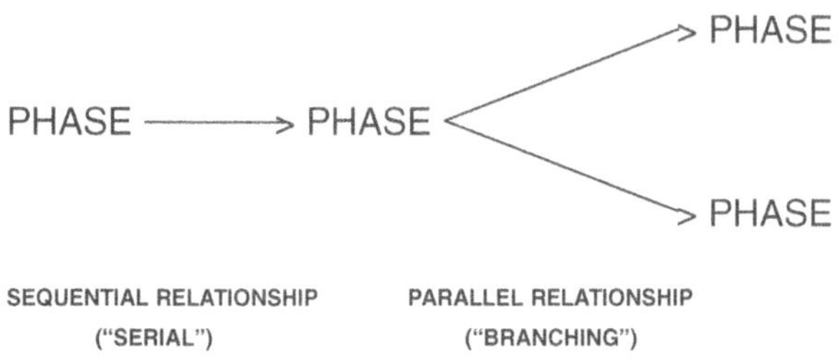

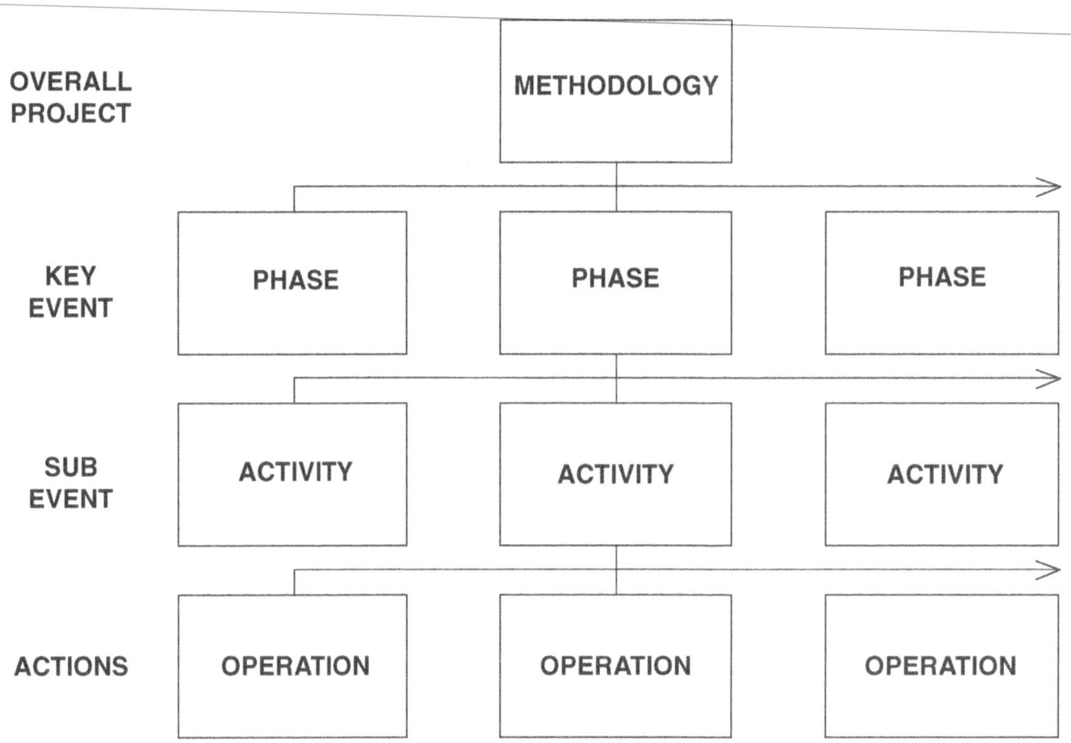

Each phase, activity and operation of a methodology should produce a reviewable result (work product) to substantiate completion of assignments. Otherwise, a methodology becomes a meaningless series of tasks.

Precedent relationships can be defined between worksteps in the same level of the methodology structure. This means:

1. PROJECT-TO-PROJECT relationships.

2. PHASE-TO-PHASE relationships.

3. ACTIVITY-TO-ACTIVITY relationships.

4. OPERATION-TO-OPERATION relationships.

This brings up two points:

1. Progression between project worksteps at the same level cannot proceed until the subordinate levels are fulfilled. This means you cannot move from one project to another until all of the phases from the first project have been performed; nor can you move from one phase to another until all of the activities from the first phase have been performed; nor can you move from one activity to another until all of the operations from the first activity have been performed.

2. You cannot define lower level worksteps until you have first defined the higher levels. In other words, you must define phases before you define activities, before you define operations.

The one exception to this is a PHASE-TO-PROJECT relationship where a separate project can be activated pending completion of a phase. This can be demonstrated by separate "PRIDE"-ISEM and "PRIDE"-DBEM projects:

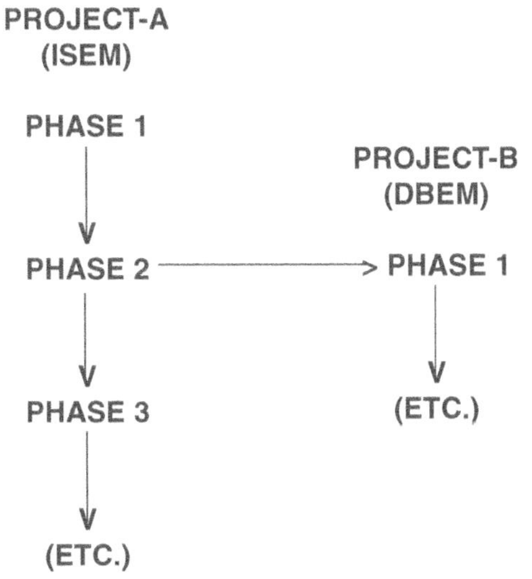

NOTE: Although Project-to-Project and Phase-to-Project Relationships are permitted, they are uncommon. Most projects will only show inner dependencies (phase-to-phase, activity-to-activity, operation-to-operation).

Although "branching" (parallelism) can occur at any level in the methodology, the project manager will typically find less need for branching at the lower levels of the methodology structure. This means phases are more apt to branch than operations. Most operational steps within an activity are performed serially (sequentially).

To expedite the development of methodology structures, "PRIDE" includes a "Methodology Definition Worksheet" which is used to define the Work Breakdown Structure and precedent relationships. To illustrate:

LEVEL-1: DECOMPOSING A METHODOLOGY INTO PHASES

PRIDE®

METHODOLOGY DEFINITION WORKSHEET

NAME: Building Project

PD- 0 0 0 1 0 CONTROL NUMBER

PAGE 1 OF 1

PREPARED BY Mick	APPROVED BY John	DATE April 1

TYPE:		IDENTIFIER:
METHODOLOGY	BUILDING CONSTRUCTION	CONST
PHASE		
ACTIVITY		

WORK STEP	RESOURCE (OPT)	FUNC CODE	NAME	PRECEDENT RELATIONSHIPS
1			PLAN PROJECT	
2			PREPARE SITE	
3			LAY FOUNDATION	
4			CONSTRUCT FRAME	
5			CONSTRUCT ROOF	
6			FINISH INTERIOR	
7			FINISH EXTERIOR	
8			INSPECTION	
9			REVIEW PROJECT	

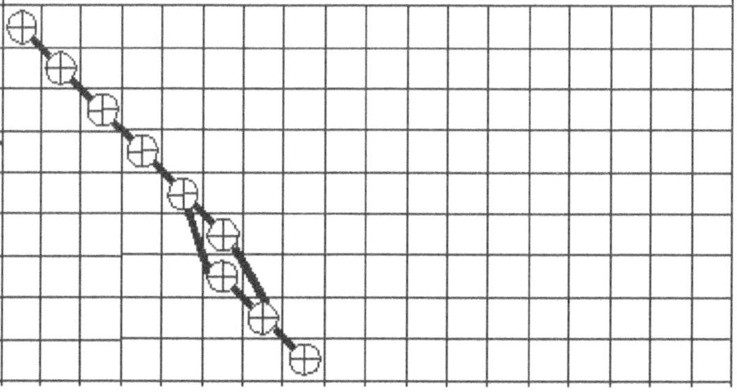

LEVEL-2: DECOMPOSING A PHASE INTO ACTIVITIES

PRIDE®

METHODOLOGY DEFINITION WORKSHEET

NAME: Building Project

PD- 0 0 0 1 0 CONTROL NUMBER

PAGE 1 OF 1

PREPARED BY Mick	APPROVED BY John	DATE April 1

TYPE:		IDENTIFIER:
METHODOLOGY	BUILDING CONSTRUCTION	CONST
PHASE	LAY FOUNDATION	3
ACTIVITY		

WORK STEP	RESOURCE (OPT)	FUNC CODE	NAME	PRECEDENT RELATIONSHIPS
A			EXCAVATE LAND	
B			PREP CONC FORM	
C			INSTALL SITE UTIL	
D			ROUGH PLUMBING	
E			ROUGH ELECTRIC	
F			POUR CONCRETE	
G			INSPECT	

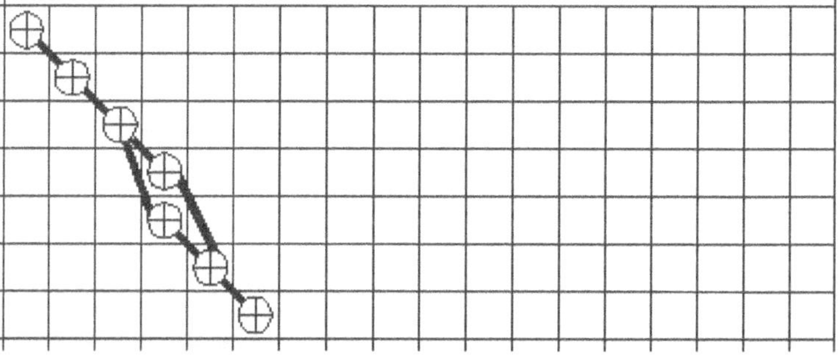

LEVEL-3: DECOMPOSING AN ACTIVITY INTO OPERATIONS

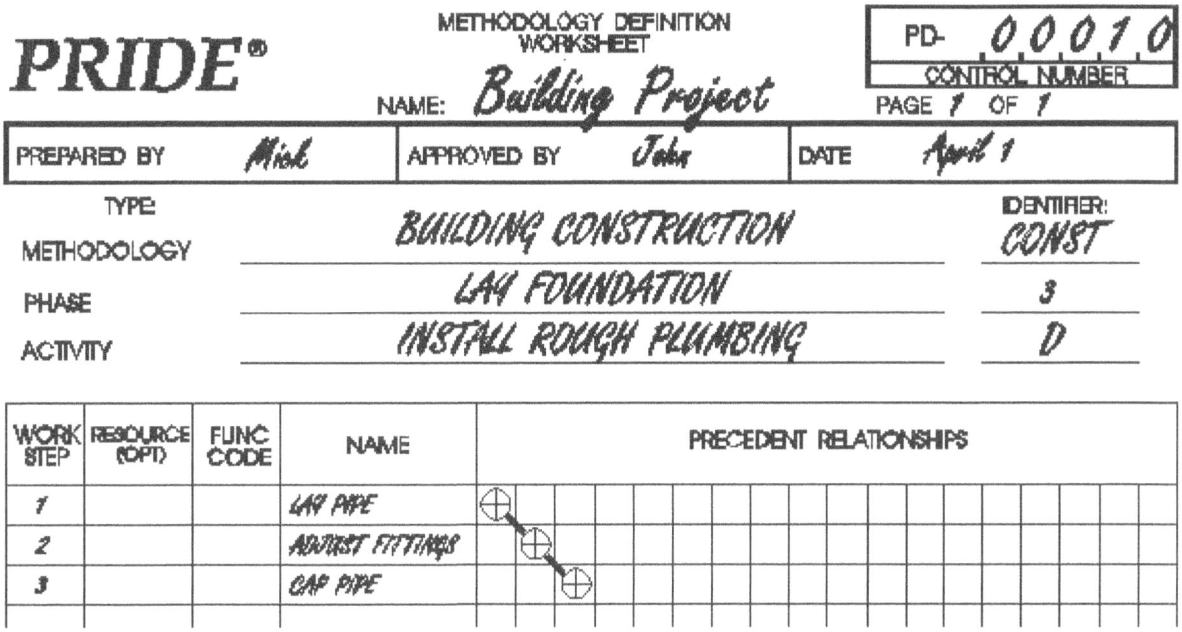
PRIDE®

METHODOLOGY DEFINITION WORKSHEET

NAME: Building Project

PD- 00010
CONTROL NUMBER
PAGE 1 OF 1

PREPARED BY Mick	APPROVED BY John	DATE April 1

TYPE:		IDENTIFIER:
METHODOLOGY	BUILDING CONSTRUCTION	CONST
PHASE	LAY FOUNDATION	3
ACTIVITY	INSTALL ROUGH PLUMBING	D

WORK STEP	RESOURCE (OPT)	FUNC CODE	NAME	PRECEDENT RELATIONSHIPS
1			LAY PIPE	
2			ADJUST FITTINGS	
3			CAP PIPE	

To view and/or print a copy of the "Methodology Definition Worksheet" worksheet, see the "Forms" Section.

METHODOLOGY CRITERIA

In order to effectively organize a project it is important to recognize the basic elements of a methodology:

Mandatory Requirements

1. A defined Work Breakdown Structure (WBS) - consisting of a series of work steps in various levels of abstraction (e.g., Phases, Activities, Operations).

2. Defines the project functions responsible for performing the various work steps.

3. Defines the project dependencies (the precedent relationships between work steps).

Without these mandatory requirements, a methodology is illegitimate and should be referred to as something else.

Optional Requirements

1. Have a single phase to initiate a project, and a single phase to conclude it. Multiple starts and multiple ends are not desirable from a management point of view.

2. The methodology structures should be based on reviewable work products to verify completeness. If the methodology is not defined accordingly, the "Dance of the Fairies" phenomenon occurs - this is where a series of meaningless work steps are defined with no verifiable end result.

3. The methodology structures should be reusable on multiple projects.

4. Provide for both sequential and parallel project execution.

5. The methodology structures should accommodate a product structure, thereby allowing parallel processing.

Although these latter requirements are not mandatory, they are highly desirable features and have been incorporated into the methodologies in "PRIDE".

Project planning is made simpler by the existence of standard methodologies, such as "PRIDE"-EEM, ISEM, and DBEM, which include defined phases, activities, deliverables, precedent relationships and the functions to perform the work.

END SECTION

Section 28

"PRIDE" PROJECT ESTIMATING

Most estimating errors are errors of omission, not commission. It's what we forget to estimate that gets us in trouble.
- Bryce's Law

INTRODUCTION

Estimating is one of the most controversial subjects in project management. There are some people who have turned the subject into a cryptic science involving esoteric techniques bearing a close resemblance to "voodoo."

Estimating is simply the process used to determine the amount of effort and cost required to implement a project, in part or in full. It is important to acknowledge that estimating is fundamentally an effort at projecting the future. Like all projections, the more facts and information available, the better the estimate. There is a natural human tendency to avoid making estimates because estimates are commitments, and people tend to shy away from commitments, particularly when they are not sure of the facts. Nevertheless, little progress would be made if we never attempted to plan for the future.

Most estimating errors are errors of omission, not commission. It is what we forget to estimate that often leads to problems. Methodologies, with their defined structure, materially assists with eliminating some of the unknowns when estimating. They provide frameworks and structures that act as checklists for estimating. Methodologies isolate the activities that have to be performed into small enough increments, thereby minimizing the margin of error.

An estimate improves in accuracy in direct relation to the level of detail considered. A methodology defines the sequence by which parts are assembled. For example, a construction methodology identifies all of the resources of a product, such as lumber, steel, glass, etc. and how they are assembled. An IRM related methodology specifies the sequence by which data elements, records, files, inputs, outputs, processes, etc. are assembled. This provides the ability to use a "bill of materials" technique to count all of the resources in the product and develop an estimate for the project, in part or in full, based on the standards developed for completing and/or installing a resource. This is why the "PRIDE" methodologies put in emphasis on "rough designs" in the early phases of work. At this time, the Project Manager is asked to consider:

- The number and types of NEW resources to be created.

- The number and type of existing resources requiring MODIFICATION.

- The number and type of existing resources that can be RE-USED as is (no modification required).

To illustrate, in a "PRIDE"-ISEM Project (Phase 1), a complete "rough design" of the envisioned system is produced in Activity F. In Activity G, the Project Manager takes the rough design and makes the following type of assessment:

IRM RESOURCE	NEW	MODIFY	RE-USE
SYSTEM	1		
SUB-SYSTEMS	14		
ADMIN PROC	23		
COMP PROC	13		
PROGRAMS	28		
MODULES	33	10	112
INPUTS	17		5
OUTPUTS	37	13	
FILES	56	5	43
RECORDS	250	50	306
DATA ELEMENTS	60		257

This analysis of the rough design is used as the roadmap for the project (in the above example, there will have to be 14 Phases 3-7 because there are 14 sub-systems and 13 Phase 4-II & 6 for the 13 computer procedures, and at least 28 Phase 5's for the programs). It is also the basis for the project estimate. Such estimating is greatly facilitated through the use of an IRM Repository.

The concept of direct and indirect hours permits the establishment and application of estimating guidelines. To illustrate; how much direct time does it take to weld a six inch pipe? Define a data element? Define an information requirement? It would be almost impossible to establish guidelines if indirect hours were included in the guidelines. Indirect time is a part of the work environment and can vary from company to company, even from group to group. Estimating, therefore, must be accomplished using direct hours only.

TYPES OF ESTIMATES

There are fundamentally two types of estimates: Order-Of-Magnitude and detail. Both are expressed in direct hours and both are based on the resources of the system as known at that point in the methodology.

The Order-Of-Magnitude (O-O-M) is an estimate of the effort remaining to complete a project and is typically prepared by the project manager or project team. It is expressed in terms of remaining phases, not activities. The O-O-M estimate is principally used to develop a cost evaluation summary which contains such things as "return on investment" and "break even points." These figures are used by management for decisions relative to the project's economic feasibility and priority. Based on this information, management may determine to continue, discontinue or revise the project within a reasonable level of accuracy.

The Detail Estimate is an estimate of the effort required to complete the activities of an individual phase and is prepared prior to the start of the phase by the individual or individuals assigned to the phase. It is based on detail specifications defined in the preceding phase. The important difference here is that **the detail estimate is an expression of someone's personal commitment to a project** and is based on detail knowledge and study. Since it is prepared and expressed in direct hours by the individual assigned to it, that person becomes responsible for meeting the estimate. In this way, the staff is treated as professionals and held accountable for their actions. By working with direct hours, more accurate estimates are made. If the analyst has to take into account scheduling and indirect time, the estimates become meaningless.

TYPES OF ESTIMATES

	ORDER-OF-MAGNITUDE	**DETAIL**
PURPOSE	BASIS FOR PROJECT DECISIONS	EXPRESSION OF PERSONAL COMMITMENT
PREPARED BY	PROJECT MANAGEMENT OR PROJECT TEAM	INDIVIDUAL(S) ASSIGNED TO PERFORM THE WORK
WHEN PREPARED	END OF PHASE	BEGINNING OF PHASE
SCOPE	THROUGH THE END OF THE PROJECT	THROUGH THE END OF THE PHASE
LEVEL OF DETAIL	FOR ALL PHASES WITHIN A PROJECT	FOR ALL ACTIVITIES WITHIN A PHASE
SKILL LEVEL CONSIDERED	AVERAGE	UNIQUE TO WORKER

ALL ESTIMATES ARE EXPRESSED IN DIRECT TIME

Is it possible to develop an accurate estimate for an entire project from the outset? Yes and no. If the nature of the project is to modify or maintain an existing product or system, the answer is 'yes.' In this situation, the parts of the product affected are known and the level of complexity deduced. The answer is 'no' when the variables of the product are unknown. For example, it would be impossible to estimate the cost for constructing a building without first detailing requirements and preparing a rough design. The same is true in the systems field. It is a relatively simple task to estimate a systems project where the variables are known, such as the processes, inputs, outputs, files, records and data elements. However, if the resources are unknown, a preliminary phase involving requirements definition and rough design is needed. In this regard, there is virtually no difference between a systems project and a construction or manufacturing project. To try to estimate the time and cost of an entire project without first knowing all of the variables is like asking a physician to diagnose a problem without a preliminary examination. It is sheer nonsense.

In terms of estimating guidelines, there is very little difference between the two estimates. Both are based on the resources and the steps involved in the project. The fundamental difference is the O-O-M estimate is based on an average skill level and the Detail estimate is based on the specific skill level of the assigned Resource. Whereas the O-O-M estimate can use historical data of an entire group of people, the detail estimate would use only historical data of a particular person.

No matter how accurate or inaccurate the generated estimates are, the individual makes the final decision on whether to accept, revise, or reject them. This is because the estimates are intended to be an expression of personal commitment. If an estimate is missed, it is because it was prepared incorrectly, not because the guidelines are wrong. The guidelines can only advise, they are not a panacea. The person preparing the estimate must make the ultimate decision as to the realism of the figures. It should always be remembered that nothing will ever be accomplished if the individual does not assume responsibility for the estimates made. Management can make the time available, but it takes a motivated employee to take advantage of the direct time made available.

One last note: the project estimate, combined with effectiveness rate and the corporate calendar, is the principal basis for plotting project schedules (not the other way around). In other words, we formulate the estimate first, then compute the schedule second.

SAMPLE DETAIL ESTIMATE FROM "PRIDE"-ISEM, PHASE 3, SUB-SYSTEM DESIGN

```
-------------------------------------------------------------------------------------------------
OD-90410                         PROJECT CONTROL SHEET                               "PRIDE"-PM
-------------------------------------------------------------------------------------------------
OCT 18, 2004                         RELEASE 1.2.0                                      PAGE 1

                    ***************************************************
                    *       PD-00132 ORDER PROCESSING SYSTEM          *
                    ***************************************************

                    PHASE 3         SUB-SYSTEM DESIGN

                    RESOURCE        OP08 ORDER PROC. MAINTENANCE

ACT/OP/FC                                           START DATE      END DATE

A       DEFINE PROCEDURES

        PM PROJECT MANAGEMENT       EST     1.0     2004/10/31      2004/10/31
          RE-00154                  ACT     0.0
          BILKO, ERNEST             TO DO   1.0

        SE SYSTEMS ENGINEERING      EST     15.0    2004/10/31      2004/11/02
          RE-00066                  ACT     0.0
          BORG, SUE                 TO DO   15.0

B       DEFINE PROCEDURE LOGIC

        SE SYSTEMS ENGINEERING      EST     26.0    2004/11/02      2004/11/08
          RE-00066                  ACT     0.0
          BORG, SUE                 TO DO   26.0

C       PREPARE DESIGN EVALUATION

        PM PROJECT MANAGEMENT       EST     2.0     2004/11/09      2004/11/09
          RE-00154                  ACT     0.0
          BILKO, ERNEST             TO DO   2.0

        QA QUALITY ASSURANCE        EST     4.0     2004/11/09      2004/11/10
          RE-10034                  ACT     0.0
          STEIN, FRANKIE            TO DO   4.0

        SE SYSTEMS ENGINEERING      EST     2.0     2004/11/09      2004/11/10
          RE-00066                  ACT     0.0
          BORG, SUE                 TO DO   2.0

D       PHASE 3 REVIEW

        PM PROJECT MANAGEMENT       EST     1.0     2004/11/11      2004/11/11
          RE-00154                  ACT     0.0
          BILKO, ERNEST             TO DO   1.0

        SE SYSTEMS ENGINEERING      EST     1.0     2004/11/11      2004/11/11
          RE-00066                  ACT     0.0
          BORG, SUE                 TO DO   1.0
```

(continued on next page)

(continued from previous page)

TIME/COST ANALYSIS			COST	
PM PROJECT MANAGEMENT	EST	4.0	200.00	
	ACT	0.0	0.00	
	TO DO	4.0	200.00	
QA QUALITY ASSURANCE	EST	4.0	200.00	
	ACT	0.0	0.00	
	TO DO	4.0	200.00	
SE SYSTEMS ENGINEERING	EST	44.0	2200.00	
	ACT	0.0	0.00	
	TO DO	44.0	2200.00	
PHASE TOTAL	EST	52.0	2600.00	
	ACT	0.0	0.00	
	TO DO	52.0	2600.00	
PHASE SCHEDULE			START DATE	END DATE
	EST		2004/10/31	2004/11/11
	ACT			

SAMPLE ORDER-OF-MAGNITUDE (OOM) ESTIMATE FROM "PRIDE"-ISEM, PHASE 1

OD-90360 PROJECT ESTIMATE/SCHEDULE RECAP "PRIDE"-ISEM

OCT 12, 2004 RELEASE 1.2.0 PAGE 9.1

****
* PD-00132 ORDER PROCESSING SYSTEM *
****

O R D E R - O F - M A G N I T U D E

REQUESTED VERSION: FIRST

PHASE	RESOURCE	FUNCTION	HOURS	START	END
1		SYSTEM STUDY AND EVALUATION			
		DB DATA BASE ADMINISTRATION	9.0	2004/10/06	2004/10/12
		PE PROGRAM ENGINEERING	14.0	2004/10/06	2004/10/12
		PM PROJECT MANAGEMENT	23.0	2004/08/25	2004/10/17
		QA QUALITY ASSURANCE	10.0	2004/09/26	2004/10/13
		SE SYSTEMS ENGINEERING	306.0	2004/08/25	2004/10/17
		PHASE TOTAL	362.0	2004/08/25	2004/10/17
		PHASE COST-	$ 18,100.00		
2	OP	SYSTEM DESIGN			
		PM PROJECT MANAGEMENT	5.0	2004/10/18	2004/10/28
		QA QUALITY ASSURANCE	3.0	2004/10/27	2004/10/27
		SE SYSTEMS ENGINEERING	50.0	2004/10/18	2004/10/28
		PHASE TOTAL	58.0	2004/10/18	2004/10/28
		PHASE COST-	$ 2,900.00		
3	OP-00	SUB-SYSTEM DESIGN			
		PM PROJECT MANAGEMENT	4.0	2004/10/31	2004/11/09
		QA QUALITY ASSURANCE	4.0	2004/11/07	2004/11/08
		SE SYSTEMS ENGINEERING	47.0	2004/10/31	2004/11/09
		PHASE TOTAL	55.0	2004/10/31	2004/11/09
		PHASE COST-	$ 2,750.00		

(continued on next page)

(continued from previous page)

41	OP-00	ADMINISTRATIVE PROCEDURE DESIGN			
		PM PROJECT MANAGEMENT	2.0	2004/11/10	2004/11/15
		QA QUALITY ASSURANCE	2.0	2004/11/14	2004/11/14
		SE SYSTEMS ENGINEERING	15.0	2004/11/10	2004/11/15
		PHASE TOTAL	19.0	2004/11/10	2004/11/15
		PHASE COST-	$ 950.00		
42	OP-00-02	SOFTWARE ENGINEERING			
		DB DATA BASE ADMINISTRATION	16.0	2004/11/17	2004/11/21
		PE PROGRAM ENGINEERING	64.0	2004/11/10	2004/11/25
		PM PROJECT MANAGEMENT	5.0	2004/11/10	2004/11/25
		QA QUALITY ASSURANCE	5.0	2004/11/22	2004/11/23
		SE SYSTEMS ENGINEERING	3.0	2004/11/23	2004/11/25
		PHASE TOTAL	93.0	2004/11/10	2004/11/25
		PHASE COST-	$ 4,650.00		

/\

8	OP	SYSTEM OPERATION			
		QA QUALITY ASSURANCE	2.0	2005/03/21	2005/03/21
		SE SYSTEMS ENGINEERING	15.0	2005/03/20	2005/03/22
		TR IRM EDUCATION/TRAINING	18.0	2005/03/21	2005/03/24
		PHASE TOTAL	35.0	2005/03/20	2005/03/24
		PHASE COST-	$ 1,750.00		
9		ISEM EVALUATION			
		PM PROJECT MANAGEMENT	15.0	2005/04/10	2005/04/14
		QA QUALITY ASSURANCE	3.0	2005/04/13	2005/04/13
		SE SYSTEMS ENGINEERING	30.0	2005/04/05	2005/04/12
		PHASE TOTAL	48.0	2005/04/05	2005/04/14
		PHASE COST-	$ 2,400.00		

PROJECT TOTALS				
	DB DATA BASE ADMINISTRATION	40.0	2004/10/06	2004/12/21
	PM PROJECT MANAGEMENT	223.0	2004/08/25	2005/04/14
	SE SYSTEMS ENGINEERING	1530.0	2004/08/25	2005/04/12
	PE PROGRAM ENGINEERING	943.0	2004/10/06	2005/03/10
	QA QUALITY ASSURANCE	70.0	2004/09/26	2005/04/13
	TR IRM EDUCATION/TRAINING	18.0	2005/03/21	2005/03/24
	PROJECT TOTAL	2824.0	2004/08/25	2005/04/14
	PROJECT COST-	$ 141,200.00		

*** END OUTPUT ***

END SECTION

Section 29

"PRIDE" PROJECT SCHEDULING

Time lost is time lost forever; You cannot buy it back.
- Bryce's Law

INTRODUCTION

Scheduling is a distinctly separate activity from estimating even though it depends upon accurate estimates. It is the process of converting direct hour estimates into date schedules. Scheduling includes the assignment of resources to tasks in the correct sequence to meet the objectives of the project. This is a complex process and one that is almost impossible without automation. In most organizations, there are many projects, resources and environmental conditions that can influence projects and cause the need for changes in schedules.

Scheduling is an iterative process of balancing the resources available against the time requirements established for the project. There are several variables involved:

- The direct hour estimate, either O-O-M or detail.
- The methodology for the project, as defined by the key events of the project.
- The human resources to be used on the project, with their effectiveness rate, current schedule, and skills and proficiencies.
- Project priorities as established by executive management.
- A corporate calendar which notes weekends, holi days, shutdowns, etc.

As an example of how a schedule is calculated, assume that there is a professional working on an assignment with an estimate of 100 direct hours. If the individual's planned effectiveness rate is 70%, then the 100 hour estimate converts to a projected 142.85 elapsed hours required to complete the assignment. The elapsed hours in the normal business day is divided into the 142.85 and (using 8 hour work days) results in 17.85 business days required to complete the assignment. These 17.85 business days are applied to the corporate calendar which expresses weekends and holidays, also other commitments for the professional, such as vacations and other work assignments. Taking these into account, start and end dates are then determined.

If the end date is unsatisfactory to the manager, there are several possible alternatives which can be used to improve the schedule:

- Reduce the employee's indirect activities and increase their effectiveness rate.
- Schedule work on weekends and/or holidays.
- Add additional personnel or use alternative personnel.

- Increase the number of available hours in the business day (within reason of course).

- Or combinations of the above.

By understanding and adjusting the variables associated with a project, a project manager has the ability to play "what if" with a high degree of certainty.

Scheduling is an activity that is greatly influenced by the concept of effectiveness rate. By examining the use of time, both direct and indirect, realistic schedules can be prepared. The "man hour" approach mentioned earlier does not take the environmental influences into consideration and assumes an effectiveness rate of 100%. Under this approach, the sample schedule would be completed in 12.5 business days as opposed to the 17.85 days mentioned. The difference is that effectiveness rate builds reality into the schedule. It does not make use of some esoteric "fudge factor" as typically used in relation to "man hours."

Time lost, is time lost forever. You cannot buy it back. Because of this, we must carefully manage the use of resources throughout the methodology of the project. The methodology defines the project network which may involve parallel and sequential work. The longest path in the methodology which is critical to the timely completion of a project is referred to as the "critical path." A project will always have a critical path until it is completed or closed. The path will not always be constant; it may vary depending on accomplishments. As one part of a project is concluded, the critical path may change to another branch of the project. Defining and monitoring the critical path is an essential part of project management, not only in scheduling, but in planning and control also.

PROJECT SCHEDULING

DEFINITION: The process of converting direct time estimates into date schedules.

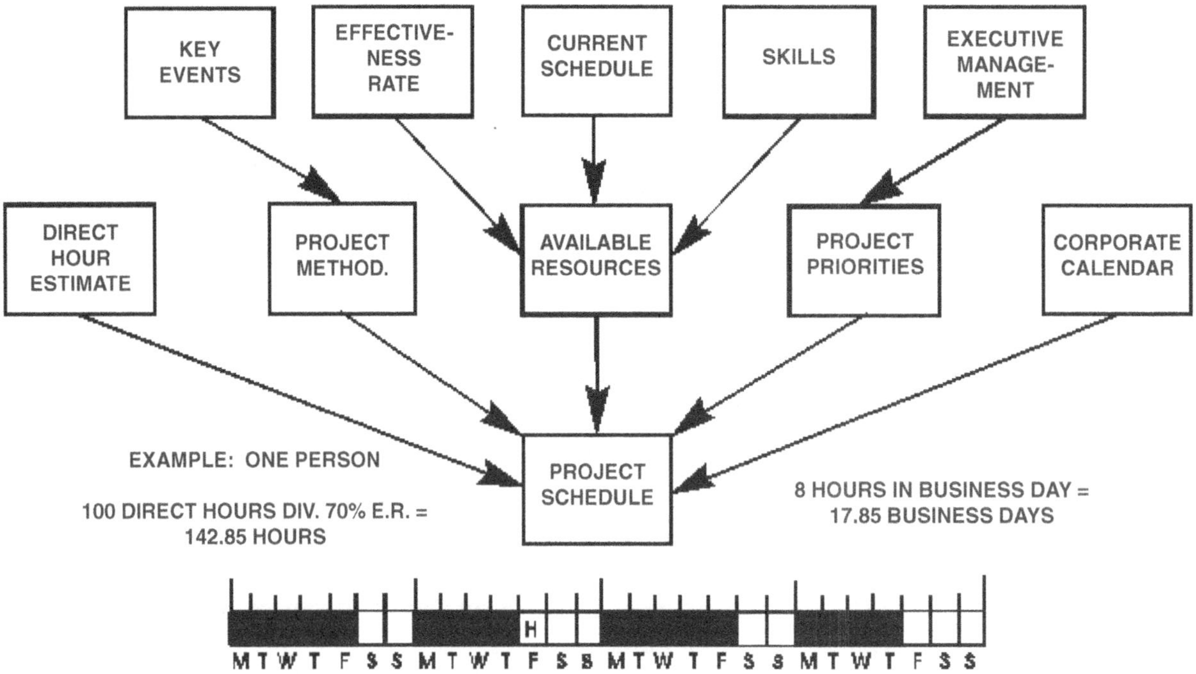

GANTT CHARTS

One tool that has always been useful for scheduling is the Gantt Chart, which was developed by Henry L. Gantt. It is used for resource allocation and project scheduling. Other names for the Gantt Chart include: planning chart, bar chart, milestone chart, histogram.

The Gantt Chart is used to plot project schedules as well as human resource allocations; to illustrate:

SAMPLE GANTT CHART USED TO PLOT PROJECT SCHEDULES

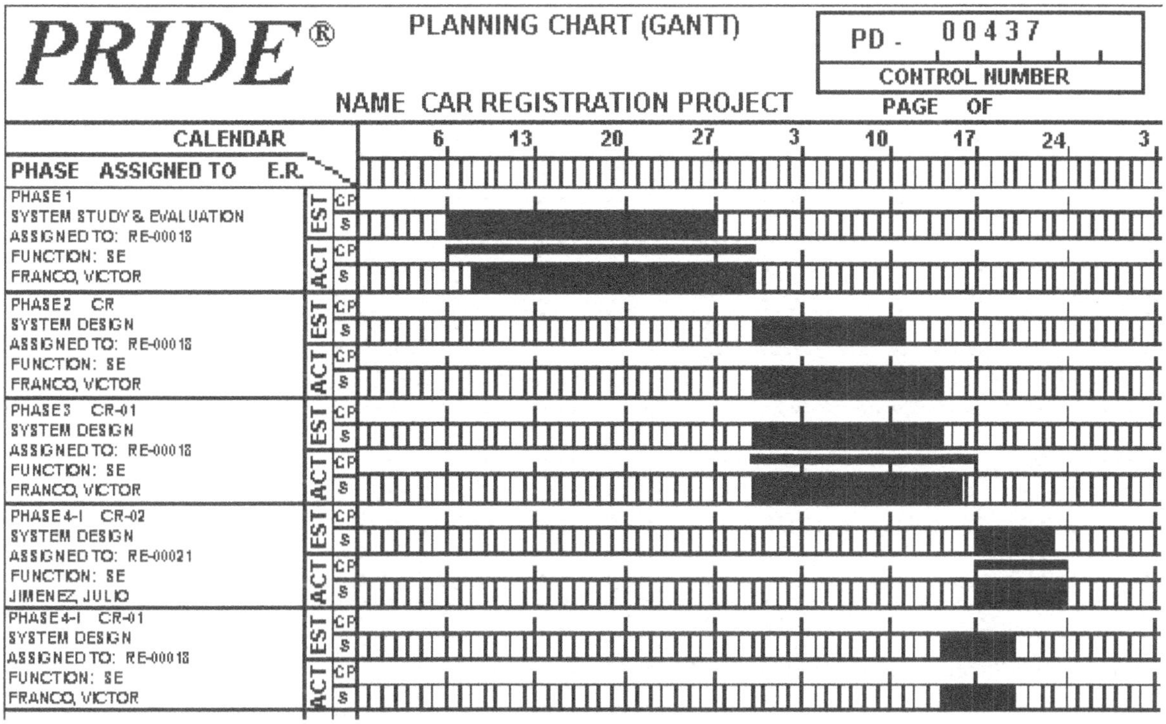

NOTE: CP represents the 'Critical Path'

SAMPLE GANTT CHART USED TO PLOT RESOURCE ALLOCATIONS

```
---------------------------------------------------------------------------------------------------
OD-90413                          RESOURCE ALLOCATION CHART                              PRIDE-PM
---------------------------------------------------------------------------------------------------
JAN 11, 2004                                                                               PAGE 1
                                   RE-00073 BROWN, IRENE

                                  JAN                               FEB
       2004 WEEK ENDING DATE      10      17      24      31        07      14      21
CORPORATE & PERSONAL UNAVAILABLES X                                                   XXXXX

PD-00132 COST CONTAINMENT SYS
PH: 3    31-08    FUNCTION: SE
SUB-SYSTEM DESIGN                 EST XXX X
HOURS/WEEK=20   FIX/VAR-V         ACT XXX

PD-00555 FINANCE SYSTEM
PH: 1             FUNCTION: SE
SYSTEM STUDY & EVALUATION         EST      XXXX XXXXX XXXXX
HOURS/WEEK=30   FIX/VAR-V         ACT

PD-00555 FINANCE SYSTEM
PH: 2             FUNCTION: SE
SYSTEM DESIGN                     EST                               XXXXX XXXXX
HOURS/WEEK=30   FIX/VAR-V         ACT

PD-00555 FINANCE SYSTEM
PH: 3    FS-01    FUNCTION: SE
SUB-SYSTEM DESIGN                 EST                                               XXXXX XXXXX
HOURS/WEEK=30   FIX/VAR-V         ACT
---------------------------------------------------------------------------------------------------
                                    RESOURCE SUMMARY
       NUMBER OF DAYS IN WORK WEEK-  5              SCHEDULED EFFECTIVENESS RATE-  .75%
       HOURS PER WORK DAY-           8.5            DIRECT HOURS PER DAY-          6.3
---------------------------------------------------------------------------------------------------
```

UNAVAILABLE HOURS PER WEEK	8.5	0.0	0.0	0.0	0.0	0.0	0.0	42.5
AVAILABLE HOURS FOR WEEK	34.0	42.5	42.5	42.5	42.5	42.5	42.5	0.0
CALCULATED DIRECT HOURS FOR WEEK	25.2	31.5	31.5	31.5	31.5	31.5	31.5	0.0
SCHEDULED HOURS FOR WEEK	26.0	30.0	32.0	30.0	31.0	27.0	28.0	0.0
UNSCHEDULED HOURS FOR WEEK	0.0	1.5	0.0	1.5	.5	4.5	3.5	0.0

Samples of schedules are included throughout the "PRIDE" methodologies and are expressed on the various Project Control Sheets and Project Estimate/Schedule Recaps. The Gantt Chart and other forms (the "Project Control Sheet" to prepare schedules for Detail estimates (also see alternate form), and; the "Project Estimate/Schedule Recap" worksheet for O-O-M estimates) can be obtained in the "Forms" section.

END SECTION

Section 30

"PRIDE" PROJECT REPORTING

Without an effective way to record and verify the use of time, all other project management activities are useless.
- Bryce's Law

INTRODUCTION

Reporting is the "Achilles' heel" of project management. If time is not reported accurately, the resulting information is of little value. Any report produced from inaccurately reported time will have little value or use.

To gain the maximum precision in reporting, time for the individual employee should be recorded on a daily basis. The time should then be posted to the various projects on a weekly basis to assure effective control.

Everyone reporting time should make a personal assessment of the remaining work to be completed. This is expressed as the amount of time "estimated to do" the work, which is different than the original estimate. Whereas the latter is used for estimating purposes, the former is used for reporting purposes and promotes the "Mini-Project Manager" concept.

"ESTIMATE TO DO"

To illustrate how "estimated to do" works, suppose an employee is scheduled to work 20 hours on an assignment during a single week. During the week, the worker reports a total of 5 hours of work performed against the assignment. The employee then considers the amount of remaining work on the assignment. The "estimated to do" will not necessarily be 15 hours (20 minus 5), although it is a possibility. After considering the amount of remaining work, the employee posts 30 hours "estimated to do." The 5 hours reported and the 30 hours "estimated to do" represents a total of 35 hours to perform the assignment, which is noticeably different than the original 20 hour estimate. Project management can easily spot this discrepancy and calculate its effect on the remainder of the project, showing a chain reaction of overruns or underruns on estimates and schedules.

ESTIMATE TO DO

Estimating the work remaining encourages each project participant to evaluate where they are and where they are going. This is the real value of project management. By estimating the remaining time for an assignment, project management can accurately pinpoint the effect of delays or accelerations to the project schedule and modify plans accordingly.

Since the project methodology is engineered to produce a reviewable result, whenever an employee indicates there is zero "estimated to do," it means they have completed the assignment and produced the appointed deliverable which can then be re-

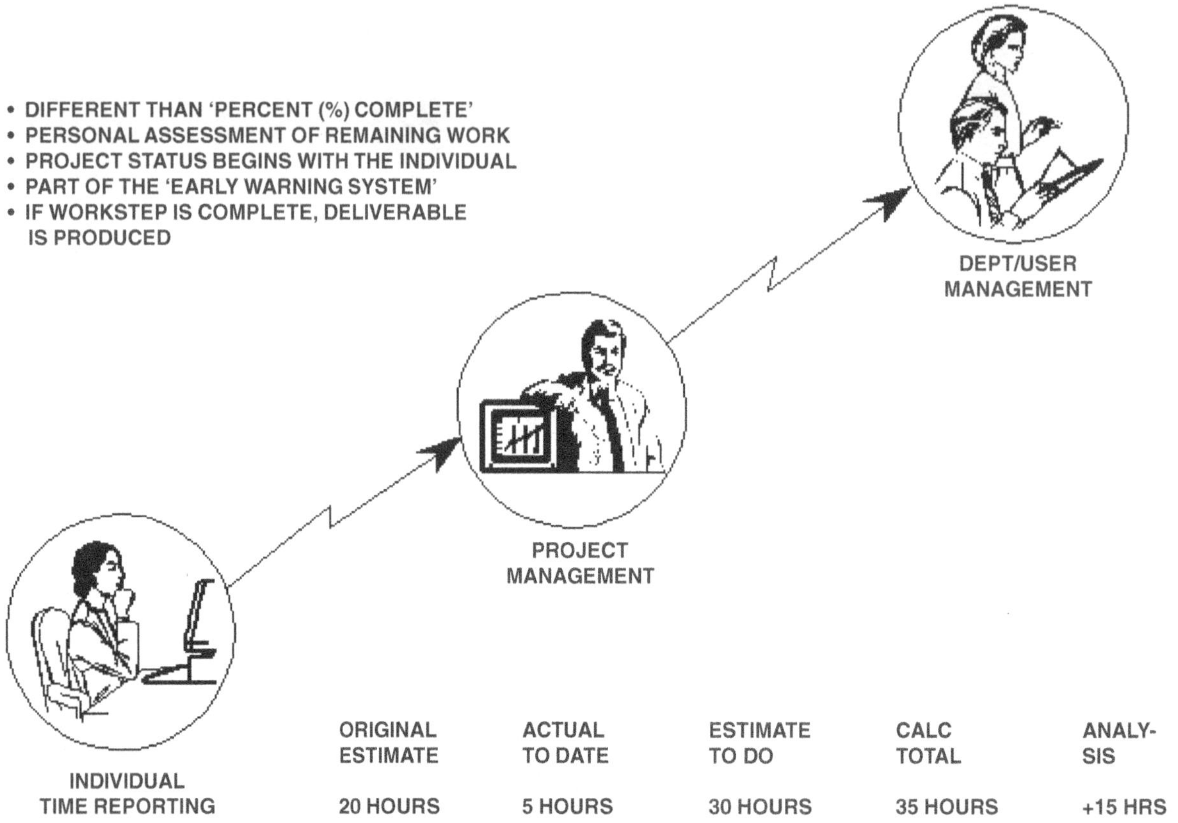

viewed. If it has not been produced, the work is not completed.

This "estimated to do" approach overcomes the "99% complete syndrome," which is an estimate of the percent of work completed. The problem with this approach is that it is less than scientific. For example, the project may be 99% complete, yet it may take another year to complete the critical 1% remaining to work on the project. Under the "estimated to do" approach, with emphasis on deliverables, there is much greater accuracy in project reporting and control.

Requiring employees to prepare estimates, report time and personally assess remaining work creates accountability and commitment. It provides for the ability to manage from the bottom-up, not just from the top-down. The data prepared by the individual employee is "rolled up" through various levels of management reports; for example:

A blank copy of the "Time Distribution Worksheet" is included in the "Forms" section.

SAMPLE TIME DISTRIBUTION WORKSHEET (MANUAL SYSTEM)

PRIDE® TIME DISTRIBUTION WORKSHEET

12 / 13 / 04
WEEK ENDING DATE

PAGE OF

RESOURCE NAME **Mick Anderson** RE- **00037**

PROJECT	RESOURCE (OPT)	PHS	ACT	OP	FUNC	D-I-U	MON	TUE	WED	THU	FRI	SAT	SUN	TOTAL	TO DO
00450	3102	3	C		SE	D	4	2						6	0
00450	3102	3	D		SE	D				7				7	0
00450	310202	42	A		PE	D					8			8	13
00450	3102	3			PM	D	1								0
00450	3103	3			PM	D	2								5
95001	MEETINGS					I		4		1	2			7	
95001	PERSONAL					I	1	2		1	1			5	
99601	VACATION					U			8					8	

		MON	TUE	WED	THU	FRI	SAT	SUN	TOTAL
UNAVAILABLE HOURS	U			8					8
DIRECT HOURS	D	7	2		7	8			24
INDIRECT HOURS	I	1	6		2	3			12
TOTAL HOURS	T	8	8	8	9	11			44

APPROVAL(S) BB

EFFECTIVENESS RATE 66 %

E.R. = DIRECT DIV. (DIRECT + INDIRECT)

IW019 (C) MBA 2004

END SECTION

Section 31

"PRIDE" PROJECT CONTROL

Projects will only be completed if people want to complete them,
- Bryce's Law

INTRODUCTION

"You only get out of project management what you put into it." Without follow-up, there is little point to managing a project. Although project management is an on-going process, it is important that project status reports be prepared and reviewed on a routine basis for consistency. A weekly time frame is perhaps the best cycle to routinely evaluate project progress. A monthly reporting cycle may not detect project errors in time for correction.

Each resource must review their work for the past period and the next succeeding period. The project manager, in turn, prepares a comparable report for the overall project. A careful review will reveal problems and suggest areas for improvement. Communication problems requiring correction will be uncovered. This review provides the opportunity to examine each individual's performance and to give recognition to accomplishments and progress. By careful attention on the part of management to these reviews, problem areas will not occur as surprises. Management will be able to take corrective action on a timely basis.

To assist in the detection of problem areas, project tolerances should be established to note when time estimates, costs and schedules will be overrun or underrun. Tolerances represent the percent allowable deviation on a project, e.g., 10%, 15%, etc. It is the relationship between "estimated to do" and project tolerances that provides for an early warning system in project management. If an employee expresses a slipped assignment and it surpasses project tolerances, then an estimate or schedule deviation is brought to the attention of the project manager. An estimate and/or schedule revision should then be prepared.

Revisions are a natural part of all projects. Any revision in estimate and schedule should be accompanied with an explanation for the change. These explanations can be used later for project auditing purposes and to provide insight for adjusting estimating guidelines.

Any time a project becomes inordinately large, it will be difficult to control. This problem usually occurs due to poor planning where the work effort is ill-defined and bottlenecks created. The temptation to try to do more than what is possible in the required time frame must be avoided. You simply cannot put two quarts of liquid into a quart bottle. Projects should be broken into smaller, more manageable pieces. If a project does not produce a significant result within one month, it will lose control.

The most important control within project management is that the responsibility for accomplishment be clearly assigned. Each professional understands this responsibility and makes personal project commitments. Given a good attitude on the part of the professional, individual responsibility promotes successful project execution with minimal supervision. Remember, projects will be accomplished only when people want to do them.

One last word on project control. It seldom occurs to people that it is just as bad to underrun a project as it is to overrun a project. Maybe it is because most people overrun them. However, underrunning a project can be equally bad when a company has specifically budgeted funds and resources for the project. An excessive underrun could represent a waste of resources, when they could have been invested more wisely elsewhere.

PROJECT CONTROL SHEETS

The Project Control Sheet included in "PRIDE" serves two purposes:

1. To prepare Detail estimates and schedules, and...

2. To report time against assignments.

A blank copy of the "Project Control Sheet" is available in the "Forms" section. See the next page for a sample as implemented under a manual system.

SAMPLE PROJECT CONTROL SHEET (MANUAL SYSTEM)

PRIDE® PROJECT CONTROL SHEET

PD - 00421 CONTROL NUMBER

TITLE TS-Φ6

PREPARED BY	DATE	APPROVED BY	DATE
TJ	2/7/06	CT	2/7/06

PHASE 3 EFFECTIVENESS RATE USED 70% ASSIGNED TO TED JONES

ACT	ACTIVITY DESCRIPTION	FUNC	ESTIMATE Hours Direct	ESTIMATE Hours Computer	ESTIMATE Schedule Start Date	ESTIMATE Schedule End Date	ACTUAL Hours Direct	ACTUAL Hours Computer	ACTUAL Schedule Start Date	ACTUAL Schedule End Date	ESTIMATE TO DO
A	DEFINE PROC	SE	14		2/7	2/9	12		2/7	2/9	Φ
B	DEFINE LOGIC	SE	16		2/9	2/15	13		2/9		12
C	PREP EVAL	SE	4		2/16	2/16					4
D	REVIEW	SE	2		2/17	2/17					2
		TOTALS	36		2/7	2/17	25		2/7		

RECAP BY FUNCTIONS (e.g., PM, SE, PE, DE, EE, DBA, etc.)	Rate Per Hour	ESTIMATE HOURS	ESTIMATE COST	ACTUAL HOURS	ACTUAL COST	(OVER RUN) / UNDER RUN HOURS	(OVER RUN) / UNDER RUN COST
PM	50	4	200				
SE	50	36	1800				
	TOTALS	40	2000				

APPROVALS CT 2/7, 2/14,

IW023A (C) MBA 2004

END SECTION

Section 32

"PRIDE" SUPPLEMENTAL NARRATIVE - ESTABLISHING A TECHNICAL LIBRARY

Every step in a methodology should produce a reviewable result in order to substantiate completion and assure a quality product.
- Bryce's Law

The purpose of this narrative is to provide guidelines for the establishment of a Technical Library.

INTRODUCTION

The concept of a Technical Library is an important part of Information Resource Management (IRM) and "PRIDE." As projects progress, documentation is produced, both formal and informal deliverables, representing project decisions and descriptions of the various information resources. This intelligence represents a considerable investment for the enterprise. Documentation must be maintained and controlled to protect this investment. It will be referenced frequently for a variety of research purposes.

The library is typically managed by a Technical Librarian who works closely with Project Administration and other "PRIDE" support functions. In many cases, both the Technical Librarian and Project Administrator are implemented by a single person.

OVERVIEW

The Technical Library is the organization's repository for all project records and approved copies of phase documentation. It may also contain other materials including computer hardware and software manuals, trade publications, text books, source program listings, standards manuals, supplies of forms, etc.

The Technical Library should be organized and patterned after a normal library, complete with procedures for cataloging documentation, check-in/check-out of documentation, etc. Any items which cannot be replaced should remain in the library and copies should be issued instead. If access to the library will be required during off hours, special provisions for controlling this access will need to be established.

The content and use of the Technical Library should not be confused with the content and use of the IRM

Repository. Most of the materials in the library are historical and will not require constant access. The IRM contains current documentation for all information resources. This documentation may be accessed as required by the staff. If program source listings are maintained in the library, as opposed to computer source code librarian aids, they may be the only material requiring constant access. During the setup of the library, this subject may be considered as a separate issue.

ORGANIZATION

The types of items to be maintained in the Technical Library include:

1. **PROJECT FILES** - filed in project number sequence. There may be sub-files for:

- **PHASE DELIVERABLES** - signed copies of all paper deliverables resulting from the methodologies, particularly:

  1. "PRIDE"-EEM Phase 1, "EEM Project Plan"
  2. "PRIDE"-EEM Phase 5, "EEM Project Evaluation"
  3. "PRIDE"-ISEM Phase 1, "System Study & Evaluation Report"
  4. "PRIDE"-ISEM Phase 9, "System Evaluation Report"
  5. "PRIDE"-DBEM Phase 1, "Data Base Study & Evaluation Report"
  6. "PRIDE"-DBEM Phase 9, "DBEM Project Evaluation Report"

- **DETAIL ESTIMATES/SCHEDULES** - as prepared by the various human resources assigned to the project.

2. **SYSTEM DOCUMENTATION** - filed according to the system identifier (XX-XX-XX-XX). This includes:

- **SYSTEM DESIGN MANUALS** - provision should also be made to file TRAINING AIDS.
- **SUB-SYSTEM DESIGN MANUALS** - provision should be made to file sub-system TEST DATA and TEST RESULTS.
- **ADMINISTRATIVE PROCEDURE MANUALS**
- **COMPUTER RUN BOOKS** - provision should be made to file computer procedure TEST DATA and TEST RESULTS.
- **ALDBM DESIGN MANUALS**
- **APDBM DESIGN MANUALS**

NOTE: System documentation should include both current and historical versions of the documentation.

3. **PROGRAM SOURCE CODE LISTINGS** - should be organized according to program number (XX-XX-XX-XX) and MD number (MD-XXXXX). Provision should be made to file program/module TEST DATA and TEST RESULTS with the source code.

Although maintaining paper copies of source code and test data is certainly feasible, it may no longer be practical for companies to do so. Instead, simple control over computer libraries with read/write privileges is often more feasible. It is not uncommon for Technical Libraries to maintain separate computers with production code and test data which is also handy for backup purposes. Under this scenario, libraries are determined by the program number and MD number. For example:

```
x:\SOURCE\CC15
x:\SOURCE\CC15\CC1502
x:\SOURCE\CC15\CC1502\CC150201
x:\SOURCE\CC15\CC1502\CC150202
x:\SOURCE\CC16
x:\SOURCE\CC16\CC1602
x:\SOURCE\CC16\CC1602\CC160202
x:\SOURCE\CC17
x:\SOURCE\CC17\CC1702
x:\SOURCE\CC17\CC1702\CC170201
x:\SOURCE\CC17\CC1702\CC170202
x:\SOURCE\IS01
x:\SOURCE\IS0102
x:\SOURCE\IS010200
x:\SOURCE\IS010201
x:\SOURCE\IS010202
x:\SOURCE\IS010203
x:\SOURCE\IS010204
x:\SOURCE\IS010208
x:\SOURCE\MODULES\MD-90001
x:\SOURCE\MODULES\MD-90002
x:\SOURCE\MODULES\MD-90003
x:\SOURCE\MODULES\MD-90004
x:\SOURCE\MODULES\MD-90005
```

Notice that Sub-System identifiers are used (CC15, CC16, CC17, IS01) as well as computer procedures (CC1502, CC1602, CC1702, IS0102) which are used to store test data and test results at their respective levels. Since modules are normally shared, they are placed in their own sub-directory ("MODULES"). However, there may be instances where it is necessary to place unique modules subordinate to a given program; for example:

x:\SOURCE\DX07\DX0702\DX070201
x:\SOURCE\DX07\DX0702\DX070201\MD-30045
x:\SOURCE\DX07\DX0702\DX070201\MD-30048

Since modules represent "re-usable" code, this approach is seldom used.

One final note, these libararies should be routinely backed-up for security and historical purposes.

4. **DATA BASE DOCUMENTATION** - filed according to the FD number and consists of:

- **ELDBM DESIGN MANUAL**
- **EPDBM DESIGN MANUAL**

NOTE: Data Base documentation should include both current and historical versions of the documentation.

5. PROJECT ADMINISTRATION FILES - summary reports filed in chronological sequence (Weekly Reports, Monthly Reports, Quarterly Reports, Annual Reports, etc.). Also, copies of Time Distribution Worksheets (Employee Time Reports) filed by each human resource in chronological order.

6. OTHER ITEMS

- Trade publications (magazines and newsletters).
- Conference Proceedings.
- Text books.
- Computer manuals (hardware and software).
- Forms (including "PRIDE" Forms).
- Audio/Video (tapes, CDs, DVDs, etc.)
- Standards Manuals

CONTROL

Of these items, determine:

1. Which must be uniquely identified and cataloged for control purposes.

2. Which must remain in the Technical Library and have only copies issued under check-in, check-out control.

3. Which may be obtained directly from the Technical Library under check-in and check-out control. Such items would include: reference books, text books, computer manuals, software manuals, audio/video cassettes, standards manuals, etc.

4. Which may be obtained from the Technical Library without any formal control procedures. This might include such things as trade publications (magazines and newsletters).

5. Which are consumable (do not have to be returned). Possible items include forms, and other perishable supplies.

CONSTRUCTION

1. Determine the types of files required for the library. This is based on the types of items (and quantity) selected for storage in the library. Consider normal shelving and filing cabinets, computer printout binders, microfilm/microfiche readers, etc.

2. Determine office equipment required for the library (e.g., tables, chairs, photocopiers, telephone, etc.).

3. Select a suitable location for the library. This is based on library size and access requirements.

4. Design an information system (with "PRIDE"-ISEM of course) to accommodate Technical Library information requirements.

END SECTION

Section 33

"PRIDE" SUPPLEMENTAL NARRATIVE - QUALITY ASSURANCE

Quality must be built into the product DURING design, not inspected in afterwards.
- Bryce's Law

The purpose of this section is to provide a framework for Quality Assurance during the execution of the "PRIDE" methodologies (EEM, ISEM, DBEM).

INTRODUCTION

It is generally agreed that the quality of a product is judged by how well it conforms to specifications. It is not a matter of how well it compares to another product of the same kind, but how well it fulfills the requirements for which it was designed. What this means is that quality is concerned with fulfilling the expectations of the customer. In this regard, quality can be described as providing a product the customer wants, when the customer wants it, and at the price the customer agreed to purchase it.

Quality Products Come From Quality Workmanship

Historically, quality was checked simply by how well the finished product conformed to specifications. This is the "black hole" approach for developing a quality product. Here, requirements are loosely formulated before they are turned over to the development staff for design. A quality inspection is performed only after the product has been installed and made operational. The process looks something like this:

This approach concentrates only on the final deliverable and not on the overall process by which the product is to be developed. By the time the final product is produced, it may be unrecognizable to the user and the project may have exceeded estimated cost and schedule. To overcome this problem, the entire development process must be precisely defined so the workers know who is to perform what, when, where, why and how. This is the intent of an assembly line or methodology which specifies the sequence of steps required to complete a project, from initiation to completion.

Quality must be built into the product DURING design. Instead of inspecting the whole product all at once after the fact, the product is reviewed in smaller, more easily examined pieces during its development, thus allowing for the early detection of defects, if any.

Each stage of product development is defined as a work station on an assembly line. At each station specific tasks are performed and deliverables produced. As the product progresses from one work station to the next, it is checked for completeness. For example, following a sequential flow of work stations (WS) from A to E, the review process looks like this:

WORK STATION	REVIEWS
WS-A	WS-E
WS-B	WS-A
WS-C	WS-A & B
WS-D	WS-A & B & C
WS-E	WS-A & B & C & D

Notice the progression of review and expertise. This means the worker at WS-B must be knowledgeable in the tasks and deliverables for both WS-A and WS-B. The worker at WS-C must be knowledgeable about WS-A, B and C. The worker at WS-D must be knowledgeable about WS-A, B, C and D. The worker at WS-E must be knowledgeable about all of the Work Stations. To close the loop, the worker at WS-A must be knowledgeable about WS-E.

In other words, there are no formal inspectors required. All of the employees participate in the development of a quality product. The benefit of this approach is that it eliminates the problem of viewing quality as the responsibility of a single person, such as an inspector. All stations work as a team and make all of the workers directly responsible for quality. Because the worker must learn about the activities at other work stations, the worker has a more comprehensive understanding of the total product and the process required to develop it. In some Japanese companies, the workers rotate from station to station, thus enabling them to experience all facets of work. From this perspective, the workers are more apt to recommend improvements in the way the product is built. The Japanese refer to this as thinking in "360 degrees."

Remember, quality is a way of life, not a burdensome overhead.

THE METHODOLOGY'S ROLE

Quality Assurance is concerned not only with the integrity of the deliverables, but also with the mechanism by which they are produced. The "PRIDE" methodologies help in this regard. They clearly identify all of the required processes (phases) and work products (deliverables). Further, through the Phase Review Checklists, they prescribe the acceptance criteria for each deliverable. This enables developers to design, build and evaluate the product with a high degree of precision.

With the methodology defining the specific benchmarks in the development process, it becomes a relatively simple matter to review the progress of the product while it is in development. Each station along the assembly line has a clear responsibility for the development of the product and defines the tangible deliverables for review.

What this highlights is that Quality Assurance is much more than an inspection type of function. It is a corporate philosophy and shared responsibility, requiring participation from personnel throughout the enterprise. It is a combined corporate responsibility, not just the responsibility of a single individual or department.

"PRIDE" subscribes to the "mini-Project Manager" concept. This is a concept that places the responsibility of the project with everyone involved with the project. Each person is viewed as a professional and an essential link in the development process. They participate in both development and review activities in their phase of work. As such, each person is responsible for developing their own estimate and schedule for their part of the project. This, of course, must be reviewed with a Project Manager who coordinates the entire process.

QA PROGRAM

There are three parts to an effective Quality Assurance program:

1. Establish the STANDARD process which defines the work environment.

2. EXECUTE according to these standards.

3. Establish QUALITY CONTROL to review, verify and enforce the standards.

STANDARDS

Standards improve communications between parties. They are an essential aspect of any Quality Assurance program in that they promote understanding and improve productivity.

The establishment of standards is a responsibility of Executive Management. Only management can determine how much quality is sufficient. Standards are used to define the formal mode of operation or work environment. They define WHO is to do WHAT, WHEN, WHERE, HOW and WHY. Executive Management must determine how much regimentation and discipline is required. Methodologies, such as those embodied in "PRIDE", are used since they delineate these considerations. All participants must have an in-depth knowledge of the methodology and a thorough understanding as to how their duties and responsibilities are related. In essence a methodology is an assembly line consisting of a series of workstations. Although each station has its own purpose and objectives, together they produce a finished product.

Although "PRIDE" includes a substantial number of standards for an organization to implement, there are other standards which must be addressed. For example, there should be formal standards for the preparation of a cost/benefit analysis (such as mathematical formulas), interviewing, programming, etc. "PRIDE" provides for these standards but recognizes that the techniques and tools used may vary from company to company. These should all be identified and incorporated into the methodology. It is extremely important that these standards are accurately defined relative to the methodology so that participants know precisely where and when they are to be used.

In order to improve the development process, standards should be reviewed, updated and distributed to appropriate personnel on a routine basis. In this respect, Quality Assurance should be considered as an expediter for the methodology. They should be constantly measuring and evaluating the data management work effort and looking for ways to simplify. These standards should be maintained in either a standards manual or on the computer (such as an in-house web page or "Help" file).

EXECUTION

When a standard is adopted, each person in the process must understand that they have a vital role to play in producing a quality product. This includes management, the development staff, and support personnel.

During the development process, each person at each phase in the methodology is asked to evaluate the product so far. An evaluation criteria for the product at each phase is required. To this end, "PRIDE" provides Phase Review Checklists for the staff to evaluate completeness and product integrity. At this time, participants are asked to accept the product as is, reject it, or ask that it be modified. If required, the project will either go forward as planned, iterate to the previous phases for rework, or be cancelled completely. It is important that the participants realize that these are their decisions to make. Problem areas should be reconciled by Executive Management.

Changes and constructive criticisms should be openly encouraged, not discouraged. Making changes can be irritating and frustrating. They should be done in moderation, not all at one time. This is the approach that "PRIDE" takes. If necessary, a project may have to stop and back-up to a previous phase or activity in order to move forward again. There is little point to progressing with a project if it is consciously known that it is going to have problems later on. All participants must keep in mind that the objective is to develop a quality product.

QUALITY CONTROL

There is no reason to enact legislation if it is known that it will not be enforced. For this reason, Quality Control is concerned with verifying that the development staff is adhering to standards.

This is the area that is concerned with inspecting the product for completeness and workmanship. Before each phase is closed, a final review with a neutral third party can be valuable to assure that nothing was overlooked and that the best possible work was produced. In this instance, Quality Control can halt a project for glaring omissions and poor workmanship. They can also offer constructive criticisms that can be helpful to the project.

THE QUALITY ASSURANCE ORGANIZATION

Reporting as a staff position to the senior IRM executive, such as a Chief Information Officer (CIO), is the Quality Assurance group. Instead of being regarded as a clerical burden, the QA group should be viewed as industrial engineers to the IRM assembly lines, constantly looking for new and improved ways to expedite the development and control of information resources. In this capacity, they are continually looking for new techniques and tools to be used in the IRM methodologies.

The IRM engineer will have the triple role of standards analysis, inspector and technology advisor. Project Administration polices project management and works closely with the technical librarian who maintains project and IRM related documentation, both current and historical. The IRM training coordinator provides a curriculum to continually sharpen IRM related skills.

At this point, you might believe that creating an IRM quality assurance group is creating additional overhead. Just the reverse will occur; it will reduce overhead and bring development costs down, while the quality of the product will soar. In most organizations, the systems analysts and programmers do this work now (at least they are supposed to do it). By centralizing the function, it can be handled more effectively and quickly by a dedicated group. As a result, the systems analysts and programmers can concentrate on their primary functions.

This is not to suggest the burden of quality assurance rests with one department or group. Far from it. Each employee is responsible for quality.

END SECTION

TYPICAL IRM QUALITY ASSURANCE ORGANIZATION

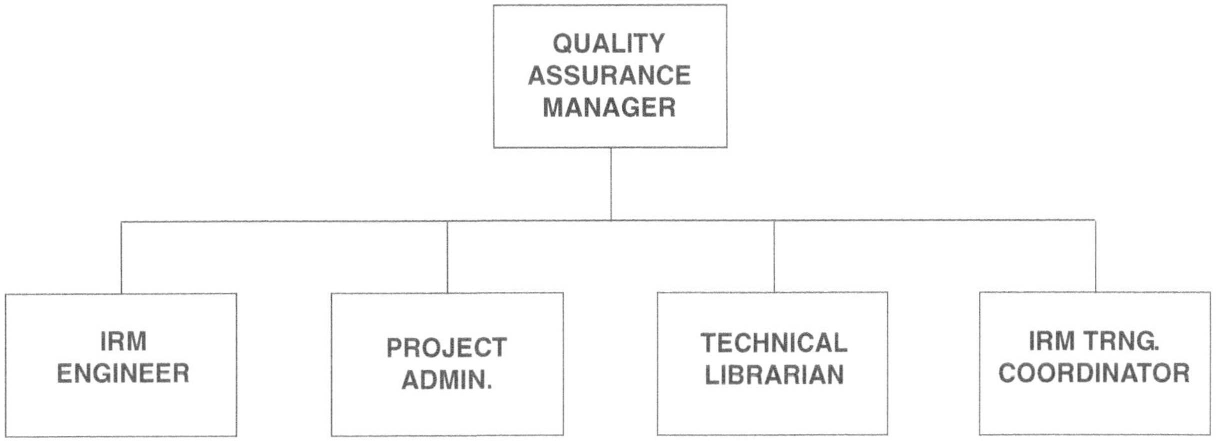

Section 34

"PRIDE" SUPPLEMENTAL NARRATIVE - ESTABLISHING AN IRM REPOSITORY

IRM is the view of the enterprise from 50,000 feet.
- Bryce's Law

The purpose of this section is to describe the role of a "Repository" in a development organization and how to implement a repository, using either a manual or automated approach.

WHAT IS A REPOSITORY?

As should be obvious from studying the "PRIDE" methodologies, the cohesive bond behind all development efforts is the "IRM Repository" (aka, "IRM" or "Repository"), a system to inventory and control all of the resources in an enterprise needed to produce information. As developers progress through the "PRIDE" methodologies, information resources are defined and cross-referenced to other resources in the repository. In this way, the repository becomes a manifest of all information resources, thereby providing the means to share and re-use resources. The IRM is also a useful documentation tool; the various resource definitions should contain sufficient intelligence to generate reports, draw graphics (e.g., flowcharts), and interface with third-party development aids (e.g., DBMS, application generators, etc.).

The repository concept has been a part of "PRIDE" since its inception in 1971. Unlike classic data dictionary/directories that support CASE tools and DBMS packages, the "PRIDE" approach views a repository as a "bill-of-material processor" (BOMP) to catalog and cross-reference ALL information resources, not just those associated with the computer; a much more global perspective of an enterprise's information resources.

RESOURCE DEFINITION

To implement a repository, it will be necessary to create the means to uniquely define each resource and link it to other resources. This can be implemented two ways:

1. Manually - using the resource worksheets accompanying "PRIDE".

2. Computer assisted - using commercial packages or to develop such software in-house. Today's Data Base Management Systems (DBMS) provide the

necessary facilities to develop a computer assisted IRM, be it on the mainframe, mid-range, or PC level. To this end, the "PRIDE" resource worksheets provide a convenient template for creating such a repository.

Under the "PRIDE" approach, all resources are given a unique eight digit alphanumeric part number to uniquely identify it, along with a name and descriptive text. Most resources, follow a "Part Number" numbering convention consisting of a two digit prefix denoting the type of resource, followed by an alphanumeric control number to identify the individual description. For example:

NUMBER	**RESOURCE TYPE**	**USED TO DEFINE..**
FE-XXXXX	Functional Entity	Enterprises, Business Functions
OE-XXXXX	Organizational Entity	Jobs/Positions
RE-XXXXX	Resource Entity	Employees, Consultants, Machines
SD-XXXXX	Skill Description	Knowledge, Experience, Talent, Feature (for equip)
MI-XXXXX	Modification/Improve.	Business Objectives, Problems, Work Requests
PD-XXXXX	Project Description	Projects (any size or type)
IR-XXXXX	Information Requirement	Information related or non-info related (equip specs)
MD-XXXXX	Module Description	Reusable program code (including subroutines and software "objects")
DD-XXXXX	Data Description	Data Elements, and field entries.
FD-XXXXX	File Description	Physical Files (manual & computer), and Objects (Logical Files), and Data Bases
ID-XXXXX	Input Description	Screens, Documents, Verbal
OD-XXXXX	Output Description	Screens, Reports, Audio
RD-XXXXX	Record Description	Records, Views (Logical Records), Transactions, Maps, Panels, Messages, and Call Parameters.

Unlike the other resources, System resources follow a product structure numbering convention reflecting the four levels in a system hierarchy (a basic "PRIDE" concept):

XX	System number	LEVEL 1
XXXX	Sub-System number	LEVEL 2
XXXXXX	Procedure number (Administrative or Computer)	LEVEL 3
XXXXXXXX	Program or Operational Step number	LEVEL 4

The XX-XX-XX-XX convention can be modified as required; for example: XXX-X-XX-XX, XX-X-XXX-XX, etc.

Identifying information resources using this control number approach is a much more effective means of identification than relying on naming conventions that are prone to redundancy. (See section below on Logical Attributes). Further, control numbers and names provide the means to establish meaningful logs by each resource type; for example:

SAMPLE LOG OF DATA DESCRIPTIONS (DD) SORTED BY RESOURCE NAME

OD-90355	RESOURCE LOG BY NAME		"PRIDE"-IRM
SEP 3, 2004	RELEASE 1.2.0		PAGE 1

NUMBER	DATA DESCRIPTION NAME	TYPE	STATUS
DD-00008	ACCOUNT NUMBER	INDI - PRIM	ACTIVE
DD-00043	ADDRESS TYPE	INDI - PRIM	ACTIVE
DD-00019	ALTERNATE IDENTIFICATION	DESC - PRIM	DEVELOPMENT
DD-00104	AMOUNT PAID ON INVOICE	QUAN - PRIM	ACTIVE
DD-00528	ANNUAL QUOTA FOR SALES FORCE	QUAN - GEN	DEVELOPMENT
DD-00524	ANNUAL QUOTA FOR SALES REGION	QUAN - GEN	DEVELOPMENT
DD-00537	ANNUAL QUOTA FOR SALES REP	QUAN - PRIM	DEVELOPMENT
DD-00173	ASSEMBLY STATION ID CODE	INDI - PRIM	DEVELOPMENT
DD-00014	BACK ORDER NUMBER	INDI - PRIM	ACTIVE
DD-00055	BACK-ORDERS FOR WEEK	QUAN - GEN	ACTIVE
DD-00196	BANK ACCOUNT NUMBER	DESC - PRIM	DEVELOPMENT
DD-00011	BILLING NUMBER	INDI - PRIM	ACTIVE
DD-00187	BUSINESS OBJECTIVE NUMBER	INDI - PRIM	DEVELOPMENT
DD-00030	BUSINESS TITLE	DESC - PRIM	ACTIVE
DD-00024	CITY	DESC - PRIM	ACTIVE
DD-00146	CITY INCOME TAX PERCENT	QUAN - PRIM	DEVELOPMENT
DD-00141	CITY INCOME TAX WITHHELD	QUAN - GEN	DEVELOPMENT
DD-00047	COMMISSION ASSIGNMENT NUMBER	INDI - PRIM	DEVELOPMENT
DD-00536	COMMISSION PERCENTAGE RATE	QUAN - PRIM	DEVELOPMENT
DD-00540	COMMISSION PRODUCT AMOUNT	QUAN - GEN	DEVELOPMENT
DD-00172	COMPETITOR ID NUMBER	INDI - PRIM	DEVELOPMENT
DD-00042	CONTACT NUMBER	INDI - PRIM	ACTIVE
DD-00188	CORPORATE POLICY NUMBER	INDI - PRIM	DEVELOPMENT
DD-00027	COUNTRY	DESC - PRIM	ACTIVE
DD-00025	COUNTY	DESC - PRIM	ACTIVE
DD-00133	CREDIT LIMIT	QUAN - PRIM	ACTIVE
DD-00195	CREDIT REQUEST NUMBER	INDI - PRIM	DEVELOPMENT
DD-00134	CURRENT CUSTOMER BALANCE	QUAN - GEN	ACTIVE
DD-00001	CUSTOMER NUMBER	INDI - PRIM	DEVELOPMENT
DD-00046	D-U-N-S NUMBER	DESC - PRIM	ACTIVE
DD-00016	DATE	QUAN - PRIM	ACTIVE
DD-00050	DELIVERY PRIORITY	DESC - PRIM	ACTIVE
DD-00004	EMPLOYEE NUMBER	INDI - PRIM	DEVELOPMENT

368 DD RESOURCE(S)

* * * END OF DD LOG * * *

SAMPLE LOG OF DATA DESCRIPTIONS (DD) SORTED BY RESOURCE CONTROL NUMBER

OD-90354 RESOURCE LOG BY CONTROL NUMBER "PRIDE"-IRM

SEP 3, 2004 RELEASE 1.2.0 PAGE 1

NUMBER	DATA DESCRIPTION NAME	TYPE	STATUS
DD-00001	CUSTOMER NUMBER	INDI - PRIM	DEVELOPMENT
DD-00002	VENDOR NUMBER	INDI - PRIM	ACTIVE
DD-00003	PRODUCT NUMBER	INDI - PRIM	ACTIVE
DD-00004	EMPLOYEE NUMBER	INDI - PRIM	DEVELOPMENT
DD-00005	POLICY NUMBER	INDI - PRIM	DEVELOPMENT
DD-00006	PART NUMBER	INDI - PRIM	DEVELOPMENT
DD-00007	EQUIPMENT NUMBER	INDI - PRIM	DEVELOPMENT
DD-00008	ACCOUNT NUMBER	INDI - PRIM	ACTIVE
DD-00009	GOVERNMENT AGENCY NUMBER	INDI - PRIM	ACTIVE
DD-00010	SHIPMENT NUMBER	INDI - PRIM	ACTIVE
DD-00011	BILLING NUMBER	INDI - PRIM	ACTIVE
DD-00012	ORDER NUMBER	INDI - PRIM	ACTIVE
DD-00013	PURCHASE NUMBER	INDI - PRIM	DEVELOPMENT
DD-00014	BACK ORDER NUMBER	INDI - PRIM	ACTIVE
DD-00015	TRANSACTION NUMBER	INDI - PRIM	DEVELOPMENT
DD-00016	DATE	QUAN - PRIM	ACTIVE
DD-00017	TIME	QUAN - PRIM	DEVELOPMENT
DD-00018	NAME	DESC - PRIM	ACTIVE
DD-00019	ALTERNATE IDENTIFICATION	DESC - PRIM	DEVELOPMENT
DD-00020	PASSWORD	DESC - PRIM	DEVELOPMENT
DD-00021	SOCIAL SECURITY NUMBER	DESC - PRIM	DEVELOPMENT
DD-00022	POST OFFICE BOX	DESC - PRIM	ACTIVE
DD-00023	STREET ADDRESS	DESC - PRIM	ACTIVE
DD-00024	CITY	DESC - PRIM	ACTIVE
DD-00025	COUNTY	DESC - PRIM	ACTIVE
DD-00026	STATE/PROVINCE	DESC - PRIM	ACTIVE
DD-00027	COUNTRY	DESC - PRIM	ACTIVE
DD-00028	ZIP/POSTAL CODE	DESC - PRIM	ACTIVE
DD-00029	MAIL DROP/STOP	DESC - PRIM	ACTIVE
DD-00030	BUSINESS TITLE	DESC - PRIM	ACTIVE
DD-00031	INPUT PARAMETER	DESC - PRIM	DEVELOPMENT
DD-00032	TEXT	DESC - PRIM	DEVELOPMENT
DD-00033	TELEPHONE NUMBER	DESC - PRIM	ACTIVE
DD-00034	FACSIMILE NUMBER	DESC - PRIM	ACTIVE
DD-00035	TELEX NUMBER	DESC - PRIM	ACTIVE
DD-00036	QUANTITY ORDERED	QUAN - PRIM	ACTIVE
DD-00037	QUANTITY PURCHASED	QUAN - PRIM	ACTIVE
DD-00038	QUANTITY BILLED	QUAN - PRIM	ACTIVE
DD-00039	QUANTITY SHIPPED	QUAN - PRIM	ACTIVE
DD-00040	QUANTITY BACK ORDERED	QUAN - PRIM	ACTIVE
DD-00041	STATUS	DESC - PRIM	ACTIVE
DD-00042	CONTACT NUMBER	INDI - PRIM	ACTIVE

/\

368 DD RESOURCE(S)

* * * END OF DD LOG * * *

SAMPLE LOG OF SYSTEM RESOURCES SORTED BY NAME

OD-90355 RESOURCE LOG BY NAME "PRIDE"-IRM

SEP 3, 2004 RELEASE 1.2.0 PAGE 1

NUMBER	SYSTEM RESOURCE NAME	BOOK/TYPE	STATUS
OP-00-05	ACCEPT/VALIDATE ORDER	COMPUTER	ACTIVE
AP	ACCOUNTS PAYABLE SYSTEM	A/P	DEVELOPMENT
AR	ACCOUNTS RECEIVABLE SYSTEM	A/R	DEVELOPMENT
OP-04-11	ANALYZE WEEKLY ORDER STATUS	MANUAL	ACTIVE
BL	BILLING SYSTEM	SYS-10	DEVELOPMENT
OP-04-09	BURST/DISTRIBUTE ORDER STATUS	MANUAL	ACTIVE
OP-04-07	COMPUTER GENERATE ORDER STATUS	COMPUTER	ACTIVE
OP-05-05	COMPUTER PRODUCE DAILY ORDERS	COMPUTER	ACTIVE
OP-04-05	CONVERT ORDER STATUS REQUEST	KEYING	ACTIVE
OP-06	CREDIT CHANGE REQUEST	TOPS	ACTIVE
OP-07-05-01	CUSTOMER DETAIL SHEET PROGRAM	COMPUTER	ACTIVE
OP-00	CUSTOMER ORDER INQUIRY	TOPS	ACTIVE
OP-07	CUSTOMER RECORD MANAGEMENT	TOPS	ACTIVE
CS	CUSTOMER SERVICE	SYSTEM	DEVELOPMENT
OP-05	DAILY ORDER PRODUCTION	TOPS	ACTIVE
OP-06-03	ELECTRONIC CREDIT REQUEST	OFFICE AUT	ACTIVE
GL	GENERAL LEDGER SYSTEM	FINANCE	DEVELOPMENT
OP-07-05	GENERATE CUSTOMER DETAIL SHEET	COMPUTER	ACTIVE
OP-05-05-01	GENERATE DAILY ORDERS	COMPUTER	ACTIVE
HR	HUMAN RESOURCE SYSTEM	PERSONNL	DEVELOPMENT
IC	INVENTORY CONTROL	SYSTEM	DEVELOPMENT
MN	MANUFACTURING	SYSTEM	DEVELOPMENT
OP-00-05-01	ORDER PROCESSING PROGRAM	COMPUTER	ACTIVE
OP-00-03	ORDER PROCESSING SESSION	KEYING	ACTIVE
OP-08-05-01	ORDER-PROC DB MAINTENANCE	COMPUTER	ACTIVE
OP-08	ORDER-PROC FILE MAINTENANCE	TOPS	ACTIVE
OP-08-03	ORDER-PROC FILE UPDATE SESSION	KEYING	ACTIVE
PA	PAYROLL SYSTEM	PAYROLL	DEVELOPMENT
OP-05-03	PREP DAILY ORDERS REQUEST	KEYING	ACTIVE
OP-07-03	PREPARE CUSTOMER DETAIL REQST.	KEYING	ACTIVE
OP-04-03	PREPARE ORDER STATUS REQUEST	MANUAL	ACTIVE
OP-05-07	REVIEW DAILY ORDERS	MANUAL	ACTIVE
OP-07-07	REVIEW/REVISE CUSTOMER DETAILS	MANUAL	ACTIVE

/\/

239 SYSTEM RESOURCE(S)

*** END OF SYSTEM LOG ***

SAMPLE LOG OF SYSTEM RESOURCES SORTED BY CONTROL NUMBER

OD-90354 RESOURCE LOG BY CONTROL NUMBER "PRIDE"-IRM

SEP 3, 2004 RELEASE 1.2.0 PAGE 1

NUMBER	SYSTEM RESOURCE NAME	BOOK/TYPE	STATUS
AP	ACCOUNTS PAYABLE SYSTEM	A/P	DEVELOPMENT
AR	ACCOUNTS RECEIVABLE SYSTEM	A/R	DEVELOPMENT
BL	BILLING SYSTEM	SYS-10	DEVELOPMENT
CS	CUSTOMER SERVICE	SYSTEM	DEVELOPMENT
GL	GENERAL LEDGER SYSTEM	FINANCE	DEVELOPMENT
HR	HUMAN RESOURCE SYSTEM	PERSONNL	DEVELOPMENT
IC	INVENTORY CONTROL	SYSTEM	DEVELOPMENT
MN	MANUFACTURING	SYSTEM	DEVELOPMENT
OP	TRUNDLE ORDER PROC. SYS (TOPS)	TOPS	ACTIVE
OP-00	CUSTOMER ORDER INQUIRY	TOPS	ACTIVE
OP-00-03	ORDER PROCESSING SESSION	KEYING	ACTIVE
OP-00-05	ACCEPT/VALIDATE ORDER	COMPUTER	ACTIVE
OP-00-05-01	ORDER PROCESSING PROGRAM	COMPUTER	ACTIVE
OP-04	WEEKLY ORDER STATUS	TOPS	ACTIVE
OP-04-03	PREPARE ORDER STATUS REQUEST	MANUAL	ACTIVE
OP-04-05	CONVERT ORDER STATUS REQUEST	KEYING	ACTIVE
OP-04-07	COMPUTER GENERATE ORDER STATUS	COMPUTER	ACTIVE
OP-04-07-01	WEEKLY ORDER STATUS PROGRAM	COMPUTER	ACTIVE
OP-04-09	BURST/DISTRIBUTE ORDER STATUS	MANUAL	ACTIVE
OP-04-11	ANALYZE WEEKLY ORDER STATUS	MANUAL	ACTIVE
OP-05	DAILY ORDER PRODUCTION	TOPS	ACTIVE
OP-05-03	PREP DAILY ORDERS REQUEST	KEYING	ACTIVE
OP-05-05	COMPUTER PRODUCE DAILY ORDERS	COMPUTER	ACTIVE
OP-05-05-01	GENERATE DAILY ORDERS	COMPUTER	ACTIVE
OP-05-07	REVIEW DAILY ORDERS	MANUAL	ACTIVE
OP-06	CREDIT CHANGE REQUEST	TOPS	ACTIVE
OP-06-03	ELECTRONIC CREDIT REQUEST	OFFICE AUT	ACTIVE
OP-07	CUSTOMER RECORD MANAGEMENT	TOPS	ACTIVE
OP-07-03	PREPARE CUSTOMER DETAIL REQST.	KEYING	ACTIVE
OP-07-05	GENERATE CUSTOMER DETAIL SHEET	COMPUTER	ACTIVE
OP-07-05-01	CUSTOMER DETAIL SHEET PROGRAM	COMPUTER	ACTIVE
OP-07-07	REVIEW/REVISE CUSTOMER DETAILS	MANUAL	ACTIVE

239 SYSTEM RESOURCE(S)

* * * END OF SYSTEM LOG * * *

In addition to control numbers and names, each resource should allow the inclusion of alternate identification fields for customers with other in-house numbering conventions.

LOGICAL ATTRIBUTES

Although resource names may be used to identify redundant resource definitions, it is not necessarily the most reliable method for duplication identification. Instead, "PRIDE" recommends the "Logical Attribute" technique. Logical Attributes are a set of synonym words (pre-defined keywords governed by an administrator; normally consisting of nouns and adjectives) used to classify each resource. For example, consider how "Employee Number" and "Social Security Number" would be classified as data elements:

DD-00124 EMPLOYEE NUMBER
Attributes: NUMBER
IDENTIFICATION
EMPLOYEE
PERSONNEL

DD-00237 SOCIAL SECURITY NUMBER

Attributes: NUMBER
IDENTIFICATION
EMPLOYEE
GOVERNMENT
FICA

If both data descriptions were defined by only the first three attributes (NUMBER, IDENTIFICATION, and EMPLOYEE) the IRM Repository would identify them as redundant resource definitions. However, since both resources have additional attributes, each resource is viewed as unique and distinct. The logical attribute words used are unique to each enterprise and represents the vocabulary of the business.

"PRIDE" users have found this logical attribute technique to be a much more flexible and effective way for identifying resources than the "Key-Word-In-Context" (KWIC) approach, employed by some data dictionaries, which demands rigid and unfriendly naming conventions.

The principal by-product of the logical attribute technique is the ability to quickly search the IRM Repository for a particular resource. For example, in order to find DD-00124, "EMPLOYEE NUMBER," we might select a search argument consisting of the logical attribute words NUMBER, IDENTIFICATION, PERSONNEL (sequencing of the search argument does not matter). This search argument would result in finding...

```
DD-00124 EMPLOYEE NUMBER - meets 100%
of our requirements.

DD-00237 SOCIAL SECURITY NUMBER - meets
66% of our requirements.
```

This means an effective repository should provide a flexible means to search for information resources; either by control number, name, or by Logical Attributes.

RESOURCE RELATIONSHIPS

In addition to being able to identify and control each resource, a repository must be able to link and cross-reference resources in a meaningful way. This cross-reference capability alone is perhaps the most powerful aspect to a repository. A single resource may be used by hundreds of other resources. By maintaining these linkages, which can be extensive, a user can measure how the change of a single resource can effect other resources (e.g., an "Impact Analysis" showing resource relationships).

The following table represents the types of superior/subordinate relationships that the repository must accommodate:

SUPERIOR/SUBORDINATE RESOURCE RELATIONSHIPS

SUPERIOR/SUBORDINATE RESOURCE RELATIONSHIPS

S U P E R I O R   R E S O U R C E S

SUBORDINATE RESOURCES	FE	OE	RE	SD	MI	PD	IR	SY	SS	PC	OP	PG	MD	ID	OD	FD	RD	DD
FE	X																	
OE	X	X																
RE		X				X												
SD	X		X															
MI	X	X																
PD		X			X													
IR	X	X			X	X												
SY	X	X			X	X	X											
SS	X	X			X	X	X	X										
PC	X	X			X	X	X		X									
OP	X	X			X	X	X			X								
PG	X	X			X	X	X			X								
MD												X	X					
ID	X	X			X		X	X	X	X	X	X	X	X	X	X		
OD	X	X			X		X	X	X	X	X	X	X	X	X	X		
FD	X	X			X		X	X	X	X	X	X	X	X	X	X		
RD												X	X	X	X	X	X	
DD							X							X	X	X	X	X

NOTES:

1. In addition to the normal control prefixes used (FE, OE, PD, DD, etc.), the table uses the following codes to indicate system resources:

- SY - System
- SS - Sub-System
- PC - Procedure (Administrative or Computer)
- OP - Operational Step (within an Administrative Procedure)
- PG - Program

2. Work Breakdown Structures (WBS) and Precedent Relationships between work steps should be expressed on the PD resource.

INTERFACES TO DEVELOPMENT AIDS

Because of the vast corporate intelligence contained within it, the resource repository is the focal point of all development activities. Virtually all planning, design and programming aids will need to interface with it, as well as project management software. Because of this, the repository must have an "open architecture" allowing it to interface with a variety of tools. An "Import/Export Facility" is required to upload and download resource specifications to such things as CASE tools, program generators, data dictionaries, DBMS packages, word/text processors/editors, financial planning aids, graphics packages, Project Management systems, etc.

BUILDING A REPOSITORY

When building a computer assisted IRM Repository, each Resource should be defined in terms of:

1. **Identification** - include control number (as mentioned above), name and, alternate identification (to accommodate in-house codes).

2. **Logical Attributes** - to uniquely identify each resource and detect duplicates.

3. **Characteristics** - describing the properties of the resource (see the "PRIDE" Resource Worksheets for specifics). Be sure to add validation rules and default values wherever possible.

4. **Descriptive** Text - to describe each resource and produce reports (interfaces to text editors/word processors are highly desirable).

5. **Program Labels** - primarily for data resources (DD, RD, FD, ID, OD). Remember, one data resource may have multiple program labels for different programming languages; e.g., "Customer Number" may be "CUST-NO" in COBOL, "cust_numb" in C, "CUST" in Assembler, etc.

6. **Subordinate Relationships** - linking subordinate resources to the resource. For example, an RD "Record" will have one or more subordinate DD "Data Definitions" subordinate to it.

7. **"Where-Used" Relationships** - showing where the resource is used by other resources. For example, an RD "Record" may be in the "Subordinate Relationship" set of an Input (ID), Output (OD), File (FD), or linked to other Records (RD).

The IRM Repository should then have facilities to:

• Input resource definitions. To this end, the Resource Worksheets as contained in the "PRIDE" Forms section makes a convenient template to design input screens. Web page based "forms" make a convenient means to collect data.

• Produce reports, including inventory control reports to analyze/monitor information resources (as shown above), to quickly search for resources (by name, number, or logical attributes), and to generate "PRIDE" deliverables (e.g., text reports, flowcharts, matrices, etc.).

• To interface with third-party products, either to import data or to export data for such things as a Project Management system, an application development aid (e.g., code generator), DBMS, spreadsheets, etc.

The IRM Repository should be viewed as the hub of all development efforts and accessed by all development personnel, executives and managers, and support personnel. As such, consideration should be given to the Repository's access security, read/write privileges, file sharing, and backup/restore procedures.

LOCATION

Since the intent of the repository is to promote sharing and re-using resources, it must be placed in a location accessible by anyone creating, updating or referencing information resources. This could potentially represent hundreds of people. The challenge becomes finding a centralized location that is convenient to everyone, without jeopardizing security.

If implemented manually, the "PRIDE" resource worksheets should be filed in numeric sequence in a filing cabinet. If automated, the repository can either be located on a mainframe and accessed through "dumb" terminals or on a file server accessible through networks.

The "PRIDE" vendor does not recommend maintaining multiple repositories on separate platforms. Although it is certainly feasible to set up such a configuration, keeping the repositories synchronized would be more trouble than it is worth.

SUMMARY

Sharing and re-using resources is at the heart of a corporate-wide IRM strategy. In order to facilitate this, it is necessary to create a "repository" to uniquely identify, define and cross-reference ALL information resources, not just the computer or programming portions. Creating such a repository can be accomplished either manually (using the "PRIDE" resource worksheets), by purchasing a software product, or by creating such a mechanism internally. Regardless, the repository is vital to the successful implementation of the "PRIDE" methodologies and to Information Resource Management in general.

For more information on the resource worksheets, see the "Forms" section.

END SECTION

Section 35

"PRIDE" SUPPLEMENTAL NARRATIVE - "PRIDE" IMPLEMENTATION

It is one thing to enact legislation, quite another to enforce it.
- Bryce's Law

The purpose of this section is to provide guidance in the implementation of "PRIDE" in a company.

INTRODUCTION

Never underestimate the implementation of "PRIDE" since it represents a change of attitude and perspective that is uncommon in the typical programming shop. This is why we refer to it as *"Software for the finest computer - the Mind."* Because "PRIDE" is first and foremost a unique management philosophy, it is not installed in the same manner as hardware or computer software. We have seen many approaches to the implementation of "PRIDE" over the years; some successful, some disastrous. The disastrous implementations are those where a "Dictator" approach is taken and "PRIDE" is jammed down everyone's throat. This will only work as long as the dictator remains in power. It is typically abandoned shortly thereafter. The more successful implementations have been those where the responsibility for "PRIDE" is placed on several key people in the organization, thereby giving the appearance that "PRIDE" is the will of the company and not just one individual.

STEP 1 - ESTABLISH A PROJECT

The first step in installing "PRIDE" is to first establish a project for this purpose. This can either be done using an automated Project Management system or using the manual forms as included in "PRIDE"-PM. Use of a Project Management system is extremely useful for keeping the project on schedule and within costs.

Key to the startup of the project is the appointment of a "PRIDE" Coordinator who will act as the Project Manager for the implementation of "PRIDE". Considerable thought should go into the selection of this person. The "PRIDE" Coordinator should be respected by the development staff as well as management; should work well with people, but more importantly; must be results oriented.

STEP 2 - ESTABLISH SUPPORT TEAM

A Support Team is assembled who will be assigned tasks in this project. One of the principal reasons for forming a Support Team is to share the responsibility for implementing "PRIDE" throughout the company. Again, this conveys the image that "PRIDE" is the will of the company, not just a single person.

Selecting members for the Support Team is critical. During the implementation process, they will have high visibility and will become the in-house experts in the use of "PRIDE" products. As such, the persons selected must be able to speak with authority and command respect. The people typically involved in the implementation of "PRIDE" include:

• "PRIDE" Coordinator - the person selected for this key assignment must have a management background.

• Enterprise Resource Manager - this will be the person primarily concerned with "PRIDE"-EEM related matters.

• Systems Resource Manager - this will be the person primarily concerned with "PRIDE"-ISEM related matters.

• Data Resource Manager - this will be the person primarily concerned with "PRIDE"-DBEM related matters.

• Quality Assurance Manager - the person who will be concerned with the development and enforcement of all IRM related standards.

• Training Coordinator - the person who will be concerned with providing educational services for the company.

• Project Administrator - the person primarily responsible for installing and administering the Project Management system.

• Technical Librarian - the person responsible for maintaining all "PRIDE" related documentation, e.g., phase deliverables, and project documentation.

This does not mean implementing "PRIDE" requires enormous resources. Depending on the "PRIDE" methodologies to be installed, certain people may not be involved. Also, some members of the team may share responsibilities (such as Project Administration/Technical Librarian). Participation in the support team is not necessarily a full time job especially if the work is evenly distributed between members of the team.

It is important that a unique mix of both managers and staff from various areas participate in the Support Team in order to give the project effectiveness, credibility, and balance. Junior people may be useful for establishing the mechanics of the product, but it will require managers to set standards, promote the use of the methodologies, and handle political issues.

One of the first steps by the Support Team is to become conversant in "PRIDE" themselves. This can be accomplished by reviewing the "PRIDE" documentation and by attending a "PRIDE" training course.

STEP 3 - DEVISE STRATEGY

In essence, the Support Team will be fulfilling the role of "Industrial Engineering" as found in a manufacturing facility. Under this scenario, they will be studying the "PRIDE" methodologies and determine:

• Supplemental tools and techniques to be used throughout the methodologies. From this, standards are devised.

• How to implement an IRM Repository, either manually or with automated assistance.

• How to implement a Project Management system either manually or with computer software. This includes defining standard "Indirect" and "Unavailable" assignments for posting such time. Also, the reporting cycle and the mechanisms of Project Management are defined.

• The necessary management infrastructure to support "PRIDE." This specifically includes the development of a Quality Assurance organization which includes the Technical Library and Project Administration functions.

• Training requirements - for developers, support functions, as well as management and end-users.

Perhaps the biggest decision to be made at this point is an implementation strategy whereby the company either installs "PRIDE" all at once or takes an evolutionary approach where key projects are selected for the initial use of "PRIDE" (a sort of "snowball" effect). The latter approach is probably the most effective for getting started.

STEP 4 - INITIATE PLAN

During this stage, the Support Team will initiate the IRM Repository and Project Management system, implement the necessary support infrastructure, execute their training plan, and begin to use "PRIDE." During the first few projects, pay particular attention to how "PRIDE" is used and look for problem areas. Here, the Support Team becomes a SWAT team to correct problem areas as quickly as possible. The intent is to gain momentum and perfect the use of the product (which will become an on-going goal).

After "PRIDE" is installed, encourage forums (perhaps after normal working hours) where the mechanics of "PRIDE" are discussed with the staff. Such forums promote self-improvement.

Whenever you have any questions about "PRIDE" or require training, please do not hesitate to contact MBA, the vendor.

END SECTION

Section 36

"PRIDE" FORMS

Just because it is on a PRIDE form,
doesn't mean it was prepared using PRIDE.
- Bryce's Law

This section includes the forms used throughout the "PRIDE" methodologies (EEM, ISEM, DBEM) and Project Management system. "PRIDE" forms consist of worksheets to document information resources, and forms for Project Management purposes. The worksheets are used to interview users, organize notes, and act as a data entry vehicle for an IRM Repository. The Project Management forms are used to plan, estimate, schedule, and report on project activities. They are also used as a means to collect data for a Project Management system.

These forms are also available from the web version of the product at MBA's corporate web site.

If you are going to use the "PRIDE" forms in-house, you may add your own corporate logo to the form but you must maintain the "PRIDE" logo and trademark, along with the copyright notation at the bottom of the form.

RESOURCE WORKSHEETS

- Functional Entity (FE) Worksheet
- Organizational Entity (OE) Worksheet
- Human/Machine Resource (RE) Worksheet
- Skill Description (SD) Worksheet
- Objective/Modification-Improvement (MI) Worksheet
- Project Description (PD) Worksheet
- Information Requirement (IR) Worksheet
- System Worksheet
- Sub-System Worksheet
- Computer Procedure Worksheet
- Program Worksheet
- Module Description (MD) Worksheet
- Playscript Worksheet
- Input/Output Description (ID/OD) Worksheet
- File Description (FD) Worksheet
- Record Description (RD) Worksheet
- Data Description (DD) Worksheet
- Matrix Worksheet
- Control Log

PROJECT MANAGEMENT FORMS

- General "PRIDE" Documentation Form
- (Project Description (PD) Worksheet)
- Methodology Definition Worksheet
- Project Control Sheet
  - Project Control Sheet - alternate form to also post "Actual" time.
- Project Estimate/Schedule Recap Worksheet
- Planning Chart (Gantt)
- Time Distribution Worksheet
- Cost/Benefit Analysis Worksheet

PRIDE®

FUNCTIONAL ENTITY WORKSHEET

FE- ______ CONTROL NUMBER

NAME

PREPARED BY	SECURITY	DATE
LOGICAL ATTRIBUTES		

ENTERPRISE TYPE FE

○ INTERNAL ○ EXTERNAL

BUSINESS PURPOSE/CHARTER

TYPES OF PRODUCTS/SERVICES OFFERED

FORM OF MANAGEMENT/GOVERNMENT

DEFINE HIGHEST LEVEL FUNCTIONS

FE-	FE-	FE-
PRODUCE INCOME	ADMINISTER RESOURCES	PRODUCE PRODUCTS/SERVICES

DEFINE HIGHEST LEVEL POSITIONS

OE-

OE-

JOB TITLE/POSITION

BUSINESS FUNCTION TYPE FE

○ INTERNAL ○ EXTERNAL

CIRCLE AREA OF BUSINESS AFFECTED

PRODUCE INCOME — ADMINISTER RESOURCES — PRODUCE PRODUCTS/ SERVICES

CHECK THE FUNCTIONAL LEVEL IN THE ENTERPRISE

○ POLICY
○ CONTROL
○ OPERATIONAL

BUSINESS PURPOSE/SCOPE OF FUNCTION

ACTIONS/DECISIONS/RESPONSIBILITIES

DEFINE OE'S IMPLEMENTING THIS FUNCTION

OE-

OE-

JOB TITLE/POSITION

OBJECTS (FD), REQUIREMENTS (IR) & SYSTEMS

IRM NUMBER NAME

IW001 (C) MBA 2004

(USE REVERSE SIDE TO SKETCH A FUNCTION CHART)

PRIDE®

ORGANIZATIONAL ENTITY WORKSHEET

OE-
CONTROL NUMBER

NAME

PREPARED BY	SECURITY	DATE
LOGICAL ATTRIBUTES		

CHARACTERISTICS

○ INTERNAL ○ EXTERNAL

ORGANIZATION CODE/NUMBER ____________

TELEPHONE NUMBER ____________

LOCATION/MAIL DROP ____________

TOLERANCES (FOR PROJECT MANAGEMENT)

WEEKLY DEVIATION OF TIME REPORTING ____ %

DAILY DEVIATION OF TIME REPORTING ____ %

PROJECT SCHEDULE OVERRUNS ____ %

PROJECT ESTIMATE OVERRUNS ____ %

BUSINESS PURPOSE/SCOPE (WHAT BUSINESS FUNCTIONS DOES IT SUPPORT?)

____

____

____

LIST SPECIFIC DUTIES AND RESPONSIBILITIES

____

____

____

____

RESOURCE(S) IMPLEMENTING OE

NUMBER	NAME	NUMBER	NAME
RE -		RE -	
RE -		RE -	

IRM RESOURCES USED (IR/ID/OD/FD/SYSTEM)

NUMBER	NAME	NUMBER	NAME

USE REVERSE SIDE TO SKETCH AN ORGANIZATIONAL CHART

IW002 (C) MBA 2004

PRIDE® HUMAN/MACHINE RESOURCE WORKSHEET

RE - ___ CONTROL NUMBER

NAME

PREPARED BY	SECURITY	DATE
LOGICAL ATTRIBUTES		

TYPE OF RESOURCE:

- ○ HUMAN
- ○ MACHINE

TIME REPORTING (FOR PM):

- ○ MUST
- ○ MAY
- ○ CANNOT

EMPLOYEE/MACHINE NUMBER: ____________

DATE OF HIRE ______

DAYS IN WORK WEEK ______

HOURS IN WORK DAY ______

SCHED EFFECT RATE ______ %

HOURLY RATE ______

TELEPHONE NUMBER: ____________

YEARS OF EDUCATION ______

POSITION PROFICIENCY ______

OVERALL PROFICIENCY ______

LAST EVALUATED ______

E-MAIL

RESOURCE PROFILE

____

____

____

SKILLS & PROFICIENCIES (9 HIGH TO 1 LOW)

NUMBER	NAME	PROFICIENCY	LAST EVALUATED
SD-			
SD-			
SD-			
SD-			
SD-			

OE (POSITION) & PD (PROJECT) ASSIGNMENTS

NUMBER	NAME	FIXED/VARIABLE	PROJECT FUNCTION
OE-			
PD-		F V	
PD-		F V	
PD-		F V	
PD-		F V	

PRIDE® SKILL DESCRIPTION WORKSHEET

SD -
CONTROL NUMBER

NAME

PREPARED BY	SECURITY	DATE
LOGICAL ATTRIBUTES		

SKILL TYPE:

- O HUMAN RELATED/TANGIBLY MEASURED
- O MACHINE RELATED/TANGIBLY MEASURED
- O HUMAN RELATED/INTANGIBLY MEASURED
- O MACHINE RELATED/INTANGIBLY MEASURED

SKILL DESCRIPTION (WHAT IS IT?)

SKILL DEVELOPMENT (WHAT EDUCATION/TRAINING/EXPERIENCE IS REQUIRED?)

HOW IS THIS SKILL EVALUATED OR TESTED?

WHAT BUSINESS FUNCTION(S) DOES THIS SKILL RELATE TO?

NUMBER	NAME	REQUIRED PROFICIENCY
FE-		
FE-		
FE-		

IW004 (C) MBA 2004

PRIDE®

OBJECTIVE/
MODIFICATION-IMPROVEMENT
WORKSHEET

MI -
CONTROL NUMBER

NAME

PREPARED BY	SECURITY	DATE
LOGICAL ATTRIBUTES		

TYPE OBJECTIVE:

- O MODIFICATION-IMPROVEMENT
- O MAINTENANCE (CORRECTIONS)
- O NEW DEVELOPMENT

CHARACTERISTICS:

DATE RECEIVED ______________

DATE REQUIRED ______________

PRIORITY WEIGHT ______________

SCOPE OF OBJECTIVE (STATE THE BUSINESS PROBLEM/OPPORTUNITY AND THE AREAS OF THE BUSINESS AFFECTED - PLEASE BE AS SPECIFIC AS POSSIBLE - USE REVERSE SIDE IF NECESSARY)

FE/OE USER SPONSORS

NUMBER	NAME	USER SHARE
		%
		%
		%

PROJECT(S) IMPLEMENTING REQUEST

NUMBER	NAME
PD-	
PD-	
PD-	

IRM RESOURCES DIRECTLY AFFECTED (IR/ID/OD/FD/SYSTEMS)

NUMBER	NAME	NUMBER	NAME

IW005 (C) MBA 2004

ATTACH EXAMPLES AS REQUIRED

PRIDE®

PROJECT DESCRIPTION WORKSHEET

PD -
CONTROL NUMBER

NAME

PREPARED BY	SECURITY	DATE
LOGICAL ATTRIBUTES		

TYPE PROJECT:
- O DIRECT
- O INDIRECT
- O UNAVAILABLE

CHARACTERISTICS:
- DATE REQUIRED ____________
- PLANNED INSTALLATION ____________
- COST TOLERANCE ____________

- O P-EEM
- O P-ISEM
- O P-DBEM
- O SPECIAL
- O OTHER ________

PROJECT SCOPE (STATE THE BUSINESS PROBLEMS/OPPORTUNITIES AND THE AREAS OF THE BUSINESS AFFECTED - PLEASE BE AS SPECIFIC AS POSSIBLE - USE REVERSE SIDE IF NECESSARY)

USER SPONSORS

NUMBER	NAME	USER SHARE
OE-		%
OE-		%
OE-		%

IRM RESOURCES DIRECTLY AFFECTED (IR/SYSTEMS)

NUMBER	NAME

ASSIGNED HUMAN RESOURCES

NUMBER	NAME	FIXED/VARIABLE	FUNCTION	HOURLY RATE	WKLY HOURS
RE-		F V			
RE-		F V			
RE-		F V			
RE-		F V			
RE-		F V			

IW006 (C) MBA 2004 USE REVERSE SIDE TO LIST REQUIRED PHASES/ACTIVITIES/OPERATIONS

PRIDE® INFORMATION REQUIREMENT WORKSHEET

IR - ____ CONTROL NUMBER

NAME

PREPARED BY	SECURITY	DATE
LOGICAL ATTRIBUTES		

TYPE REQUIREMENT:

- O INFORMATIONAL
- O NON-INFO RELATED

TYPE INFORMATION:

- O POLICY
- O CONTROL
- O OPERATIONAL

TIMING (FOR INFORMATION ONLY)

FREQUENCY ____________

OFFSET ____________

RESPONSE TIME ____________

BUSINESS PURPOSE (TO ACCOMPLISH WHAT?)

____

____

ACTIONS AND/OR BUSINESS DECISIONS (LIST 'WHO' WILL DO 'WHAT' WITH THE INFORMATION)

1. ____________
2. ____________
3. ____________
4. ____________

BENEFITS (LIST BOTH TANGIBLE & INTANGIBLE

1. ____________
2. ____________

SUPPORTING IRM RESOURCES

NUMBER	NAME

REQUIRED DATA ELEMENTS

NUMBER	NAME
DD-	
DD-	
DD-	
DD-	
DD-	
DD-	

IW007 (C) MBA 2004

LIST FE/OE 'RECEIVERS; - 'ORIGINATORS' - 'PARTICIPATORS' ON REVERSE SIDE

PRIDE® SYSTEM WORKSHEET NAME

	SYSTEM ID NUMBER

PREPARED BY	SECURITY	DATE
LOGICAL ATTRIBUTES		

BUSINESS PURPOSE

DESCRIPTION (SPECIFY THE INFORMATION REQUIREMENTS AND BUSINESS FUNCTIONS SERVED BY THE SYSTEM. THIS TEXT SHOULD CORRESPOND WITH THE SYSTEM CONCEPT DIAGRAM).

SUB-SYSTEMS

NUMBER	NAME	FREQUENCY	OFFSET	RESPONSE TIME

LIST PERTINENT FE/OE USERS OF THE SYSTEM ON THE REVERSE SIDE

PRIDE® SUB-SYSTEM WORKSHEET

NAME

SYSTEM	SUB-SYSTEM

PREPARED BY	SECURITY	DATE
LOGICAL ATTRIBUTES		

FE's/OE's AFFECTED BY SUB-SYSTEM

NUMBER	NAME	RESPONSIBILITY=
		RESPONSIBILITY=
		RESPONSIBILITY=
		RESPONSIBILITY=

EXECUTE-APPROVE-REVIEW-PARTICIPATE

TIMING

FREQUENCY ____________

OFFSET ____________

RESPONSE TIME ____________

COMES FROM	INPUTS
	ID-
	ID-
	ID-
	ID-
	ID-

BUSINESS PURPOSE

STEPS IN EXECUTION

TEST CRITERIA
(DEFINE TEST PLAN ON REVERSE SIDE)

OUTPUTS	GO TO
OD-	
OD-	
OD-	
OD-	
OD-	

FILES USED

NUMBER	NAME	C	U	R	VOLAT.	HIT RATIO
FD-					%	%
FD-					%	%
FD-					%	%
FD-					%	%

PREPARE ROUGH DESIGN OFPROCEDURES ON REVERSE SIDE

PRIDE®

COMPUTER PROCEDURE WORKSHEET

SYS	SUB-SYS	PROC

NAME

PREPARED BY	SECURITY	DATE
LOGICAL ATTRIBUTES		

FE's/OE's AFFECTED BY PROCEDURE

NUMBER	NAME	EXECUTE-APPROVE-REVIEW-PARTICIPATE
______	______	RESPONSIBILITY=______
______	______	RESPONSIBILITY=______
______	______	RESPONSIBILITY=______

CHARACTERISTICS

ALTERNATE ID ______
OFFSET ______
RESPONSE TIME ______

COMES FROM | INPUTS

______ ID-
______ ID-
______ ID-
______ ID-
______ FD-

INPUT TRANSACTION FILE

BUSINESS PURPOSE

STEPS IN EXECUTION

TEST CRITERIA
(DEFINE TEST PLAN ON REVERSE SIDE)

OUTPUTS | GO TO

OD- ______
OD- ______
OD- ______
OD- ______
FD- ______

OUTPUT DATA FILE

FILES USED

NUMBER	NAME	C	U	R	FROM	GOES TO
FD-						
FD-						
FD-						
FD-						

PREPARE ROUGH DESIGN OF PROGRAM JOB STREAM ON REVERSE SIDE

PRIDE®

PROGRAM WORKSHEET

NAME

SYS	SS	PRC	PGM

PREPARED BY	SECURITY	DATE
LOGICAL ATTRIBUTES		

FE's/OE's AFFECTED BY PROGRAM

NUMBER	NAME	EXECUTE-APPROVE-REVIEW-PARTICIPATE
______	______	RESPONSIBILITY=______
______	______	RESPONSIBILITY=______
______	______	RESPONSIBILITY=______

CHARACTERISTICS:

PROC OFFSET ______
RESPONSE TIME ______

ALTERNATE ID ______
LANUAGE ______
MEMORY ______

COMES FROM — INPUTS

ID- ______
ID- ______
ID- ______
ID- ______
FD- ______

BUSINESS PURPOSE

______
______
______

STEPS IN EXECUTION

______
______
______
______
______
______

TEST CRITERIA
(DEFINE TEST PLAN ON REVERSE SIDE)

OUTPUTS — GO TO

OD- ______
OD- ______
OD- ______
OD- ______
FD- ______

FILES USED

NUMBER	NAME	C	U	R	FROM	GOES TO
FD-						
FD-						
FD-						
FD-						

PREPARE ROUGH DESIGN OF PROGRAM JOB STREAM ON REVERSE SIDE

PRIDE®

MODULE DESCRIPTION WORKSHEET

NAME

MD - ____ CONTROL NUMBER

PREPARED BY	SECURITY	DATE
LOGICAL ATTRIBUTES		

TYPE OF MODULE

○ PERFORMED SEPARATELY

○ CALLED BY OTHER MODULES

IF CALLED, SPECIFY CALL PARAMETERS

RESPONSE TIME ____

ALTERNATE ID ____

PROGRAM LANGUAGE ____

MEMORY ____

RD- ____ ____ ____

RD- ____ ____ ____

RD- ____ ____ ____

ENTRY POINT/TEXT

MODULE/SUBROUTINE PURPOSE

PROCESSING LOGIC

SUBORDINATE MODULES

MD- ____ ____ ____

MD- ____ ____ ____

MD- ____ ____ ____

ENTRY POINT/TEXT

IRM RESOURCES AFFECTED (ID/OD/FD/RD)

NUMBER	NAME	OPEN	CLOSE	C	U	R

IW012 (C) MBA 2004 PREPARE MODULE DESIGN/LOGIC ON REVERSE SIDE

PRIDE® PLAYSCRIPT WORKSHEET

NAME

SYS	SUB-SYS	PROC

PREPARED BY	SECURITY	DATE
LOGICAL ATTRIBUTES		

TYPE PROCEDURE:

- ○ MANUAL
- ○ OFFICE AUTOMATION
- ○ KEYING

TIMING

OFFSET: ____________

RESPONSE TIME: ____________

BUSINESS PURPOSE

____

____

____

SETUP (LIST ID/OD/FD USED)

NUMBER	NAME	C	U	R	COMES FROM	GOES TO

OPERATIONS

ACTOR	STEP NO.	INSTRUCTIONS (ALWAYS BEGIN WITH A VERB, NEVER WITH AN 'IF')

PRIDE®

INPUT/OUTPUT DESCRIPTION WORKSHEET

D -
ID/OD CONTROL NUMBER

NAME

PREPARED BY	SECURITY	DATE
LOGICAL ATTRIBUTES		

TIMING:

FREQUENCY ____________

OFFSET ____________

RESPONSE TIME ____________

PAGINATION:

CONTINUOUS

NONE

OTHER ______________

MEDIUM (PAPER, CRT, AUDIO, ETC.)

____

FORM/SCREEN NUMBER:

____

BUSINESS PURPOSE

____

____

DESCRIPTION (HOW PRODUCED, FORMAT, SECTIONS, PRINT SEQUENCE, ETC.)

____

____

____

____

____

SORT/ACCESS KEY(S)

NUMBER	NAME	NUMBER	NAME
DD-		DD-	
DD-		DD-	

IRM RESOURCES USED (RD/DD/ID/FD/OD)

NUMBER	NAME	NUMBER	NAME

IW014 (C) MBA 2004

PREPARE A COMPLETE EXAMPLE ON THE REVERSE

PRIDE® FILE DESCRIPTION WORKSHEET

FD - ____ CONTROL NUMBER

NAME

PREPARED BY	SECURITY	DATE
LOGICAL ATTRIBUTES		

TYPE OF FILE:

- O APPLICATION LOGICAL
- O ENTERPRISE LOGICAL
- O ENTERPRISE PHYSICAL
- O APPLICATION PHYSICAL

PHYSICAL FILE CHARACTERISTICS:

CHARACTERISTICS:	MEDIA:	ORGANIZATION:
O FACT/PRIMARY	O DISK	O SEQUENTIAL
O EVENT/WORKING	O TAPE	O INDEX SEQUENTIAL
RESIDENCE:	O CD	O RANDOM
O COMPUTER	O PAPER	O DBMS
O MANUAL	O ______	O ______

RECORD FORM:	RECORD LENGTH:	NUMBER OF RECORDS:	BLOCKING FACTOR:
O FIXED	MINIMUM ______	MINIMUM ______	______
O VARIABLE	AVERAGE ______	AVERAGE ______	RETENTION:
O UNDETERMINED	MAXIMUM ______	MAXIMUM ______	______

BUSINESS PURPOSE (WHAT OBJECT DOES THE LOGICAL FILE REPRESENT? WHAT TYPE OF DATA IS MAINTAINED IN THE PHYSICAL FILE? DESCRIBE FILE TO FILE RELATIONSHIPS)

____

____

KEY(S) TO FILE

NUMBER	NAME
DD-	
DD-	
DD-	

PROGRAM LABELS FOR PHYSICAL FILE

LANGUAGE	LABEL

IRM RESOURCES USED (RD/DD/ID/OD/FD)

NUMBER	NAME	NUMBER	NAME

IW015 (C) MBA 2004 PREPARE FILE DESIGNS ON REVERSE. ATTACH DDL AS REQUIRED.

PRIDE® RECORD DESCRIPTION WORKSHEET

RD - ______ CONTROL NUMBER

NAME

PREPARED BY	SECURITY	DATE
LOGICAL ATTRIBUTES		

TYPE OF RECORD:

- O APPLICATION LOGICAL
- O ENTERPRISE LOGICAL
- O ENTERPRISE PHYSICAL
- O APPLICATION PHYSICAL

USER SOURCE:__________

CHARACTERISTICS:

- O IDENTIFICATION VIEW (LOGICAL)
- O CHARACTERISTICS VIEW (LOGICAL)
- O RELATIONSHIP VIEW (LOGICAL)
- O STORAGE RECORD
- O INPUT TRANSACTION
- O PRINT MAP
- O SCREEN PANEL
- O CALL PARAMETER
- O __________

PHYSICAL FORM:

- O FIXED
- O VARIABLE
- O UNDETERMINED

RECORD LENGTH:

MINIMUM ______

AVERAGE ______

MAXIMUM ______

DESCRIPTION (WHAT DATA DOES THIS RECORD REPRESENT? WHAT DATA IS MAINTAINED IN THIS RECORD?)

____

____

____

KEY(S) TO RECORD (BASIC GROUPING)

NUMBER	NAME
DD-	
DD-	
DD-	

PROGRAM LABELS FOR PHYSICAL RECORD

LANGUAGE	LABEL

DATA ELEMENTS USED

NUMBER	NAME	NUMBER	NAME
DD-		DD-	
DD-		DD-	
DD-		DD-	
DD-		DD-	

IW016 (C) MBA 2004

DEFINE RD-RD RELATIONSHIPS ON THE REVERSE SIDE

PRIDE® DATA DESCRIPTION WORKSHEET

DD - ___ CONTROL NUMBER

NAME

PREPARED BY	SECURITY	DATE
LOGICAL ATTRIBUTES		

BUSINESS DEFINITION (PROVIDE A WEBSTER/OXFORD DICTIONARY STYLE DEFINITION, AVOID ENTERING VALIDATION RULES AND PROGRAM LABELS)

____

____

____

TYPE DATA

○ INDICATIVE - IT IDENTIFIES... ____ AN OBJECT ____ A VIEW

○ DESCRIPTIVE - IT DESCRIBES... ____ A FACTUAL OBJECT ____ AN EVENT OBJECT ____ GENERATED

○ QUANTITATIVE - IT QUANTIFIES... ____ A FACTUAL OBJECT ____ AN EVENT OBJECT ____ GENERATED

IF DESCRIPTIVE OR QUANTITATIVE DATA, SPECIFY...

DD-____ ____________ TO REPRESENT THE PRIMARY 'OBJECT'

DD-____ ____________ TO REPRESENT SECONDARY/FOREIGN 'OBJECTS' OR A 'VIEW' IDENTIFIER

DD-____ ____________

FORM

○ PRIMARY - SPECIFY THE USER SOURCE ____________

○ GROUP - SPECIFY DATA DEPENDENCIES BELOW

○ GENERATED - SPECIFY REQUIRED DD's & CALCULATIONS BELOW

DD-____ ____________ ____________

DD-____ ____________ ____________

EXPRESS FORMULAS & CALCULATIONS HERE

CLASS:	JUSTIFICATION:	FILL CHARACTER:	MODE:	NUMERIC DATA:
○ ALPHABETIC	○ LEFT	○ BLANK	○ CHARACTER	PRECISION ____
○ NUMERIC	○ RIGHT	○ ZERO	○ FIXED POINT	SCALE ____
○ ALPHANUMERIC	○ AROUND DECIMAL	○ ____	○ BINARY	UNIT OF MEAS ____
○ SIGNED NUMERIC	○ ____	VOID STATE:	○ PACKED DECIMAL	BASE ____
○ VARIABLE STRING	LENGTH:	○ BLANK	○ ____	
○ ____	____	○ ZERO		
		○ ____		

 EXPRESS VALIDATION/EDITING RULES ON THE REVERSE SIDE

PRIDE® MATRIX WORKSHEET

CONTROL NUMBER

TITLE:

PAGE OF

PREPARED BY	DATE	APPROVED BY	DATE

PRIDE®

CONTROL LOG

NAME

LOG TYPE

NUMBER	NAME	TEXT

IW020 (C) MBA 2004

PRIDE®

TITLE

CONTROL NUMBER

PREPARED BY	APPROVED BY	DATE

PRIDE®

METHODOLOGY DEFINITION WORKSHEET

NAME

PD -
CONTROL NUMBER

PREPARED BY	APPROVED BY	DATE

TYPE:		IDENTIFIER:
METHODOLOGY	____________________	__________
PHASE	____________________	__________
ACTIVITY	____________________	__________

WORK STEP	RESOURCE (OPT)	FUNC CODE	NAME	PRECEDENT RELATIONSHIPS																	

USE THIS SYMBOL TO INDICATE OFF-PAGE CONNECTION

PRIDE®

PROJECT CONTROL SHEET

PD -

CONTROL NUMBER

TITLE

PREPARED BY	DATE	APPROVED BY	DATE
METHODOLOGY: EEM ISEM DBEM		TYPE ESTIMATE (CIRCLE): DETAIL O.O.M.	

IRM RESOURCE	PHASE	ACT	OP	FUNC	ASSIGNED TO	RATE	HRS	COST	START DATE	END DATE
					RE-					
					RE-					
					RE-					
					RE-					
					RE-					
					RE-					
					RE-					
					RE-					
					RE-					
					RE-					
					RE-					
					RE-					
					RE-					
					RE-					
					RE-					
					RE-					
					RE-					
					RE-					
					RE-					
					TOTALS					

PRIDE® PROJECT CONTROL SHEET

PD - ___________
CONTROL NUMBER

TITLE

PREPARED BY	DATE	APPROVED BY	DATE

PHASE ___________ EFFECTIVENESS RATE USED ___________ ASSIGNED TO ___________

ACTIVITY			ESTIMATE				ACTUAL				ESTIMATE TO DO
			Hours		Schedule		Hours		Schedule		
ACT	DESCRIPTION	FUNC	Direct	Computer	Start Date	End Date	Direct	Computer	Start Date	End Date	
	TOTALS										

RECAP BY FUNCTIONS (e.g., PM, SE, PE, DE, EE, DBA, etc.)	Rate Per Hour	ESTIMATE		ACTUAL		(OVER RUN) / UNDER RUN	
		HOURS	COST	HOURS	COST	HOURS	COST
	TOTALS						

APPROVALS ___________

IW023A (C) MBA 2004

PRIDE®

PROJECT ESTIMATE/
SCHEDULE RECAP

NAME

PD -
CONTROL NUMBER

PAGE OF

PREPARED BY	APPROVED BY	DATE

ORDER-OF-MAGNITUDE (OOM) ESTIMATE

PHASE	RESOURCE (OPT)	NAME	DIRECT HOUR ESTIMATES BY PROJECT FUNCTION										COST	START DATE	END DATE
		TOTALS													

IW025 (C) MBA 2004

PRIDE® PLANNING CHART (GANTT)

NAME

PD -

CONTROL NUMBER

PAGE OF

PHASE	ASSIGNED TO	E.R.			CALENDAR
			EST	CP	
				S	
			ACT	CP	
				S	
			EST	CP	
				S	
			ACT	CP	
				S	
			EST	CP	
				S	
			ACT	CP	
				S	
			EST	CP	
				S	
			ACT	CP	
				S	
			EST	CP	
				S	
			ACT	CP	
				S	
			EST	CP	
				S	
			ACT	CP	
				S	
			EST	CP	
				S	
			ACT	CP	
				S	
			EST	CP	
				S	
			ACT	CP	
				S	
			EST	CP	
				S	
			ACT	CP	
				S	
			EST	CP	
				S	
			ACT	CP	
				S	
			EST	CP	
				S	
			ACT	CP	
				S	

PRIDE® TIME DISTRIBUTION WORKSHEET

/ /
WEEK ENDING DATE

PAGE OF

RESOURCE NAME	RE-

PROJECT	RESOURCE (OPT)	PHS	ACT	OP	FUNC	D-I-U	MON	TUE	WED	THU	FRI	SAT	SUN	TOTAL	TO DO

UNAVAILABLE HOURS	U								
DIRECT HOURS	D								
INDIRECT HOURS	I								
TOTAL HOURS	T								

APPROVAL(S) ______________________

EFFECTIVENESS RATE ________%

E.R. = DIRECT DIV. (DIRECT + INDIRECT)

PRIDE®

COST/BENEFIT ANALYSIS WORKSHEET

PD -
PROJECT NUMBER
IRM CONTROL NUMBER

TITLE:

PREPARED BY	DATE	APPROVED BY	DATE

STAFFING ANALYSIS LIST DEPARTMENTS BELOW	CURRENT STAFFING	PROJECTED STAFFING	DIFFERENCE
ANNUAL SAVINGS/EXPENSES **LIST DEPARTMENTS BELOW**	**ANNUAL SAVINGS**	**ANNUAL EXPENSES**	**NET**
ONE-TIME SAVINGS/EXPENSES **LIST DEPARTMENTS BELOW**	**ANNUAL SAVINGS**	**PROJECT EXPENSES**	**NET**

INTANGIBLE BENEFITS

INVESTMENT EVALUATION

BREAK EVEN POINT =

ANNUAL RETURN ON INVESTMENT (R.O.I.) = _________ %

IW021 (C) MBA 2004

EXPRESS YOUR CALCULATIONS ON THE REVERSE SIDE OF THIS FORM

Section 37

"PRIDE" CHRONOLOGY

The first on-line, real-time, interactive, data base system was double-entry bookkeeping which was developed by the merchants of Venice in 1200 A.D.
- Bryce's Law

The "PRIDE" Methodologies for IRM have evolved into a substantial body of work. This did not happen over night. Considerable thought and work went into the product. The real-world input from our customers were invaluable for important modification/improvements to it. For this we give our thanks. "PRIDE" set many precedents over the years, and as such, the company wants to set the record straight.

FOR THE RECORD

1971

MBA is founded and introduces "PRIDE" (PRofitable Information by DEsign - through phased planning and control) for designing and building Information Systems. "PRIDE" sets the following precedents:

- The world's first commercially available methodology for developing systems (the term "Methodology" was coined in the industry by "PRIDE").
- First to apply engineering/manufacturing concepts to the development of Information Systems, e.g., Standard System Structure, Layered Documentation (Blueprinting), standard and reusable parts (data resources).
- First Structured Systems Design Technique (Chronological Decomposition).
- First Data Dictionary/Directory.

"PRIDE" also includes Documentation and a Project Management System, all integrated into a single product.

The first "PRIDE" customer was the Marion Power Shovel Company of Marion, Ohio (08/18/1971).

1974

MBA introduces "PRIDE"-LOGIK (Logical Organizing and Gathering of Information Knowledge) an automated Data Dictionary/Directory system. It is designed to complement the Data Management and Documentation techniques of "PRIDE." The product is the first of its kind in the world. It is originally developed to run under IBM MVS & VM, Honeywell GCOS, and Burroughs MPE computers. Over time it is migrated to many other configurations including: CDC Cyber Series, UNIVAC 100 Exec, Hewlett Packard 3000 MPE, ICL George III, Hewlett Packard 3000 MPE, Data General Eclipse AOS, DEC VAX/VMS, and IBM AS/400 and OS/2.

1976

MBA establishes representation in Japan, the start of a long relationship in the land of the rising sun.

1977

"PRIDE" becomes the first commercial methodology enacted into law (State of Minnesota).

1979

The Automated Design Facility (ADF) is developed by MBA as a productivity tool which is used to automatically design Information Systems and logical Data Bases. It is the world's first "Systems Generator" (not to be confused as a "Program Generator"). Based on Information Requirements, ADF automatically designs a system into its sub-systems, procedures, and programs, as well as design the logical data base and suggests physical files. By accessing "PRIDE"-LOGIK, ADF promotes the sharing and re-use of data resources.

Also in December, MBA wins a highly contested trade secret lawsuit with AMF-Harley Davidson Division and Arthur Young & Company in Milwaukee, Wisconsin. A jury determines that "PRIDE" is a unique product and that MBA took all of the necessary steps to protect it, thus making it a trade secret. They also determine that Arthur Young was guilty of trying to misappropriate the product.

1981

MBA enhances "PRIDE"-LOGIK and renames the product, the Information Resource Manager (IRM) and the name "PRIDE"-LOGIK is dropped. Among the enhancements are new components (resources) to represent business models and information requirements. As such, it is no longer referred to as a Data Dictionary/Directory but rather a Repository.

MBA demonstrates program generation by developing an interface from the IRM to the System-80 COBOL program generator. Between ADF and the IRM's interface to program generators, MBA successfully demonstrated 100% systems generation.

Several other interfaces are developed by MBA and its customers for program generation and DBMS support.

1982

MBA announces "PRIDE"-ASDM, Automated Systems Design Methodology. "PRIDE"-ASDM is a major development effort by MBA and includes Automated Instructional Materials (AIM). AIM represents all of the "PRIDE" documentation available on-line, the first such product to do so.

MBA is recognized by its customers for improving the productivity of systems development in Japan.

1984

"PMC2" (Project Management - Command & Control) is introduced by MBA. It represents the automation of the Project Management activities in "PRIDE."

1985

MBA relocates its offices to Palm Harbor, Florida.

1986

MBA renames "PRIDE" as the "PRIDE"-Information Systems Engineering Methodology (ISEM). The company also unbundles and restructures its product line consisting of manual and automated combinations of ISEM, IRM, and PMS (formerly "PMC2").

1987

MBA introduces "PRIDE"-Data Base Engineering Methodology (DBEM). DBEM introduces the concepts of "objects", 4 data base models, and "Data Taxonomy."

Also, ADF is enhanced and renamed ASE (Automated Systems Engineering). The facility now includes standard templates for engineering business processes (sub-systems and procedures).

1988

MBA introduces "PRIDE"-Enterprise Engineering Methodology (EEM). EEM introduces the techniques of "Enterprise Decomposition" and "Enterprise Information Strategy" (EIS)

MBA develops automated support for "PRIDE"-EEM in the form of "PRIDE"-CAP (Computer Aided Planning) to model and communicate priorities, group information requirements and objectives, and to perform an Organization Analysis (the first such product to do so).

MBA publishes its book, *"THE IRM REVOLUTION: BLUEPRINT FOR THE 21ST CENTURY"* (ISBN: 0962118907), which quickly makes the top ten list in Japan.

1990

MBA unveils the "PRIDE"-Information Factory, an OS/2 based implementation of the "PRIDE" product line.

1993

MBA enters into agreement with the American Management Association (AMA) to conduct a series of IRM Executive Seminars

1994

MBA announces "PRIDE"-PC, a PC implementation of its Automated Instructional Materials (AIM).

1996

MBA and "PRIDE" celebrates its 25th birthday.

2004

MBA puts the "PRIDE" Methodologies for IRM in the public domain via the Internet.

2006

MBA and "PRIDE" celebrates its 35th anniversary and publishes this book.

2011

MBA and "PRIDE" celebrates 40 years. Company is absorbed into its parent, M&JB Investment Company.

END SECTION

Section 38

BRYCE'S LAWS

A compilation of our axioms on IRM and management.

Productivity = Effectiveness X Efficiency

Information = Data + Processing

There is nothing more unproductive than to build something efficiently that should not have been built at all.

Organizations progress when the impact of good actions and decisions outweighs the impact of poor actions and decisions.

The solution to improved productivity is not better software but rather a better understanding of business and its objectives.

IRM is the view of the enterprise from 50,000 feet.

We must apply the same discipline, organization and automation that we recommend for other parts of the company.

Technology alone will not solve our problems, only effective management will.

No amount of elegant programming or technology will solve a problem if it is improperly specified or understood to begin with.

If anything in life is constant, it is change. The only constant involved with information is that it is seldom static.

If an information requirement is stated improperly to begin with, then everything else that follows will be incorrect.

The only way that information systems communicate, both internally and externally to other systems, is through shared data.

The more we understand about the external entities affecting our business, as well as our own internal operating limitations, the better we can compete.

Data is the glue that holds systems together.

Data is stored, Information is produced.

A data element has only one logical definition, but may be represented physically in many different ways.

Group Data tells us a lot about what objects are important to an enterprise.

Data Resource Management is a neutral third party who represents the enterprise's overall interests, not just a single application.

Quality must be built into the product during design, not inspected in afterwards.

Enterprises with identical missions will also be identical in terms of logical structure.

Never embark on a journey without knowing your destination.

Whereas logical information resources will remain relatively static, the physical resources will change dynamically.

100% of your design documentation is contained in the specifications of your information resources.

Information is a perishable commodity; it only has value at a particular point in time.

Information is highly volatile in that it is greatly influenced by external factors, such as government, economics, competition, customers, etc.

An elegant solution to the wrong problem solves nothing.

The day a company goes into business is the day when its information systems are born.

Only when the systems engineer can walk in the moccasins of the user does the engineer have a right to design a system for the user.

Information is for people, not for the computer.

An information system is a product that can be engineered and manufactured like any other product.

Systems =/= Computers
Systems =/= Software
Systems =/= Projects

Q: How many interpretations of systems development are there?
A: How many analysts and programmers have you got?

All information systems have the same structure. In manufacturing terms, it is known as a "four-level bill of material."

Systems are designed by 'explosion' and implemented by 'implosion.'

Documentation is a working tool and a byproduct of design.

The word 're-engineering' implies something was 'engineered' in the first place, which is rarely the case.

No one has ever built a perfect system the first time, and no one ever will.

Systems are built by evolution; not by revolution. The day when a system is installed, is the day it begins to undergo change.

85% of all systems development work is modifications/improvements.

Good Systems Design + Good Programming = Great Systems
Good Systems Design + Bad Programming = Good Systems
Bad Systems Design + Good Programming = Bad Systems
Bad Systems Design + Bad Programming = Chaos

How a system is implemented is of little importance if it solves the problem effectively.

Forgetting the human-being during design will cause the human-being to forget the system at time of startup; it will be DOA, Dead On Arrival.

Systems are logical, programming is physical.

Programming is a translation function, going from human understandable specifications to machine readable instructions.

Good specifications will always improve programmer productivity far better than any programming tool or technique.

Whenever you see a ratio of 1:4 analysts:programmers you will find systems analysis being performed at the wrong time and by the wrong person.

Beware of your "firefighters," they are probably your chief arsonists.

If we built bridges the same way we build systems in this country, this would be a nation run by ferryboats.

There are very few true artists in computer programming, most are just house painters.

Systems will fail more for the lack of administrative procedures than well written computer procedures.

A program without any form of transaction serves no useful business purpose.

The first on-line, real-time, interactive, data base system was double-entry bookkeeping which was developed by the merchants of Venice in 1200 A.D.

Successful screen design is based on how well the developer knows both the user and the data.

All organizations have a data base; some are managed, most are not.

A Data Base should naturally evolve over time and synchronize with all Information Systems.

Most organizations use a DBMS as nothing more than an elegant file access method. Consequently, the opportunity to share data and integrate systems is lost.

You must first plant the seeds in order to harvest the crop. Unfortunately, most companies tend to eat the seed and then there is no crop to harvest.

There is only one problem with common sense; it's not very common.

A methodology is nothing more than an assembly line that produces a finished product.

Every step in a methodology should produce a reviewable result in order to substantiate completion and assure a quality product.

Without a road map, you might be driving in circles.

Project management is only possible with an effective methodology.

Having a Project Management system without a methodology is like attaching a speedometer to an orange crate; it measures nothing.

The number of lines of communications grow exponentially based on the number of people involved in a project.

Systems do not have a "life cycle." They may go on forever if kept viable with change. The only thing that has a "life cycle" is a project which has a beginning for planning, a middle for execution, and an end for review.

It is one thing to enact legislation, quite another to enforce it.

Project management is a philosophy of management, not a tool or technique.

Most estimating errors are errors of omission, not commission. It is what we forget to estimate that gets us into trouble.

An estimate improves in accuracy in relation to the level of detail considered.

A project always has a critical path until it is completed or closed. The path can vary depending on accomplishments.

A project requires a methodology, but a methodology does not require a project.

It is just as bad to underrun a project as it is to overrun a project.

If we lived in a perfect world, there would not be a need for managers; projects would be executed on time and within cost. However, the reality is, we live in an imperfect world.

Managers do not create problems, they solve problems.

Manage from the bottom up; not just from the top down; this creates personal commitment and accountability.

A project will only be accomplished if the individuals performing the work want to do it.

We accomplish projects through people.

Manage more, supervise less.

The only good business relationship is where both parties benefit.

Employees should be treated as professionals and held accountable for their actions.

All companies have a culture. In order for employees to function and succeed, it is essential they understand and believe in the culture.

Culture is learned. It can be taught and enforced.

The ethics of a business are whatever the top-dog says they are.

Time lost, is time lost forever. You cannot buy it back.

You cannot capitalize on your workers' talents if you do not know their skills and proficiencies.

You cannot put two quarts of liquid into a one quart bottle. If you try, you will lose a quart.

An expert is someone who lives more than 50 miles out of town and wears a tie to work.

"We never have enough time to do things right." Translation: *"We have plenty of time to do things wrong."*

Just because it is on a "PRIDE" form, doesn't mean it was prepared using "PRIDE."

The "man" in the word "management" refers to "mankind." Don't forget it.

Do not try to apply a band-aid when a tourniquet is required to stop the bleeding.

All of your hard work, regardless of how well it is intended, is for naught if it results in a pile of rubbage.

If there is no governing science supporting it, use of the term 'engineering' is fraudulent and misleading.

You cannot treat a patient if he doesn't know he is sick.

A man's trustworthiness is measured by the number of keys he holds.

Most children are raised by amateurs, not professionals.

Never trust a person who doesn't have at least one known vice (e.g., drinking, smoking, swearing).

Business is about people, not just numbers.

Management is more of a benevolent dictatorship as opposed to a democracy.

Unless someone is looking for an excuse to duck a work assignment, nobody wants to attend an inconsequential meeting.

I have never encountered a technical problem that couldn't be conquered with a little imagination, some concentrated effort, and a lot of good old-fashioned management.

Don't watch the clock, watch the product or service to be produced.

Lawsuits primarily benefit the attorneys and nobody else.

A policy is written to protect a company from those who break the rules, not from those who follow them.

Technology without Management is Madness.

You eat elephants one spoonful at a time.

You simply can't build anything of substance without a good set of blueprints.

If you are not pissing someone off, you are probably not doing your job.

If the mind really is the finest computer, then there are a lot of people out there who need to be rebooted.

If they do not have an appreciation of whence we came, I doubt they will have an appreciation of where we should be going.

In this industry, we tend to worry about the wrong things. This is like rearranging the deck chairs on the Titanic.

Its never lonely at the top of an IT organization, primarily because the IT Director is never there.

Forget about today, build for tomorrow.

Social intercourse is a two way street. Make sure you are driving on the right side.

Do not criticize unless you can offer an alternative. Turn something negative into something positive.

"Bullshit" is the most versatile word in the English language. It can be applied in just about any business situation. Frankly, we do not use it enough.

There is more to building a team than buying new uniforms.

Youth is our only true vacation in life, and our most unappreciated.

In every person's life, you must eat at least one spoonful of dirt.

END SECTION

Section 39

GLOSSARY OF TERMS

4GL - initials for "Fourth Generation Language" (See listing).

Access Key - the data element(s) which are used to identify and search through a physical file.

Access Method - the technique used to access data on a physical file device. Types of access methods include sequential, index sequential, hierarchical structure, network structure, relational structure, etc.

Accessibility - a timing consideration used in "Information Driven Design" for specifying when and where data is to be retrieved to produce information for delivery to end user destinations. Accessibility answers the question, "Can I get to the data when I need to?" How data is accessed is a function of output. (See Availability, and Output).

Administrative Procedure - a detailed set of human processable operating instructions required to accomplish a specific process within a sub-system. Administrative procedures detail manual steps along with explanations and examples. They are procedures to be used for such things as performing routine or clerical assignments, making decisions, report distribution, input preparation, or for using "office automation" related equipment. (See Sub-System and Operational Step).

ANSI - initials for American National Standard Institute; standards institution.

ASCII - American National Standard Code for Information Interchange - standard control characters and graphic characters (as used in a simple text file).

Availability - a timing consideration used in "Information Driven Design" for specifying how and when data is collected from various sources within an enterprise. Availability answers the question, "Is the data there when I need it?" How data is made available is a function of input. (See Accessibility, and Input).

Available Time - time when an employee is available for work. Available time is divided into two categories: Direct and Indirect Time (see listings).

Basic Grouping - refers to the indicative data elements used to uniquely identify a view (logical record). It also represents a dependency between data elements in a particular context (it is how they are "grouped" into separate views). For example, in a Customer Object, it is used to segregate address data, from credit data, from customer contact data, etc. Perhaps it is easier to think of the "basic grouping" as the key to a logical record (not a physical record). Because the intent is to uniquely identify data, the "basic grouping" consists only of "Indicative" data (not "Descriptive" or "Quantitative"). There may be up to two parts in a single basic grouping:

PRIMARY BASIC GROUPING - Since views are used to describe objects, they must all be defined with the one data element used to uniquely identify the overall object; this will be a primary/indicative/object-oriented data element. The Primary Basic Grouping will be the basis to sort views into objects (e.g., all Product related Records will be put in the Product File, all Customer related Records in the Customer File, etc.).

SECONDARY KEYS are used to either distinguish Characteristic Views or Relationship Views.

Batch - a method of processing that involves the processing of several transactions and does not necessarily require instantaneous response. Batch processing is especially suited to entering or retrieving large volumes of data from the enterprise data base. (See Interactive, Processing Method, and Transaction).

Benefit - something that contributes to an improvement in condition or promotes profitability for a given enterprise, either tangibly or intangibly. When writing benefits, begin with a transitive verb, such as: Improve, Maximize, Minimize, etc. Be sure to substantiate your claim; do not simply say "Improved cash flow." Instead, say, "Improved cash flow by reducing inventory overhead."

Bill of Materials - an itemized account of all of the parts of a product and their relationships in order to assemble a complete product. In IRM terms, the bill of materials for information includes systems, data and business resources.

Bill of Materials Processing (BOMP) - a mechanized way of tracking all of the resources in a product.

BLOG - slang expression referring to a web page devoted to topics relating to a specific subject (e.g., News, Essays, etc.).

BOMP - initials for "Bill Of Materials Processing" (see listing)

Break Even Point - a point in time where cost savings match accumulated development expenses. Typically calculated as:

BEP = Investment divided by Average Annual Savings.

Business Process - (see Sub-System).

Byte - a group of eight adjacent binary digits treated as a unit and often represent a single character.

C - a procedural programming language. (See Procedural Languages).

C++ - variation of the C programming language oriented to Object Oriented Programming (see listing).

CAD - initials for "Computer Aided Design" (see listing).

CAM - initials for "Computer Aided Manufacturing" (see listing).

CASE - initials for "Computer Aided Software Engineering." Computer tools that provide support for software engineering and manufacturing activities. Their perspective is on software only.

Chief Information Officer (CIO) - an executive level officer who represents the chief information broker/architect/strategist of a company. Ideally, the CIO reports to the Chief Executive Officer (CEO) and maintains a lateral working relationship with the Chief Operating Officer (COO) and Chief Financial Officer (CFO).

Client/Server Computing - computing technique for processing data between a "client" computer and a file "server."

COBOL - acronym for COmmon Business Oriented Language; a procedural programming language (see listing).

Command File - a PC file (BAT or CMD) containing basic computer operating commands.

Command Line or Prompt (aka DOS prompt) - the ability to type a command directly to the computer (as opposed to through a GUI interface).

Compiler - a program translator that converts source code into executable programs. (See Source Code).

Computer Aided Design (CAD) - computer software/hardware used to assist in the design of a product.

Computer Aided Manufacturing (CAM) - computer software/hardware used to assist in the manufacturing of a product, such as industrial robots.

Computer Procedure - a detailed set of machine processable instructions required to accomplish a specific process within a sub-system. A computer procedure consists of one or more computer programs linked together and executed in a prescribed sequence. (see Sub-System and Program).

Control, Project - one of the five basic activities of Project Management; it is concerned with monitoring project activities in order to realize the goals of a project, this includes time estimates, date schedules, and project costs.

Cost/Benefit Analysis - an analysis of the costs or expenses incurred by a project in comparison to the benefits derived from implementing the project.

CPU - Central Processing Unit (see Microprocessor).

Critical Path - the longest path in a project methodology which is critical to the timely completion of a project or part of a project. A project always has a critical path until it is completed or cancelled. The path may vary according to accomplishments.

Critical Path Method (CPM) - a project planning and scheduling technique which was developed in the mid-1950's by Morgan R. Walker of the E.I. DuPont de Nemours Company Engineering Department of Wilmington, Delaware and James E. Kelley of Sperry Univac. It is an approach for defining the structure of a project. It establishes a network of project phases with nodes to indicate start and stop points. Critical project activities are determined and are used as a point of reference for scheduling. The technique was originally developed for linear programming and was called "network analysis." (See PERT).

CRT - initials for Cathode Ray Tube; the computer's monitor/terminal screen.

Data Base - all of the data necessary to meet the information requirements of an enterprise, regardless of where used or how stored. A DBMS is not a prerequisite for data base. A common misconception is that an organization's data base is only that data stored on direct access devices by the computer. Actually, the data base may reside in many physical forms (e.g., tape, microfiche, disk, paper, cards, etc.). By this definition, all enterprises have a data base. The where and how of storage is dictated by the time requirements for information. Because information needs vary over time and as a result of changing conditions, the data base should be readily expandable. This base should be both hardware, software and application independent. An integrated data base environment is one in which the various parts of the data base are available for shared use and coordination within the environment.

Data Base Engineering - The discipline of developing the corporate data base to accommodate the processing requirements of all systems. The objectives of Data Base Engineering are:

1. To produce a RELIABLE corporate data base that satisfies all of the corporate systems and performs according to specifications.

2. To produce a corporate data base that is easy to MAINTAIN and MODIFY.

3. To physically implement the corporate data base in the most PRACTICAL, EFFICIENT, and COST EFFECTIVE manner possible.

Data Base Management System (DBMS) - computer software used to store and retrieve data. Ideally, the intent of the DBMS is to share and re-use data between computerized systems.

Data Element - the digital representation of a fact or an event. It is the raw material required to produce information. There are three types of data elements: descriptive, indicative, and quantitative (see respective listings). A descriptive data element is used to describe names, routings, addresses, text, etc. An indicative data element is used to uniquely identify objects, in part or in full, and usually consists of identification numbers and codes. Quantitative data elements are numeric values, such as quantities and percentages. There are only three things that can happen to a data element: it can be created, updated, or referenced. Data elements also come in two forms, primary and generated (see listings). Primary data elements are entered into an information system by a user. Generated data elements are developed internally within the system and most often made up of quantitative data elements. Generated data elements are created based on some form of logic, such as a calculation or formula. (Also see Group Data).

DBA - initials for Data Base Administrator; a person responsible for administrating a DBMS.

DBCS - initials for Double Byte Character Set, the ability to handle oriental character sets (e.g., Japanese and Chinese alphabets).

DBMS - initials for Data Base Management System (see listing).

DDL - Data Definition Language - a language enabling the structure and instances of a data base to be defined in a human-, and machine-readable form.

Deliverable - a reviewable result produced from the execution of a work step in a methodology. In systems development, a deliverable can take the form of either a document (usually a formal or informal report or working paper), a computer file or an executable program. Its intent is to substantiate that the work has been completed according to the methodology (see Methodology).

Descriptive Data - descriptive data consists of alphanumeric characters that are not strong enough to identify an object, but convey important business facts about an object, such as names, addresses, text, codes, etc.

Detail Estimate - an estimate of the effort required to complete the activities of an individual phase and is prepared prior to the start of the phase by the individual or individuals assigned to the phase. The detail estimate is an expression of someone's personal commitment and is expressed in Direct Time and used to calculate project schedules.

Direct Time - that amount of time expended on productive work, such as following the assigned phases/activities/steps of a methodology. Project estimates are expressed in Direct time. Under the "PRIDE" concept of "Mini-Project Manager," the individual is responsible for managing Direct Time. Available Time include both Direct Time and Indirect Time.

DLL - initials for Dynamic Link Library, chunks of reusable program code that is called for and executed as required by other programs (see Subroutine).

Domain - the lowest level in the classification hierarchy (Data Taxonomy); a collection of data elements, one or more, with common characteristics. For example, "text" related data elements would be in one domain, "weights" in another, "percentages" in another, "monetary values" in another, etc. The domain also defines the standard physical characteristics and values the data may assume. (Also see Taxonomy)

Dumb-terminal - slang expression referring to a monitor and keyboard attached to a remote computer (usually a mainframe) where the actual data processing is performed. No processing actually occurs locally at a dumb-terminal. (See Intelligent-terminal).

EIS - initials for Enterprise Information Strategy (see listing).

Effectiveness - producing a desired result. Whereas efficiency is concerned with how well a task is performed, effectiveness deals with the necessity of the task itself. Effectiveness answers the question, "Are we doing the right things?" (See Efficiency and Productivity).

Effectiveness Rate - the relationship of Direct to Indirect Time (see listings); it is expressed as the percent of time spent on direct work.

E.R. = (Direct Time + Indirect Time) / Direct Time

If someone has a low effectiveness rate, this may highlight to the manager that an employee is experiencing too many interferences, such as meetings or breaks. The manager can then exercise authority and minimize or eliminate the indirect activities if it is deemed unproductive or incorrect. In this way, the manager is controlling the work environment. On the other hand, a low effectiveness rate is not always considered undesirable. For example, a manager may send employees to a training class to sharpen their skills and become more productive. The employee's effectiveness rate for this period would be very low. As such, "effectiveness rate" should never be considered a measurement of a worker's productivity.

Efficiency - producing a desired result without waste. Efficiency is concerned with how well a task is performed. Effectiveness deals with the necessity of the task itself. Efficiency answers the question, "Are we doing things right?" (See Effectiveness and Productivity).

Engineering - the planning, design and construction of an object; i.e., buildings, products, machinery, etc.

Enterprise - a defined business entity with a specific mission, whether profitable or non-profitable in intent. Enterprises take many forms, such as the conventional commercial venture, whether private or public, a government agency, a fraternal organization, even a simple household can be viewed as an enterprise. It exhibits the following characteristics:

1. It has a mission or purpose.

2. An enterprise depends on other enterprises in order to survive.
3. An enterprise is a self-governing body.
4. An enterprise is a self-contained unit.

Enterprise Decomposition - the technique for defining the logical structure of the enterprise (building a functional hierarchy). Performed in Phase 2 of "PRIDE"-EEM

Engineering Engineering - The discipline of developing logical and physical models of the business. This includes documenting and analyzing business functions, administrative relationships, and performing an organizational analysis. Based on this, Enterprise Engineering is also used to identify information requirements and establish the priorities for business objectives, along with their supporting projects. The objectives of Enterprise Engineering are:

1. To produce ACCURATE and RELIABLE models of the enterprise.

2. To produce corporate models that are easy to MAINTAIN and MODIFY.

3. To recommend the most PRACTICAL, EFFICIENT and COST EFFECTIVE means to physically organize a business.

4. To develop and maintain corporate priorities pertaining to business objectives and their supporting projects.

Enterprise Information Strategy (EIS) - a plan to develop or modify information systems synchronized with business plans. The EIS is prepared in Phase 4 of "PRIDE"-EEM.

Ergonomics - an expression referring to consideration given to the human element in systems design. During sub-system (business process) design consideration is given to the physical implementation of the system. Systems analysis must consider the human senses and intelligence level, as well as man/ machine interfaces in determining an appropriate system implementation.

Estimate to Do - the amount of time a worker he/ she believes has remaining on a given task. This is not an automated calculation but rather a human assessment of remaining work and an important part for project reporting and control. Estimate to Do is expressed in Direct Time.

Estimating, Project - one of the five basic activities of Project Management; estimating is the process used to determine the amount of effort and cost required to implement a project, in part or in full. there are two types of estimates in "PRIDE" Project Management: Detail (for a single phase) and Order-of-Magnitude (for the whole project) (see separate listings). All estimates are expressed in Direct Time.

Explosion - expression refers to the top-down decomposition of a product into its assemblies, subassemblies and parts. "Implosion" represents the reversal of this process. A product, such as an information system, is designed by explosion (top-down) and implemented by implosion (bottom-up).

Feasibility Study - normally the first phase in a project used for specifying and analyzing a business problem/opportunity and proposing to management a course of action. It is usually one of the most important phases of a project due to its concentration on the proper identification and specification of business problems and requirements.

File - a logical and/or physical grouping of one or more records. A logical file is used to represent an object used in the operation of the enterprise. A physical file is used to describe a particular way to store data. Physical files can be implemented two ways; manually or computer-assisted. Computer files can be formatted (organized) many different ways; e.g., AVI, BMP, DOC, GIF, JPG, TXT, WAV, etc. A manual file is a storage location for inputs and outputs (human-readable forms and documents). A computer file is a storage location for machine-readable records.

Flowchart - graphical representations of system and software designs expressing processing movement ("flow"). Flowcharts are used as a planning tool in a manner similar to the way architects use perspective drawings or blueprints, and the way process engineers use flow diagrams to describe processes.

Form - A document used to collect data (see Input).

Fortran - a procedural programming language (see listing).

Fourth Generation Language - a specification driven, non-procedural programming language used to interpret and manipulate data in a DBMS or some other computer file. (See Program Generator, and Report Writer).

Frequency - an attribute of timing that defines how often information is required. This timing consideration will be used to determine the time cycle of subsystems to produce the required information. Frequency specifies the specific time cycle (e.g., daily, weekly, monthly, upon request, etc.) in which the subsystem is to be processed and the number of times within the cycle this processing is anticipated to take place, (e.g., three times daily, once a week, twice a month, etc.). (See Information, Offset, and Response Time).

Function - a logical grouping of one or more responsibilities for carrying out a specific portion of the mission of the enterprise. Functions define WHAT work must be performed and WHY. When organized into a hierarchy, functions represent the logical structure of the enterprise.

Gantt Chart - a documentation technique developed by Henry L. Gantt which allows management to visualize work to be performed and scheduled in relation to a calendar. It is a useful tool for determining project schedules, as well as for monitoring actual performance. In this way, it can be used to identify the critical path of a project. Other expressions of a Gantt Chart include "Bar Chart," "Planning Chart," "Milestone Chart" or "Histogram." (See Critical Path).

Generated Data - refers to data that relies on other data elements in order to produce the necessary result. This type of data can involve elaborate calculations and algorithms (e.g., Data-1 + Data-2 = Data-3). "Net Pay," "Balance Amount" and "Percent Complete" are some examples of calculated data. (See Primary Data, and Group Data).

Group Data - a form of generated data that represents a concatenation of indicative data elements in a prescribed format. For example, "Telephone Number" is a group data element consisting of "Area Code," "Exchange Number," and "Account Number" (for example: 727/786-4567). There are many other examples of "group" data: such as credit card numbers, product identification codes, public utility account numbers, etc.

GUI - Graphical User Interface - industrial term referring to the graphical screen format of an operating system.

Illustrative Examples - sample inputs and outputs produced prior to software design so that a user can visualize its layout and content. For this purpose, illustrative output/input uses sample data values (not X, Z, 9, etc.). Prototyping tools are useful for preparing such examples.

Impact Analysis - an analysis of the impact of a proposed change to a part to the sub-assemblies, assemblies, products and storage areas that are affected by the part. An impact analysis results in a bill of materials of all of the resources affected by the proposed change. This listing can then be used for project planning, estimating and scheduling purposes. In this way, it is used for "what if" analysis purposes. In IRM terms, an impact analysis results in a bill of materials of information resources affected by the change to one resource. (See Bill Of Materials Processing).

Implosion - expression refers to the bottom-up assemblage of parts to build a product. "Explosion" represents an opposite process. A product, such as an information system, is designed by explosion (top-down) and implemented by implosion (bottom-up).

Indicative Data - data used to uniquely identify an object in part or in full. "Uniqueness" is an inherent property of indicative data so that it can be used to clearly differentiate occurrences of an object. This is why control numbers and codes are typically used as indicative data, as opposed to names. Names can be too vague. For example, there may be more than one employee named "John Smith." Without some form of qualifier, it is virtually impossible to distinguish one "John Smith" from another. Consequently, an "Employee Number" is assigned to uniquely identify each employee.

Indirect Time - time that does not contribute to productive work, such as non-project related meetings, training, breaks, etc. Indirect Time is used to help calculate project schedules. Under the "PRIDE" concept of "Mini-Project Manager," the manager is responsible for managing Indirect Time. Available Time include both Direct Time and Indirect Time.

Information - the understanding or insight gained from the processing and/or analysis of data. Information is created as a result of the collection, processing and analysis of data in a prescribed manner. Information supports specific business related actions and decisions. The accuracy of information depends on the validity and completeness of the data and the processing logic used. There are three types of information: policy, control and operational. Policy information is used to establish policies and corporate direction. Control information is used by middle management to implement policy decisions and control corporate operations. Operational information is used by employees in the daily operations of the business, such as processing orders, payroll, and shipping products.

Information Requirement - a specific need for information in order to perform actions and decisions related to the business of the enterprise.

Information Resource Management (IRM) - the function of controlling and coordinating all of the resources required to produce information, such as data, systems, and business resources. The concept of IRM is analogous to Materials Resource Planning (MRP) as used in manufacturing.

Information System - an organized and methodical approach for the collection, storage and retrieval of data necessary to produce information in a timely basis to satisfy business objectives. It is a product that can be engineered and manufactured like any other product. Because of this, a system has a hierarchical structure consisting of one or more sub-systems; consisting of one or more administrative procedures and one or no computer procedures; each administrative procedure consists of operational steps, and each computer procedure consists of one or more programs. Systems vary in size depending on the scope of information required and are often arbitrarily established. Systems communicate to each other through shared data. An information system may or may not be automated; a manual system may be more cost-effective. An information system is often erroneously referred to as a "data processing system," "software system," or a "computer system." (See System, and Sub-System).

Input - human-intelligible medium that is used to collect data to store in a data base.

Interactive - a method of processing which refers to the interaction and instantaneous responsiveness that a person requires in order to solve a problem. An individual works "interactively" through a problem by submitting transactions and obtaining results that possibly require more transactions until the problem is answered. This can be applicable to both manual and automated information systems. (See Batch, Processing Method, and Transaction).

I/O - Input/Output

IRM - initials for Information Resource Management.

ISO - International Standards Organization - standards organization responsible for defining computer related standards.

JAD - initials for Joint Application Development; technique used in programming where the developer works closely with the end-user to design a program. Using application development aids such as a prototyping tool and program generators, the programmer prepares and reviews the layout of a screen or a report for the user before preparing the final executable program.

Java - programming language and trademark of Sun Microsystems, Inc. Java provides the ability to write a program that can execute on many different operating systems ("write once, run anywhere").

LAN - Local Area Network - allowing multiple computers within a limited geographical area to communicate and share computer resources (see WAN).

Legacy Systems - slang expression referrings to an older information system currently in use by an enterprise.

Mainframe - refers to a large computer with far more capacity and processing speed than a PC. Because of this, mainframes are commonly used as file servers (see Server).

Maintenance - this activity is concerned with correcting errors or defects in a product. It is intended to correct the problem and make the product operate according to specifications. In most organizations, this represents approximately 5% to 10% of the systems development work effort. (See Modification/ Improvement).

Materials Resource Planning (MRP) - a manufacturing related activity concerned with managing the materials required to produce products. This includes collecting, inventorying, and distributing parts in the most cost-effective and efficient means possible. The concept of Information Resource Management is analogous to MRP (see IRM).

Materials Management - to receive, inventory and distribute parts to the assembly lines for building products. It is also concerned with the standardization and reuse of parts for various products.

MBA - initials for M. Bryce & Associates, the devlopers of the "PRIDE" methodologies for IRM.

Methodology - one or more phases of work to be executed in a prescribed manner. The methodology denotes a project's sequence of execution or network. All projects have a structure; some are based on key events to be completed and others are based on the structure of a product to be built. All projects have a beginning for planning, a middle for execution and an end for review. The beginning phase of most projects is performed through some form of feasibility study. The ending phase is usually an evaluation phase. Each phase, activity and operational step within a methodology must produce a reviewable deliverable to substantiate adherence to the methodology. (See Deliverable).

Microprocessor - the main processing unit or chip in a PC, sometimes referred to as chip or CPU.

"Mini-Project Manager" - "PRIDE" Project Management concept. Under this approach, people are treated like professionals and are expected to act as such in return. It emphasizes less supervision and more personal management by the individual worker. Employees must learn to supervise themselves. Although, management still provides leadership and direction over assignments, employees participate in the estimating, scheduling and control process.

Modification/Improvement (Mod/Imp) - the activity required to change or enhance an existing product. For most organizations, this will represent 85% to 90% of their systems development work effort.

Modeling - to develop a preliminary design of a product based on product specifications. The model is used to evaluate the product's viability and also serves as the pattern to complete the design.

Module - compilable program source code consisting of one or more subroutines written in the same programming language. It is not executable by itself. (See Subroutine).

MRP - initials for Materials Resource Planning (see listing).

New Systems Development - process of designing, developing and implementing new information systems for an enterprise based on a specific need. This represents progress to an enterprise in the form of totally new applications. (See Modification/Improvement and Maintenance).

Object Code - executable program (see Source Code).

Object Oriented Programming - a design technique for writing programs. Object Oriented Programming (sometimes referred to as "OOPS") stresses the reuse of software by classification. (See Structured Programming).

Objective - a goal for the enterprise to achieve whether strategic, tactical, or mandatory in nature. An objective can be used to call for new development, modify or improve an existing condition (mod/imp), or to maintain or correct something. One or more objectives can be grouped into a project. An objective may relate to one or more information requirements.

Offset - a timing consideration used to denote when sub-systems, procedures or program are to begin their cycles (i.e., hour, day, week, etc.). When a sub-system is performed "upon request" there is no offset since is not schedulable. (See Frequency, Information and Response Time).

Operating System (OS) - the basic machine software used to operate a computer. Some examples: Windows, DOS, Linux, OS/2 MVS, UNIX, Exec, etc.

Operational Step (Task) - the smallest unit of work when executing an administrative procedure. One or more operational steps make up an administrative procedure (see listing).

Order-of-Magnitude (O-O-M) Estimate - an estimate of the effort remaining to complete a project and is typically prepared by the project manager or project team. It is expressed in terms of remaining phases, not activities. It is also expressed in Direct Time and used to calculate project schedules.

Organizational Analysis - an analysis of how well the physical enterprise model matches the logical model. During this activity, functions not being fulfilled are identified and overlaps in management are highlighted (both horizontally and vertically). This provides management with the ability to make more effective use of organization resources. This activity is performed in Phase 3 of "PRIDE"-EEM.

Output - human-intelligible medium which transmits information to the user. Outputs can take many forms:

a report, a CRT display, an audio response, etc.

PC - initials for Personal Computer.

PERT - a technique for planning, estimating and scheduling projects which was developed in 1958 by the U.S. Navy Special Projects Office, Bureau of Ordnance, in cooperation with Booz, Allen and Hamilton, a management consulting firm. PERT was originally used to develop the Navy's Polaris Weapon System. It utilizes the definition of a methodology for the completion of a project. A PERT network depicts the precedent relationships that exists between various phases, activities or operations, so that critical activities can be identified and the appropriate scheduling and resource allocations can be made. PERT was originally derived from the Critical Path Method (CPM). Like CPM, PERT also expresses nodes in the network and tries to express critical activities in the development of a project. The difference between PERT and CPM is basically that CPM uses one estimate and schedule, while CPM uses three: minimum, maximum and probable. (See Critical Path Method).

Phase - a prescribed set of one or more activities normally performed serially to accomplish specific project work. One or more phases make up a project. The sequence by which they are executed indicates the project structure (methodology) and usually has "deliverables" associated with them.

Planning, Project - one of the five basic activities of Project Management; project planning is concerned with establishing the scope of a project, along with the work breakdown structure and precedent relationships, and allocating resources to performing the work.

Playscript - a technique developed by Les Matthies used for writing administrative procedures. It is derived from a script for a play and describes the "actors" and their "actions" (operational steps). Computer programming procedural languages use derivatives of Playscript (e.g., COBOL, FORTRAN). (See Administrative Procedure and Operational Step).

Positions - a prescribed set of duties and responsibilities; another name is "job." Positions are implemented by human/machine resources.

Precedent Relationships, Project - the relationship of work steps within a project denoting dependencies in execution (sequential execution and parallel path ("branching")).

PRIDE - M&JB's product line. Acronym for PRofitable Information by DEsign - through phased planning & control. "PRIDE" is the registered trademark of M&JB Investment Company.

Primary Data - primary data refers to data in its virgin state; as introduced to the system from an external source (such as a person or department). "Source" defines who is responsible for entering the data to a system, and who has ultimate authority for the definition of the data element. (See Generated Data).

Procedural Languages - programming languages which make use of procedural statements, such as COBOL, Fortran, C, Basic. Procedural languages are considered "third generation" (as opposed to "Fourth Generation Languages" - see listing) and were ultimately derived from the Playscript procedure technique (see listing).

Procedure - the detailed set of steps necessary to accomplish specific work. The steps consist of instructions either in human or machine readable terms that direct the performance of particular processing functions. (See Administrative Procedure and Computer Procedure).

Process Diagram - a diagram used in engineering to show the flow of a process or work flow. A flowchart is another form of process diagram. (See Flowchart).

Processing Method - an expression used to describe the approach taken for the implementation of information systems; including such methods as interactive, on-line, real-time, transaction, batch, data and time driven. The method selected depends upon the derived logical processes and is based upon the analysis of timing and economics required to meet the information needs. The basic processing constructs include: sequence (consecutive progression), iteration (repetition until a condition is met), and choice (selection of path). The use of transactions is also an inherent part of all processing. (See Batch, Interactive, and Transaction).

Product Structure - a hierarchy that defines various levels of detail in a product's structure, from general to specific. This hierarchy defines the assemblies, sub-assemblies, and parts of the product. This becomes the basis for the design and development of the product.

Productivity - producing the correct results in the most efficient means possible. It is measured by effectiveness ("Are we doing the right things?") and efficiency ("Are we doing things right?"). Productivity = Effectiveness X Efficiency.

Program - a set of computer-executable instructions that perform a step within a computer procedure. Programs can be linked together to form a computer procedure. A program may be subdivided into modules if so desired.

Program Generator - an application development aid for producing program source code, usually for a procedural language such as COBOL. It should not be mistaken as a systems generator, but rather as a software generator. (See Fourth Generation Language, and Report Writer).

Programmer/Programming (see Software Engineering)

Project - a scope of work consisting of one or more phases. A project is an application of the material and human resources to a specific objective through the execution of a prescribed sequence of events. All projects have a structure which usually consists of a beginning (for planning), a middle (for execution), and an end (for review).

Project Administration - a support function used to validate project data and adminster a Project Management system.

Project Management - the function of applying resources to a defined goal and obtaining this goal within time and cost objectives. Project management is people management oriented, not clerical or administratively oriented. A Project Manager is sometimes called a "project leader," "team leader," "foreman," or "contractor."

Project Management system - an information system used to support project management activities. The system does not make decisions, people do, based on information received from the system.

Project Network - refers to a project's work breakdown structure; this specifically includes the phases, activities and tasks of the project, along with the precedent relationships between them (denoting the sequence of project execution). (See Methodology and Precedent Relationships).

Project Plan - a defined plan for executing a project. A Project Plan should specify: 1. The methodology for the project, including the phases, activities, and steps required to execute the project, along with their sequencing, and; 2. The human and machine resources required to perform the work, including both internal and external resources.

Project Scope - the statement of a particular business problem and/or opportunity (one or more objectives) to be addressed by a project and the organizational entities that will be affected, both directly and indirectly.

Prototyping - to develop an actual physical archetype of a product to some scale. The prototype is used for testing in order to visualize and evaluate the performance of the product and to make recommendations for improvement prior to the final design. Prototyping is used in the preparation of illustrative examples (See Modeling and Illustrative Examples).

Quality Assurance (QA) - a function concerned with building quality into a product during design, not inspecting for quality after the fact. Quality assurance is a shared responsibility and not a single job description in a company. Quality is judged by how well the product conforms to specifications.

Quantitative Data - data dealing with numeric values that are either calculated or are calculable. Measurements and computations are typical examples: "Net-Pay," "Quantity Ordered," "Elapsed Time," "Percent of Gross," etc. It is sometimes difficult to differentiate Quantitative data from Indicative data. Indicative data will often use numeric values for identification purposes, such as "Invoice Number," "Purchase Order Number," "Customer Number," etc. However, it would be a mistake to use these numbers for quantitative purposes (aside from counting the number of occurrences). (See Descriptive Data and Indicative Data).

RAD - initials for Rapid Application Development; design technique for programming involving the use of special application development tools (CASE) to expedite the development of a program (not an Information System).

Record - a collection of one or more data elements. (See Data Element, and File).

Referential Integrity - Rules for inserting, changing, and deleting data from a DBMS. This is an important consideration for selecting a DBMS in order to safeguard the integrity of the data.

Report Writer - an application development aid used to produce special reports based on an analysis of computer based data. (See Information Center, Fourth Generation Language, and Program Generator).

Reporting, Project - one of the five basic activities of Project Management; this includes the reporting of time, effort, and money towards project work.

Resource - a re-usable source of supply to produce something. Examples include human, financial, material, and information resources. To maximize the efficient and effective use of resources, they must be classified in order to share them and eliminate unwanted redundancy, and controlled in order to receive, store and distribute them properly.

Response Time - an attribute of timing that defines the maximum processing time allowable from the start of the execution (offset) to when it must be performed. Response time can be expressed in seconds, minutes, hours, days, etc. (See Frequency, Information, and Offset).

Return On Investment (ROI) - the ratio of projected cost savings versus amount invested. Typically calculated as:

ROI% = (Average Annual Savings divided by Investment) X 100

Rough Designs - preliminary designs of a product which are used for estimating, scheduling and planning purposes. Rough designs are a useful aid for determining the complexities and dimensions of a product. (See System Concept Diagram).

Scheduling, Project - one of the five basic activities of Project Management; scheduling is a distinctly separate activity from estimating even though it depends upon accurate estimates. It is the process of converting direct hour estimates into date schedules. Scheduling includes the assignment of resources to tasks in the correct sequence to meet the objectives of the project.

Skills - specific knowledge or talents as developed by education and/or experience. Proficiency denotes level of skill.

Software (computer) - machine processable instructions for operating a specific computer. There are two aspects to computer software: "system software" which is concerned with the basic operational needs required to operate computer hardware (i.e., operating systems, communication utilities, performance monitors, editors, compilers, etc.), and; "application software" which is concerned with having the computer perform specialized tasks to assist users in their work effort.

Software Engineering - a subset of information systems engineering which is concerned with the analysis of computer processing specifications, designing programs that will be the most appropriate solution to satisfying these specifications, defining the machine operating instructions for executing the programs, and successfully producing executable object code for a computer. The objectives of Software Engineering are:

1. To produce reliable programs that will perform according to specifications.

2. To produce programs that are easy to maintain and modify.

3. To physically implement programs in the most practical, efficient, and cost effective manner possible.

Source Code - the ASCII text used to actually compile a program. Source code is written according to the rules of a specific language (e.g., COBOL, Java, C, Pascal, Basic, etc.). (See Object Code).

Source, Data - the point of origin for a specific data element. A data element source can come from either a user area or, if generated, from some calculation or formula.

SQL - initials for Structured Query Language; a query language used to manipulate data in a DBMS.

Structured Programming - an expression used to categorize organized program design techniques. Structured programming adheres to a set of prescribed rules and standards used to produce a program. Its purpose is to provide programs that are easy to maintain and modify. It is simply one alternative to software engineering. Other alternatives include a 4GL, report writers, or program generators. (See Object Oriented Programming).

Sub-Assembly - a portion of an assembly used to bring together parts to build a product. In IRM terms, an assembly is represented by a sub-system and a sub-assembly is represented by procedures. (See Assembly, and Information System).

Sub-System - a business process or "work flow" within an information system which consists of inputs, outputs, and files used to produce specific information within a particular time frame. When initiated, sub-systems are executed to their logical conclusion and consist of one or more administrative procedures and not more than one computer procedure. The procedures can be executed serially or concurrently depending on flow or process in logic. Sub-systems relate to each other through the data base.

Subroutine - the smallest unit of program source code which can be invoked at program execution time by other subroutines (within the same module or other modules) to perform tasks. Its attributes include: it may or may not be re-usable or re-entrant to itself (it should have only one entry point); it is not compilable by itself; it is language dependant and can call or be called (by other subroutines). Depending on program language used, a subroutine may be referred to as a "function" or a "section."

System - an orderly arrangement, grouping or combination of one or more elements which form a whole. Its purpose is to accomplish specific goals or objectives. It operates routinely in a specific and predictable manner. (See Information System).

System Concept Diagram - the purpose of the System Concept Diagram is to communicate to users the concepts and approaches that will be used to meet their information needs. This diagram is based on the rough designs of the system, which in turn are based on the information requirements. Although there are no formal standards for this diagram, it should show who/what/when and how the system will perform. (See Rough Design).

System Life Cycles - Expression commonly used in the field to refer to a methodology or a project. Unfortunately, it is incorrect. Systems do not have life cycles. They will go on forever if kept viable with change. Projects, which are scopes of work, have life cycles. They have a beginning for planning, a middle for execution, and an end for review.

System Software - concentrates on the basic operational needs of the computer hardware and is represented by the operating system. The expression itself, "System" Software, is perhaps a misnomer in that it has nothing to do with Information Systems. Instead, it is only concerned with operating the computer.

Systems Engineering - The discipline of specifying user information requirements, designing a logical system that can produce the required information, determining the most appropriate way to physically process data, and to successfully implement the physical system. The objectives of the Systems Engineer are:

1. To produce a RELIABLE information system that will satisfy user information requirements and perform according to specifications.

2. To produce an information system that is easy to MAINTAIN and MODIFY.

3. To physically implement an information system in the most PRACTICAL, EFFICIENT, and COST EFFECTIVE manner possible.

Taxonomy - A Data Taxonomy is a hierarchical structure that separates data into specific classes of data based on common characteristics. The taxonomy represents a convenient way to classify data to prove that it is unique and without redundancy. This includes both primary and generated data elements. (Also see Domain).

Transaction - an exchange event from one object to another. A transaction always has some form of action associated with it. For output reporting, "request," "display," "print," "extract," "search," etc. are common transactions. For input, "new," "add," "change," "delete," "update," "charge," "credit," "debit," "deposit," etc. are typical transactions. (See Batch, Interactive, and Processing Method).

Unavailable Time - time when an employee is not available for work; e.g., vacations, leave, sickness, etc. Unavailable Time is considered (and recorded) in project scheduling.

User Friendly - an expression referring to the ease of use/processing of a system.

Utility - referring to a type of computer program used to manage a computer; e.g., sort, clipboard viewer, etc.

WBS - Work Breakdown Structure - a hierarchical structure consisting of one or more phases of work in a project. A phase represents a major "key event" or milestone in the project. Each phase consists of one or more activities representing "sub-events" required to meet the milestone. Each activity consists of one or more operational steps or tasks representing the individual actions to be taken in the project. The "PRIDE" methodologies represent Work Breakdown Structures.

WYSIWYG - What You See Is What You Get - term used in reference to the layout and appearance of a web page using a software tool to build the web page.

Y2K - reference to Year 2000 (2K)

END SECTION

Section 40

INDEX

Please note this index does not include any reference to the *"Glossary of Terms"* (Section 39). Also, please observe the convention used for pagination: SECTION-PAGE

Examples:
3-16 (Section 3 - Page 16)
12-4 (Section 12 - Page 4)

A

B

C

D

E

F

Q

R

S

T

U

V

W

END SECTION

Made in United States
Cleveland, OH
05 November 2026

47669886R00198